International
REVIEW OF
Neurobiology
Volume 90

International
REVIEW OF
Neurobiology

Volume 90

Transcranial Sonography in Movement Disorders

EDITED BY

DANIELA BERG
Department of Neurodegeneration
Hertie-Institute for Clinical Brain Research and German Center
for Neurodegenerative Diseases (DZNE)
University of Tübingen
Hoppe-Seyler-Straße 3
72076 Tübingen
Germany

and

UWE WALTER
Department of Neurology
University of Rostock
Gehlsheimer Str. 20
18147 Rostock
Germany

ELSEVIER

AMSTERDAM • BOSTON • HEIDELBERG • LONDON
NEW YORK • OXFORD • PARIS • SAN DIEGO
SAN FRANCISCO • SINGAPORE • SYDNEY • TOKYO
Academic Press is an imprint of Elsevier

AP

Academic Press is an imprint of Elsevier
360 Park Avenue South, New York, NY 10010-1700
525 B Street, Suite 1900, San Diego, California 92101-4495, USA
32 Jamestown Road, London NW1 7BY, UK

This book is printed on acid-free paper.

For information on all Elsevier publications
visit our website at www.elsevierdirect.com

ISBN: 978-0-12-381330-5

Printed and bound in the United Kingdom
Transferred to Digital Printing, 2011

CONTENTS

Part II TRANSCRANIAL SONOGRAPHY IN PARKINSON'S DISEASE

Transcranial Sonography in Relation to SPECT and MIBG

Yoshinori Kajimoto, Hideto Miwa and Tomoyoshi Kondo

Diagnosis of Parkinson's Disease—Transcranial Sonography in Relation to MRI

Ludwig Niehaus and Kai Boelmans

Early Diagnosis of Parkinson's Disease

Alexandra Gaenslen and Daniela Berg

Transcranial Sonography in the Premotor Diagnosis of Parkinson's Disease

Stefanie Behnke, Ute Schröder and Daniela Berg

Pathophysiology of Transcranial Sonography Signal Changes in the Human Substantia Nigra

Kay L. Double, Gabrielle Todd and Stephen R. Duma

Transcranial Sonography for the Discrimination of Idiopathic Parkinson's Disease from the Atypical Parkinsonian Syndromes

Angela E. P. Bouwmans, Annemarie M. M. Vlaar, Karin Srulijes, Werner H. Mess and W. E. J. Weber

Transcranial Sonography in the Discrimination of Parkinson's Disease Versus Vascular Parkinsonism

Pablo Venegas-Francke

TCS in Monogenic Forms of Parkinson's Disease

Kathrin Brockmann and Johann Hagenah

Part III TRANSCRANIAL SONOGRAPHY IN OTHER MOVEMENT DISORDERS AND DEPRESSION

Transcranial Sonography in Brain Disorders with Trace Metal Accumulation

Uwe Walter

Transcranial Sonography in Dystonia

Alexandra Gaenslen

Transcranial Sonography in Essential Tremor

Heike Stockner and Isabel Wurster

Transcranial Sonography in Restless Legs Syndrome

Jana Godau and Martin Sojer

Transcranial Sonography in Ataxia

CHRISTOS KROGIAS, THOMAS POSTERT AND JENS EYDING

Transcranial Sonography in Huntington's Disease

CHRISTOS KROGIAS, JENS EYDING AND THOMAS POSTERT

Transcranial Sonography in Depression

MILIJA D. MIJAJLOVIC

PART IV FUTURE APPLICATIONS AND CONCLUSION

Transcranial Sonography-Assisted Stereotaxy and Follow-Up of Deep Brain Implants in Patients with Movement Disorders

UWE WALTER

Conclusions

DANIELA BERG

CONTRIBUTORS

Numbers in parentheses indicate the pages on which the authors contributions begin.

Stefanie Behnke (93), Department of Neurology, University Hospital of the Saarland, Homburg/Saarland, Germany

Daniela Berg (81, 93, 287), Department of Neurodegeneration, Hertie Institute of Clinical Brain Research and German Center of Neurodegenerative Diseases (DZNE), Tübingen, Germany

Kai Boelmans (63), Department of Neurology, University Hospital Hamburg-Eppendorf, Hamburg, Germany

Angela E.P. Bouwmans (121), Department of Neurology, University Hospital Maastricht, Maastricht, the Netherlands

Kathrin Brockmann (157), Department of Neurodegeneration, Hertie Institute of Clinical Brain Research and German Center of Neurodegenerative Diseases (DZNE), Tübingen, Germany

Kay L. Double (107), Neuroscience Research Australia and the University of New South Wales, Sydney, NSW, Australia

Stephen R. Duma (107), Neuroscience Research Australia and the University of New South Wales, Sydney, NSW, Australia

Jens Eyding (217, 237), Department of Neurology, St. Josef-Hospital, Ruhr University Bochum, Germany

Alexandra Gaenslen (81, 179), Department of Neurodegeneration, Hertie Institute of Clinical Brain Research and German Center of Neurodegenerative Diseases (DZNE), Tübingen, Germany

Jana Godau (199), Department of Neurodegeneration, Hertie Institute of Clinical Brain Research and German Center of Neurodegenerative Diseases (DZNE), Tübingen, Germany

Johann Hagenah (157), Department of Neurology, University of Lübeck, Ratzeburger Allee 160, Lübeck, Germany

Heiko Huber (35), Department of Neurodegeneration, Hertie Institute of Clinical Brain Research and German Center of Neurodegenerative Diseases (DZNE), Tübingen, Germany

Yoshinori Kajimoto (49), Department of Neurology, Wakayama Medical University, Kimiidera, Wakayama, Japan

Tomoyoshi Kondo (49), Department of Neurology, Wakayama Medical University, Kimiidera, Wakayama, Japan

Christos Krogias (217, 237), Department of Neurology, St. Josef-Hospital, Ruhr University Bochum, Germany

Werner H. Mess (121), Department of Clinical Neurophysiology, University Hospital Maastricht, Maastricht, the Netherlands

Milija D. Mijajlovic (259), Institute of Neurology, Clinical Center of Serbia; School of Medicine University of Belgrade, Serbia

Hideto Miwa (49), Department of Neurology, Wakayama Medical University, Kimiidera, Wakayama, Japan

Ludwig Niehaus (63), Department of Neurology, Otto-von-Guericke University, Magdeburg, Germany; Department of Neurology, Castle Hospital Winnenden, Winnenden, Germany

Thomas Postert (217, 237), Department of Neurology, St. Vincenz-Krankenhaus, Paderborn, Germany

Ute Schröder (93), Department of Neurology, University Hospital of the Saarland, Homburg/Saarland, Germany

David Školoudík (7), Department of Neurology, Faculty of Medicine and Destristry, Palacky University Olomouc, Olomouc, Czech Republic

Martin Sojer (199), University of Innsbruck, Innsbruck, Austria

Karin Srulijes (121), Department of Neurodegeneration, Hertie Institute of Clinical Brain Research and German Center of Neurodegenerative Diseases (DZNE), Tübingen, Germany

Matthew B. Stern (3), Parkinson's Disease and Movement Disorders Center, University of Pennsylvania, Philadelphia, PA, USA

Heike Stockner (189), Medical University Innsbruck, Anichstrasse, Innsbruck

Gabrielle Todd (107), Discipline of Physiology, School of Medical Sciences, The University of Adelaide, Adelaide, SA, Australia

Pablo Venegas-Francke (147), Hospital Clínico, Universidad de Chile, Santiago de Chile

Annemarie M.M. Vlaar (121), Department of Neurology, Sint Lucas Andreas Hospital, Amsterdam, the Netherlands

Uwe Walter (7, 167, 275), Department of Neurology, University of Rostock, Rostock, Germany

W.E.J. Weber (121), Department of Neurology, University Hospital Maastricht, Maastricht, the Netherlands

Isabel Wurster (189), Department of Neurodegeneration, Hertie Institute of Clinical Brain Research and German Center of Neurodegenerative Diseases (DZNE), Tübingen, Germany

PREFACE

SUMMARY OF CONTENT—TCS IN THE DIAGNOSIS OF MOVEMENT DISORDERS

Transcranial B-Mode sonography (TCS) is a relatively young method which has evolved as an important diagnostic tool for various movement disorders over the past 15 years.

Due to the underlying physical principles supplementary information to other neuroimaging methods is given. The most widely recognized application of TCS is the diagnosis and differential diagnosis of parkinsonian syndromes. Many different groups have confirmed that "hyperechogenicity of the substantia nigra"—the hallmark for Parkinson's disease—can serve as a sensitive and fairly specific marker for this frequent neurodegenerative movement disorder. Importantly, there is increasing evidence that substantia nigra hyperechogenicity in yet healthy subjects may disclose nigrostriatal vulnerability. And indeed, the first longitudinal studies on healthy subjects with this echofeature support the assumption that it may contribute to defining high-risk groups for Parkinson's disease (PD).

In this volume we will provide detailed information about the ultrasound procedure and some guidelines for the scanning procedure. Besides its specific value with respect to easy applicability, noninvasiveness, common availability, and inexpensiveness also the limitations of the method will be discussed. We will focus on PD and its differential diagnoses. Moreover, we will review its clinical application in dystonia, essential tremor, restless legs syndrome, ataxia, Huntington's disease, and depression also in relation to other neuroimaging methods. As far as known pathophysiological mechanisms leading to changes in the reflection of ultrasound waves will be discussed. Moreover, we will give an outlook on important future applications—e.g., the monitoring of electrodes in deep brain stimulation.

DANIELA BERG
UWE WALTER

PART I

INTRODUCTION

INTRODUCTORY REMARKS ON THE HISTORY AND CURRENT APPLICATIONS OF TCS

Matthew B. Stern

Parkinson's Disease and Movement Disorders Center, University of Pennsylvania, Philadelphia, PA, USA

Transcranial B-mode ultrasonography (TCS) is a noninvasive imaging technique that was initially developed for evaluating cerebrovascular disorders, particularly arterial stenosis and occlusion. Since the mid-1990s, however, TCS has evolved to enable visualization of brain parenchymal structures, and, more recently, it has been applied to Parkinson's disease (PD) and other movement disorders. This is the first comprehensive monograph on TCS in movement disorders and brings together experts from around the world to review the technique, diagnostic utility, and other potential applications (including preclinical detection) of TCS in these disorders.

Becker and colleagues (1995) first observed hyperechogenicity of the substantia nigra in PD using TCS. His observations were the result of an initial intraoperative scan of a PD patient undergoing a brain tumor resection and fueled subsequent attempts to reproduce his findings transcranially. Thus began the productive collaboration with Dr. Daniela Berg beginning in the Department of Neurology at Würzburg and the ongoing studies of TCS in PD. As ultrasound resolution improved, more quantitative measurements were employed (measuring the area of hyperechogenic signals within the substantia nigra). Berg's study published in 2001 not only confirmed earlier observations of hyperechogenicity in PD, but also showed the abnormality in 91% of PD patients and larger areas of hyperechogenicity contralateral to the most affected side (Berg *et al.*, 2001). Interestingly, the authors noted that this finding was not necessarily specific for PD, as 2 of the 30 control patients in the first study on PD patients also had the abnormality. The subsequent development of PD in one of these control subjects raised the possibility that nigral hyperechogenicity by TCS might actually be a marker for eventual PD and contribute to PD's preclinical detection. This led to a larger, cross-sectional assessment of over 400 healthy subjects in whom 9% had SN hyperechogenicity and many of these individuals had reduced flourodopa uptake by positron emission tomography (PET) (Berg *et al.*, 1999), raising the possibility that some normal individuals with SN hyperechogenicity might already have a functional deficit of the nigrostriatal system and be destined to develop PD. Subsequent studies continued to suggest that TCS might be useful as a diagnostic tool both for clinical and for preclinical PD. First, Berg and co-workers (2005)

INTERNATIONAL REVIEW OF
NEUROBIOLOGY, VOL. 90
DOI: 10.1016/S0074-7742(10)90001-9

showed that the abnormal SN signal appears to be constant over time, suggesting that SN changes detected by TCS might be a trait marker of PD. Second, in a recent study, Gaenslen and colleagues (2008) followed a group of individuals with subtle parkinsonian signs and found that SN hyperechogenicity was highly predictive of the eventual diagnosis of PD. Taken together, TCS has begun to emerge as a potentially important tool in the early and potentially preclinical detection of PD and current studies are underway to further test this concept.

The ultimate utility of TCS will of course depend upon our understanding of what SN hyperechogenicity actually represents. While there is speculation that it is indicative of higher tissue iron content, which is increased in PD, this is not conclusive and other parkinsonian disorders with high SN iron content do not necessarily reveal the hyperechogenic signal. Further, similar TCS findings have been detected in patients with corticobasal degeneration (CBD) and dementia with Lewy bodies (Walter *et al.*, 2004, 2006). TCS might then be useful in differentiating PD from progressive supranuclear palsy (PSP) or multiple system atrophy (MSA) or perhaps contribute to the often challenging distinction between PSP and CBD.

TCS also appears to have value in the assessment of other movement disorders and even restless legs syndrome (RLS). Hypoechogenicity appears to distinguish RLS patients from controls (Godau *et al.*, 2008). Hyperechogenicity of the lentiform nucleus has been observed in patients with cervical dysonia (Nauman *et al.*, 1996) and may correlate with disease severity in Wilson's disease (Walter *et al.*, 2005).

TCS has distinct advantages as a diagnostic tool. It is inexpensive, noninvasive, and widely available. Any discussion of its potential uses must also acknowledge its limitations. TCS is dependent on the quality of the temporal acoustic bone window, ultrasound system, and, most importantly, the experience of the investigator. These issues are addressed thoughtfully and comprehensively in the text as are anticipated developments in TCS technology and other potential applications.

Both editors, Drs. Daniela Berg and Uwe Walter, have devoted much of their careers to elucidating the technique and specific TCS methods as they apply to movement disorders. Their studies over the past decade have fueled growing interest in TCS as evidenced by the large number of visitors to their centers to learn the technique and begin their own investigations. It is fitting that they lead the effort to assemble the body of experts represented in this volume and that they are recognized as true pioneers in the field.

References

Becker, G., Seufert, J., Bogdahn, U., Reichmann, H., and Reiners, K. (1995). Degeneration of the substantia nigra in chronic Parkinson's disease visualized by transcranial color-coded real-time sonography. *Neurology* **45**, 182–184.

Berg, D., Becker, G., Zeiler, B., Tucha, O., Hofmann, E., Preier, M., Benz, P., Jost, W., Reiners, K., and Lange K.W. (1999). Vulnerability of the nigrostriatal system as detected by transcranial ultrasound. *Neurology* **53**, 1026–1031.

Berg, D., Siefker, C., and Becker, G. (2001). Echogenicity of the substantia nigra in Parkinson's disease and its relation to clinical findings. *J. Neurol.* **248**, 684–689.

Berg, D., Merz, B., Reiners, K., Naumann, M., and Becker, G. (2005). Five-year follow-up study of hyperechogenicity of the substantia nigra in Parkinson's disease. *Mov. Disord.* **20**, 383–385.

Gaenslen, A., Unmuth, B., Godau, J., Liepelt, I., Di Santo, A., Schweitzer, K. J., Gasser, T., Machulla, H. J., Reimold, M., Marek, K., and Berg, D. (2008). The specificity of transcranial ultrasound in the differential diagnosis of Parkinson's disease: a prospective blinded study. *Lancet Neurol.* **7**, 417–424.

Godau, J., Wevers, A. K., Gaenslen, A., Di Santo, A., Liepelt, I., Gasser, T., and Berg, D. (2008). Sonographis abnormalities of brainstem structures in restless legs syndrome. *Sleep Med.* **7**, 782–789.

Nauman, M., Becker, G., Toyka, K.V., Supprian, T., and Reiners, K. (1996). Lenticular nucleus lesion in idiopathic dystonia detected by transcranial sonography. *Neurology* **47**, 1284–1290.

Walter, U., Dressler, D., Wolters, A., Probst, T., Grossmann, A., and Benecke, R. (2004). Sonographic discrimination of corticobasal degeneration vs. progressive supranuclear palsy. *Neurology* **63**, 504–509.

Walter, U., Krolikowski, K., Tarnacka, B., Benecke, R., Czlonkowska, A., and Dressler, D. (2005). Sonographic detection of basal ganglia lesions in asymptomatic and symptomatic Wilson's disease. *Neurology* **64**, 1726–1732.

Walter, U., Dressler, D., Wolters, A., Wittstock, M., Greim, B., and Benecke, R. (2006). Sonographic discrimination of dementia with Lewy bodies and Parkinson's disease with dementia. *J. Neurol.* **253**, 448–454.

METHOD AND VALIDITY OF TRANSCRANIAL SONOGRAPHY IN MOVEMENT DISORDERS

David Školoudík* and Uwe Walter†

*Department of Neurology, Faculty of Medicine and Detristry,
Palacky University Olomouc, Olomouc, Czech Republic
†Department of Neurology, University of Rostock, Rostock, Germany

Transcranial sonography (TCS) of the brain parenchyma in patients with movement and other neurodegenerative disorders has developed with increasing dynamics during the past two decades. The specific advantages of TCS are the different visualization of brain structures compared to other neuroimaging methods due to the different physical imaging principle, high-resolution imaging of echogenic deep brain structures, on-time dynamic imaging with high resolution in time, relatively low costs of technical equipment, wide availability, short investigation time, noninvasivity, mobility and bedside availability, and little corruption by patients' movements. TCS proved sensitive and reliable in

DOI: 10.1016/S0074-7742(10)90002-0

0074-7742/10 $35.00

detecting disease-specific alterations of brainstem structures and basal ganglia in various movement disorders. Here, we give an overview on the technical requirements and recommendations on the standardized application of TCS of deep brain structures in movement disorders. We discuss methodological potentials and limitations of TCS, its validity, and future developments.

I. Introduction

For many years the intact skull of adult subjects was regarded as an unsurmountable problem to investigate any intracranial structure with ultrasound, much less to get high-resolution sonographic images of the brain. However, in the year 1982, Rune Aaslid was able to detect the blood flow of intracranial brain arteries through the intact skull by applying a high-energy pulsed Doppler ultrasound system (Aaslid *et al.*, 1982). In the following decade, transcranial Doppler sonography (TCD) examination became widely used to diagnose hemodynamic changes in intracranial arteries, e.g., in cases of stenoses of brain-supplying arteries, intracranial vasospasms, or arteriovenous malformations (Aaslid *et al.*, 1986). Since the 1990s, with ongoing improvements in the ultrasound technique, including the subsequent development of transcranial color-coded sonography (TCCS), TCD has become a key diagnostic tool in the evaluation of acute and chronic cerebrovascular disorders (Bartels, 1999; Bogdahn *et al.*, 1998).

Visualization of the brain parenchyma using transcranial ultrasound (transcranial brain sonography, TCS) is even more challenging since the penetration of the skull necessitates the application of rather low ultrasound frequencies of 2–3.5 (maximum 5) MHz, which, in principle, limits the image resolution. First attempts of TCS in a clinical setting were made in the 1970s and 1980s (Berland *et al.*, 1988; Furuhata, 1989; Kamphuisen *et al.*, 1972). Since the 1990s, application of TCS for diagnostic purposes has developed with increasing dynamics. In the early 1990s, the first reports were published by Georg Becker and co-workers (1994a, 1995) demonstrating the possibility of discriminating main brain parenchymal structures sonographically in different parts of the brain, such as the pontine brainstem and the ventricular system. Measurement of ventricle widths were shown to be practical and reliable (Seidel *et al.*, 1995). Using TCS, visualization of malignant brain tumors and acute cerebral bleedings became possible (Becker *et al.*, 1992). Georg Becker was also the first to visualize an alteration of the substantia nigra in Parkinson's disease on TCS, which was not seen before with other brain imaging methods (Becker *et al.*, 1995). Meanwhile, the present-generation ultrasound systems display the brain parenchyma with improved image resolution and image homogeneity, which facilitated the development of a continuously growing number of differential diagnostic applications of TCS. Especially, TCS has proven to discriminate disease-specific patterns of deep brain structure alterations in various movement and neurodegenerative disorders (Berg *et al.*, 2008; Walter *et al.*, 2007a).

The main advantages of using TCS application in the clinical setting are as follows:

- Different visualization of brain structures compared to other neuroimaging methods due to different physical imaging principle.
- High-resolution imaging of echogenic deep brain structures (Walter *et al.*, 2008b).
- On-time dynamic imaging with high resolution in time.
- Relatively low costs of technical equipment.
- Wide availability since same transducers as for TCCS or echocardiography can be used.
- Short investigation time.
- Noninvasivity and unlimited repeatability.
- Mobility and bedside availability.
- Little corruption by patients' movements.

The main limitations of TCS compared to other imaging methods are as follows:

- Dependency on the quality of transcranial acoustic bone windows, accounting for no or only partial insonability of the brain in 10–20% of Caucasian subjects and 20–50% of Asian subjects (Bartels, 1999; Berg *et al.*, 2008; Bogdahn *et al.*, 1998).
- Limited possibility of *post-hoc* evaluation of stored images since the major deal of diagnostic information is obtained from moving images during the investigation.
- No or low assessability of brain regions near the probe (at about 0- to 2-cm distance), very near the contralateral and basal skull bone, and in the (insonable) high frontoparietal region.
- Dependency on the investigators' qualification and experience (Školoudík *et al.*, 2007; Walter *et al.*, 2007a).

Despite the limitations listed above, the assessment even of a very small brainstem structure, such as the substantia nigra, with the present-day TCS system has been proven to have a high diagnostic accuracy and excellent intra- and interindividual reproducibility provided by experienced investigators (van de Loo *et al.*, in press). Therefore, it is expected that the clinical application of TCS in the evaluation of brain disorders will further increase.

In this chapter, technical aspects and the current methodological standards of TCS application in neurodegenerative brain diseases are reviewed.

II. History of TCS in Movement Disorders

In 1995, Georg Becker and co-workers (1995) described for the first time the in vivo visualization of a disease-specific structural alteration of a brain structure in Parkinson's disease (PD): a characteristic enlargement of the echogenic signals at the anatomic area of the substantia nigra (SN) on TCS. Georg Becker called this abnormal TCS finding of SN "hyperechogenicity". By this time, there had been no way of visualizing structural changes in PD using other structural neuroimaging techniques; this limitation in diagnosis was first challenged by the finding of the typical hyperechogenicity of the SN in PD. Already in 1994, Becker and co-workers had reported a characteristic sonographic alteration of the brainstem raphe in subjects suffering from unipolar depression (Becker *et al.*, 1994b). These findings opened up a new field of TCS applications—the diagnostic assessment of brainstem and cerebellar structures and of basal ganglia in movement and affective disorders. In the following years, the characteristic patterns of changes in echogenicity of SN, brainstem raphe, red nucleus, dentate nucleus lenticular nucleus, caudate nucleus, cerebellar parenchyma, dentate nucleus, or in width of ventricular system were detected not only in various primary neurodegenerative but also in inflammatory brain diseases (Berg *et al.*, 2008; Walter *et al.*, 2007a, 2009).

Although TCS was able to detect several pathologies in patients with various movement disorders, it was met with skepticism from the scientific community for many years. For physicians, it was hard to believe that a sonographic method used mainly for detection of hemodynamic changes in intracranial arteries might show in B-mode disease-specific abnormalities that could not be shown by other well-established structural neuroimaging methods such as computed tomography (CT) or magnetic resonance imaging (MRI). Since then, thanks to confirmation and broadening of diagnostic TCS applications by several independent groups worldwide, TCS has become more widely accepted tool, especially for the diagnosis and differential diagnosis of PD and other movement disorders (Berg *et al.*, 2008). Since ultrasound systems suited for TCS are available in most hospitals, and since TCS is relatively inexpensive, and suitable to perform even in somewhat moving patients, it is worth considering as a supplementary diagnostic tool in the routine diagnostic workup of patients with movement disorders.

Development of TCS as a diagnostic tool in several neurodegenerative disorders led to the necessity of creating guidelines of TCS diagnostic procedure.

In 2004, on the occasion of the 9th Meeting of the European Society of Neurosonology and Cerebral Hemodynamics (ESNCH) in Wetzlar, Germany, the ESNCH sponsored a workshop to develop a standardized procedure for TCS in neurodegenerative diseases. As a result, the first consensus guidelines on the standardized procedure for TCS of midbrain structures, basal ganglia, and ventricles were established, and recommendations for its clinical application in the early and differential diagnosis of Parkinson's disease were mounted (Walter *et al.*, 2007a).

III. Technical Equipment and System Settings

For TCS, a high-end ultrasound system, as applied also for transcranial color-coded cerebrovascular ultrasound, equipped with a 2.0- to 3.5- (1.0 to 5.0) MHz transducer is used. By publishing working groups, several ultrasound systems have been used (Table I). It must be considered that certain measurements, e.g., of the size of a hyperechogenic area, are dependent on the applied ultrasound system and the individual system settings. System parameters such as the width of ultrasonic beam, the line density, and even the age of the probe influence the image resolution. Therefore, reference values need to be obtained (and ideally updated for the same probe every 2–3 years) separately for each ultrasound system. Hitherto, standard reference values on echogenic sizes of the SN have been published for

TABLE I
ULTRASOUND MACHINES USED BY SCIENTIFIC WORKING GROUPS FOR TCS IN MOVEMENT DISORDERS

Manufacturer	Ultrasound system
Advanced Technologies Laboratories (Washington, DC, USA)	Ultramark 3000
	Ultramark 9
Esaote (Florence, Italy)	MyLab 25
General Electric (Milwaukee, WI, USA)	Logiq 7
Philips (Eindhoven, the Netherlands)	HDI 5000
	SONOS 4500
	SONOS 5500
	IU22
Siemens (Erlangen, Germany)	Sonoline CF
	Sonoline Elegra
	Acuson Antares
	S2000
Toshiba (Tokyo, Japan)	SSH-140A
	Aplio

Source: Updated from Berg *et al.*, 2008, Walter *et al.*, 2007a; personal communications.

TABLE II
ULTRASOUND SYSTEM SETTINGS FOR TRANSCRANIAL SONOGRAPHY

Parameter	Settings
Ultrasound system	
Penetration depth	Start with 14–16 cm, reduce if needed
Dynamic range	45–50 dB
Contour amplification	Medium to high
Postprocessing function	Moderate suppression of low echo signals
Time gain compensation	Adjust manually as needed, or use the "tissue equalization" function if available
Image brightness	Adjust manually, not too high, or use the "tissue equalization" function if available
Ultrasound transducer	
Crystal/channel	As high as possible, ideally: "matrix" probe
Insonation frequency (center frequency)	2.0–3.5 MHz, preferably 3.0 MHz

the ultrasound systems Siemens Sonoline Elegra (Walter *et al.*, 2007a), Siemens Acuson Antares (Glaser *et al.*, 2006), and Philips Sonos 5500 (Mehnert *et al.*, 2010). Table II summarizes standard system settings that can be individually adapted depending on the ultrasound system applied. Usually the following ultrasound parameters are chosen: penetration depth 14–16 cm, dynamic range 45–55 dB, contour amplification medium or high. Using these parameter settings, the thermal index (TI) is about 1.9 and the mechanical index (MI) about 1.3. Image brightness and time gain compensation are visually adapted as needed, or using the "tissue equalization" button of the ultrasound system, if available (usually available with contemporary high-end systems). Postprocessing parameters should be set at moderate suppression of low echo signals. Investigation is usually performed transtemporally with preauricular position of the transducer. Other transcranial approaches used for specific questions are the foramen magnum, the transfrontal, and the transoccipital bone windows. The latter two, however, are more frequently insufficient to insonate in adults. The structures assessed at different planes and windows are detailed below.

IV. TCS Image Resolution

TCS depicts the brain structures near the midline in most detail since the focus zone of 2- to 3.5-MHz transducers is at a distance of 5–7 (4–8) cm from the contact plane of the transducer. Therefore, deep brain structures such as the mesencephalon with its echogenic substructures (i.e., SN, red nuclei, raphe) and the basal ganglia can well be assessed. While image resolution using former generation TCS systems was inferior in comparison with MRI, with an axial resolution of 0.7–1.0 mm and a lateral resolution of about 3.0 mm, with a contemporary high-end TCS the in-plane image resolution of echogenic deep brain structures in the focus zone of the transducer is currently even somewhat superior (0.6 × 1.1 mm) compared to MRI under clinical conditions (about 1.0 × 1.0 mm) (Walter *et al.*, 2008b) (Fig. 1). The reason for the limited image resolution on MRI is that even slight head movements of patients caused by breathing corrupt the maximum resolution currently achievable with MRI. This

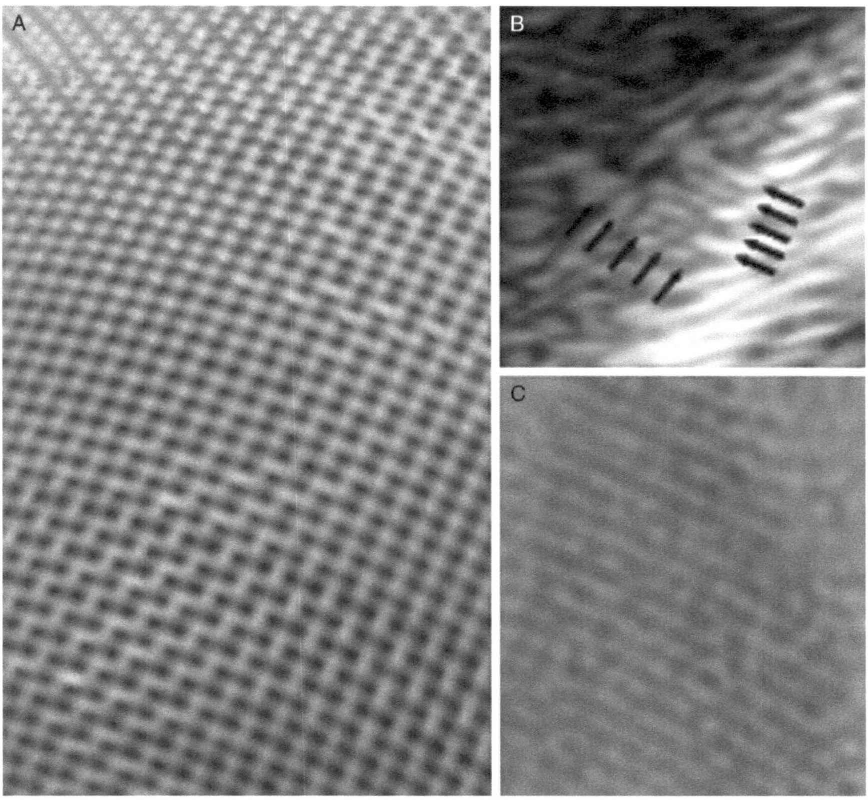

FIG. 1. Comparison of transcranial sonography (TCS) and magnetic resonance imaging (MRI) with respect to image resolution of the network structure of a kitchen strainer placed within a human skull phantom filled with ultrasound gel. (A) Photograph of the kitchen strainer consisting of polyamide threads forming a 1.1 × 0.8 mm network. (B) Using a contemporary TCS system, the network structure of the intracranially located strainer could be clearly visualized. (C) With MRI, simulating clinical conditions, the network structure was not visualized (Walter *et al.*, 2008b).

may change in future when head motion compensation technologies become commercially available. Because an on-time image is generated with TCS, and the investigator can slightly compensate for patients' head movement by keeping the probe firmly to the patients' head, the image resolution remains high even in patients with involuntary head movement due to movement disorder. This allows investigation with TCS even of those patients with hyperkinetic movement disorders who otherwise would have the need to be sedated for CT or MRI. New ultrasound technologies such as tissue harmonic imaging can improve image resolution but are more dependent on the quality of the acoustic bone window and may be applicable only to limited insonation depths (Puls *et al.*, 2000). Therefore, these technologies are not applicable in all patients, and cannot be recommended as a reference standard (Walter *et al.*, 2007a). However, for validation of slight abnormalities seen with the conventional ultrasound mode, greater contour-enhancing harmonic imaging mode can sometimes be helpful. The progress in image quality achieved with a contemporary TCS system compared to a former-generation TCS system is illustrated in Fig. 2.

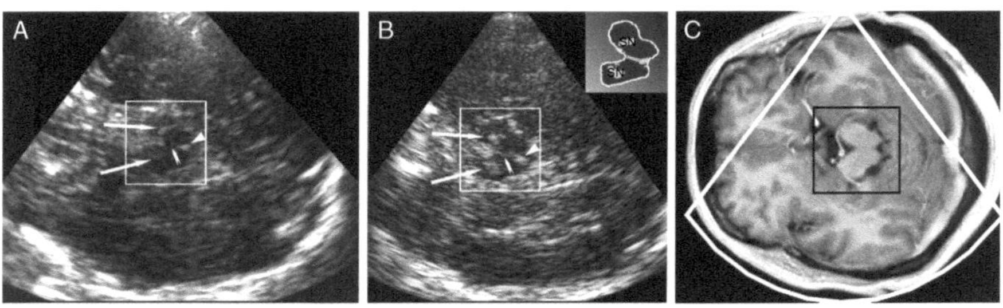

FIG. 2. Comparison of transcranial sonography (TCS) images obtained with a former-generation (Sonoline Elegra, Siemens) and a present-generation (Acuson Antares, Siemens) ultrasound system in the same patient. (A) Axial (transverse) TCS scan through the brain at midbrain level obtained with the former-generation system. The bilateral substantia nigra (arrows) is clearly displayed. On the right (upper) side, the substantia nigra shows markedly increased echogenic size ("hyperechogenicity"). The midline raphe (arrow head; triangle: aqueduct) is invisible while the highly echogenic basal cisterns surrounding the midbrain and the substantia nigra are visible. This finding is referred to as brainstem raphe hypoechogenicity. (B) Axial (transverse) TCS scan through the brain at midbrain level obtained with the present-generation system. Note that the whole scan is displayed with homogeneous image brightness, thanks to the "tissue equalization technology" function. The small substructures within the midbrain are displayed with high resolution. Compared with the former-generation TCS system, the echogenic areas of the bilaterally hyperechogenic substantia nigra are larger (arrows). Moreover, the raphe is displayed as a thin continuous line of slightly increased echogenicity (arrow head; triangle: aqueduct), making the recognition of raphe hypoechogenicity more difficult than with the former-generation system. (C) Magnetic resonance imaging (MRI) scan corresponding to the TCS images shown in (A) and (B).

V. Procedure of Standard TCS Examination in Movement Disorders

A. Patient Setting and Probe Position

For the examination, the patient is posed in a supine position, and the examiner usually sits at the head of the examination table. The transtemporal examination is performed consecutively from each side with preauricular position of the ultrasound probe. The examiner presses the ultrasound probe firmly to the temporal plane (Fig. 3A). Usually the posterior and/or the middle temporal bone windows are used. For special investigations, such as the measurement of the corpus callosum, the transfrontal approach may be used (Fig. 3B). For standardized transtemporal evaluation of deep brain structures, including the mesencephalic brainstem, the basal ganglia, and the ventricular system, standardized axial scanning planes are used as detailed below (Fig. 4). In these planes, a two-dimensional (2-D) B- (brightness-) mode image of the brain is obtained, which is composed of gray-scale speckles.

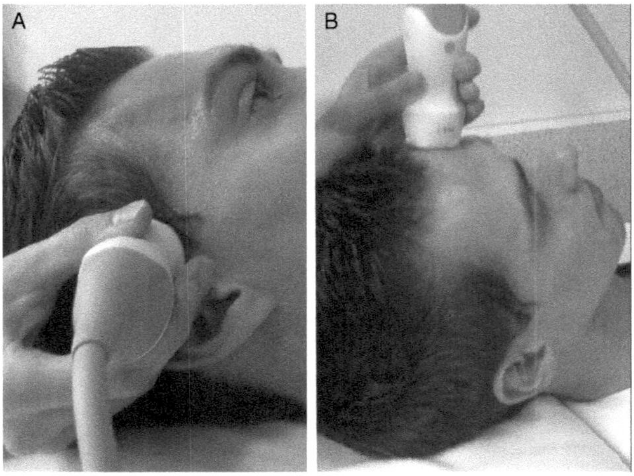

Fig. 3. Most frequently used probe positions for transcranial sonography (TCS) in adults. Note that probe position in an individual needs to be adapted for an optimal acoustic bone window. Here, the typical start positions of the probe are shown. The investigator sits behind the patients head. (A) For transtemporal investigation, the ultrasound probe is usually placed preauricularly at the temporal plane, aligning the probe in parallel to the orbitomeatal line. (B) For transfrontal investigation, the transducer is placed at a paramedian frontal position, starting with a sagittal scan.

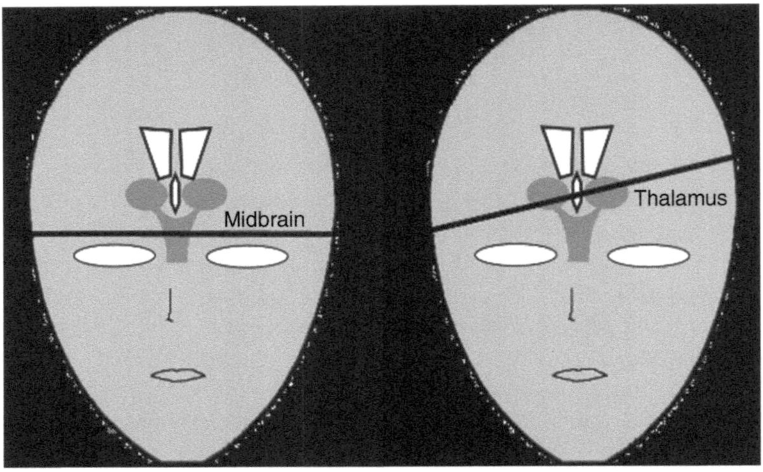

FIG. 4. Standard axial scanning planes for transcranial (transtemporal) sonography (TCS) in patients with movement disorders. The ultrasound probe is placed at the temporal plane as shown in Fig. 3A. (A) Axial scanning plane at midbrain level. At this plane, the substantia nigra, red nucleus and brainstem raphe are routinely assessed. (B) Axial scanning plane at thalamus level. At this plane, typically the third ventricle, frontal horns, thalamus, lenticular, and caudate nucleus are assessed.

B. Current and Future Methods of Echogenicity Quantification

For different structures, the intensity of their gray-scale shades and their "echogenic" extent, i.e., the extent of their echosignals, can be evaluated. A structure is usually denoted as "hyperechogenic" if the intensity of the ultrasound signal is abnormally increased compared with a reference structure, which is usually the surrounding white matter. For distinct structures, especially the SN, not only the visually assessed relative brightness but also the extent (area) of the echogenic signals of the referring structure is considered. If the planimetrically measured area of the echogenic signals of this structure exceeds a predefined cut-off value, e.g., the 90% percentile of measures in normal population, this structure (SN) is denoted as "hyperechogenic" (Walter *et al.*, 2007a). Other approaches to quantify an abnormal increase of echogenicity, such as measuring the image brightness of a region of interest (ROI) in comparison to a reference area (white matter), have not been demonstrated to be superior to planimetric area measurements. The opposite appears to be true since this approach was limited by a ceiling effect, characterized by a still increasing echogenic area despite an already stable degree of maximum image brightness of the investigated structure (Prof. Günther Seidel, personal communication, Department of Neurology, University of Luebeck, Germany). Computer-based systems for evaluation of relative brightness might overcome this problem in future. In a recent report of Schreiber *et al.* (2008), an automatic program is described for the detection and evaluation of SN. The achieved results obtained during testing suggest that this method may be helpful for the detection and analysis of brainstem TCS images. With the tested, optimum-quality images, the position of the brainstem was correctly determined and the SN was successfully measured in 76% of cases. It needs to be proven whether such a system will be helpful in a clinical setting.

Till date, it is unclear whether volume measurements of small echogenic structures will be superior to planimetric measurements. Currently, there are attempts to develop technologies for small-structure volume estimation. As soon as these technologies are commercially available, their diagnostic value needs to be assessed. Other technological approaches such as complex analysis of echo-signal characteristics of the SN, are currently under development (Prof. Günther Seidel, personal communication), might offer new insights into the pathophysiology of disease-specific changes and hopefully allow an investigator-independent assessment of SN echogenicity in future.

VI. Transtemporal Investigation

There are two scanning planes, defined by cerebral landmark structures (midbrain, thalamus), that are recommended to be applied as a standard to any diagnostic examination in patients with movement disorders, although the course of the examination may be individually adapted according to the diagnostic question (Fig. 4). Once a brain structure of interest is clearly displayed, the image should be fixed and zoomed two to fourfold for measurements. According to the current knowledge from TCS studies, the following structures are of diagnostic value in neurodegenerative disease: *SN*, *brainstem raphe*, red nucleus, cerebellum, dentate nucleus, thalamus, *lenticular nucleus*, *caudate nucleus*, widths of fourth ventricle, *third ventricle*, and of *frontal horn* of lateral ventricle. The structures highlighted in italic style should be always assessed in movement disorders if possible.

A. Structures at Midbrain Level

To display the plane of the mesencephalic brainstem, the examination starts with the identification of the hypo- to anechogenic butterfly-shaped structure of mesencephalic brainstem surrounded by the highly echogenic basal cisterns in the axial scanning plane. The mesencephalic brainstem surrounded by the highly echogenic basal cisterns can easily be delineated in 90–95% of individuals, even in those with only partially sufficient acoustic bone windows. Within the brainstem, several structures of increased echogenicity, including SN, red nucleus, the midline raphe, and the aqueduct, can be visualized (Fig. 5).

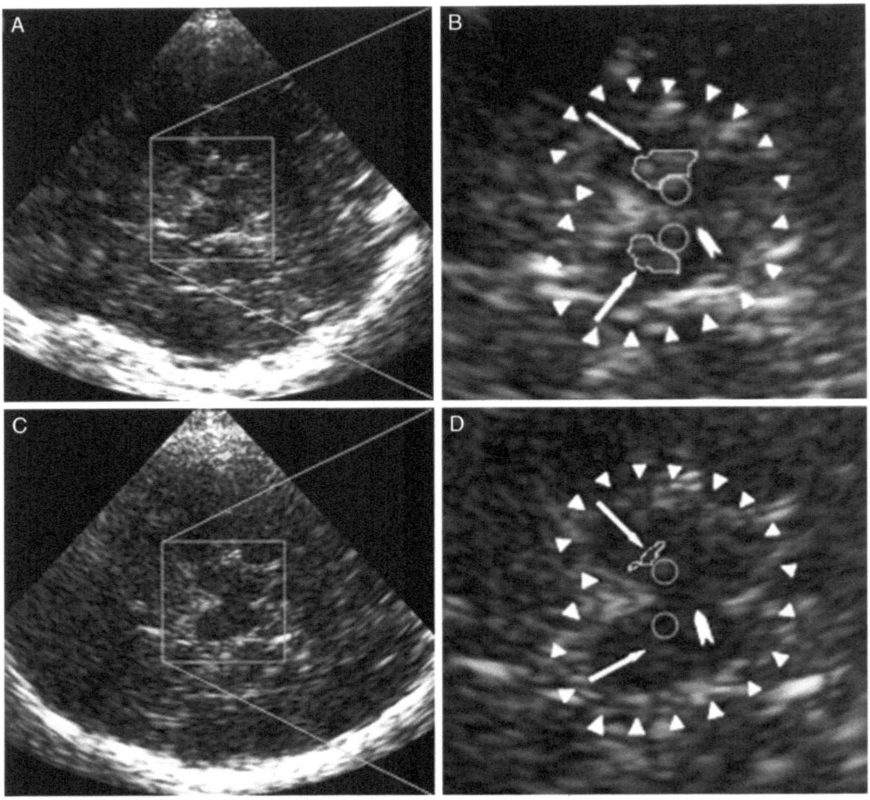

FIG. 5. Transcranial sonography (TCS) findings of midbrain structures. (A) TCS axial scan through the brain at midbrain level. It can easily be recognized that there is an increase of echogenic area ("hyperechogenicity") of the substanta nigra (SN) bilaterally while the brainstem raphe shows normal echogenicity. (B) Zoomed image corresponding to the TCS image shown in (A). The bilateral red nuclei are highlighted with circles for better recognition. The bilateral SN is surrounded for area measurements. Note that for standardized measurements only the SN ipsilateral to insonation should be measured according to consensus guidelines (Walter *et al.*, 2007a). The brainstem raphe (arrow head) is displayed as a continuous, highly echogenic line. (C) TCS axial scan through the brain at midbrain level in another subject. While the SN is displayed bilaterally with normal (low) echogenicity, the brainstem raphe is not visible despite the SN and the highly echogenic basal cisterns surrounding the midbrain are completely displayed. This finding refers to as brainstem raphe "hypoechogenicity" provided that it is seen this way on right-side and left-side transtemporal TCS. (D) Zoomed image corresponding to the TCS image shown in (C). The bilateral red nuclei are highlighted with circles for better recognition. The SN ipsilateral to insonation is surrounded for area measurement. The raphe is hypoechogenic (arrow head).

1. *Substantia Nigra*

For assessment of SN echogenicity, the patchy or tie-shaped echogenic signals superimposing the SN are visualized as clearly as possible. The assessment of SN echogenicity should always be performed ipsilaterally to insonation. This means that for bilateral SN echogenicity measurement TCS needs to be performed from both sides. Whereas quantification of SN echogenicity in terms of signal brightness is limited due to methodological problems, either semi-quantitative assessment using a scoring system (Table III) or, preferably, planimetric measurements of the SN echogenic area (Fig. 5) can be performed easily. Majority of working groups currently prefer the planimetric measurement since clear cut-off values are defined for this approach. Školoudík and co-workers (Ressner *et al.*, 2007; Školoudík *et al.*, 2007) proposed a semi-quantitative visual 5-grade scoring of SN echogenicity. A recent study of the same group (Bártová *et al.*, 2010) showed a good correlation between visually rated SN echogenicity and planimetrically measured SN area in PD patients, but this needs to be confirmed by independent groups.

By the current consensus guidelines (Walter *et al.*, 2007a), SN hyperechogenicity is defined as an increased size of the planimetrically measured extension of SN's echogenic signals compared with normal ranges. A marked SN hyperechogenicity is currently considered if the planimetrically measured echogenic area exceeds a predefined cut-off value defined by the 90% percentile of measures in normal population. A moderate SN hyperechogenicity is considered if the measured area ranges between 75 and 90% percentile of measures in normal population. For example, using the ultrasound system Sonoline Elegra (Siemens; Erlangen, Germany), SN echogenic sizes of less than 0.20 cm^2 are classified as normal since sizes of 0.20 cm^2 and above were found to represent upper 25% percentile in normal population. Sizes of 0.25 cm^2 and above, representing upper 10% percentile in normal population, are classified as markedly hyperechogenic, and sizes in-between as moderately hyperechogenic (Berg *et al.*, 1999a, 2001; Walter *et al.*, 2002, 2007a). A consensus definition of a reduced SN echogenicity,

TABLE III
SEMI-QUANTITATIVE VISUAL SCORING OF SUBSTANTIA NIGRA (SN) ECHOGENICITY AS PROPOSED BY ŠKOLOUDÍK *ET AL.* (2007)

Score	Description
Grade I	SN shows the same echogenicity as surrounding brainstem
Grade II	Very low but clearly detectable echogenic signal of SN
Grade III	Medium SN echogenicity, lower than perimesencephalic cisterns
Grade IV	SN shows the same echogenicity as perimesencephalic cisterns
Grade V	SN shows higher echogenicity than perimesencephalic cisterns

referred to as "hypoechogenicity", is still lacking. In few studies published by two groups, SN hypoechogenicity was defined as an SN echogenic area ranging below the 10th percentile of measures obtained in normal population (Godau *et al.*, 2007, 2008; Schmidauer *et al.*, 2005). Because of the limitation of image resolution when very small echogenic signals of SN are present, measures of SN echogenic sizes are normally given as sum values of bilateral measurements in case of SN hypoechogenicity. Godau *et al.* (2007) proposed bilateral sum values of less than 0.20 cm^2 obtained with the ultrasound system Sonoline Elegra (Siemens; Erlangen, Germany) as representing SN hypoechogenicity. It should be stressed that for accurate measurements of SN echogenic area, the mesencephalic brainstem should be, after image freezing, zoomed two to fourfold. Then the SN echogenic signals are surrounded manually with the cursor and finally the size of SN echogenic area is calculated automatically by the system (Fig. 5).

The examiner should be aware of a few sources of imaging artifacts that may sometimes cause difficulties in the discrimination of the SN. First, the red nucleus is sometimes highly echogenic and may be misinterpreted as SN hyperechogenicity. To overcome this problem, not only the lower but also the upper midbrain should be scanned; in case of SN hyperechogenicity, this should also appear at the upper midbrain level while RN echosignals would disappear. Another source of errors can be hyperechogenic penetrating arteries or reverberation echoes of the basal cisterns surrounding the mesencephalic brainstem.

2. *Red Nucleus*

The red nucleus is rated visually using a semi-quantitative score. Normally, the red nucleus can only be discerned on TCS by dot-like echosignals at the lateral border of the red nucleus near the SN, and/or the medial border of the red nucleus near the brainstem midline raphe. In case of a signal increase ("hyperechogenicity"), the red nucleus shows high echogenicity at its whole anatomic extension. While reduced echogenicity of red nucleus has not been described as a pathological condition, its hyperechogenicity has been associated with restless legs syndrome (Godau *et al.*, 2008). TCS assessment of red nucleus plays a minor role in the evaluation of movement disorders. But the clear differentiation between red nucleus and SN, which can sometimes be challenging, is essential for the correct assessment of SN echogenicity.

3. *Brainstem Raphe*

Echogenicity of brainstem raphe is rated semi-quantitatively using the highly echogenic red nucleus or signal intensity of the basal cisterns as reference points (Becker *et al.*, 1994b, 1997; Berg *et al.*, 1999b). In 90–95% of the normal population, the brainstem raphe is depicted as a highly echogenic line with an echogenicity identical to that of the red nucleus (Walter *et al.*, 2007d). In 5–10% of healthy population but 50–70% of depressive subjects a reduced echogenicity of brainstem raphe can be detected with TCS (Becker *et al.*, 1994b, 1997; Berg *et al.*, 1999b; Walter *et al.*, 2007d). At reduced echogenicity, the echogenic line of the brainstem raphe is interrupted or not visible, i.e., isoechogenic compared with the adjacent brainstem parenchyma (Fig. 5). A finer grading differentiating between slightly reduced echogenicity (i.e., brainstem raphe interrupted) and markedly reduced echogenicity (i.e., brainstem raphe not visible), as used in early reports (Becker *et al.*, 1994b, 1997), is no longer recommended (Walter *et al.*, 2007a). Generally, the brainstem raphe should be evaluated by bilateral TCS investigation. Brainstem raphe echogenicity should be rated as reduced only if it appears interrupted (or not visible) at both sides of investigation.

B. Structures at the Cerebellum Level (Posterior-Fossal Plane)

Visualizing the posterior fossa is started from the mesencephalic plane (midbrain level, see Section VI. A). Structures of the posterior fossa can be visualized optimally after twisting the probe with the dorsal parts downward for about 45° and tilted in an upward direction for about 10–15°. This creates a plane in which the mesencephalon and the thalamus can be seen simultaneously (Fig. 6). Then

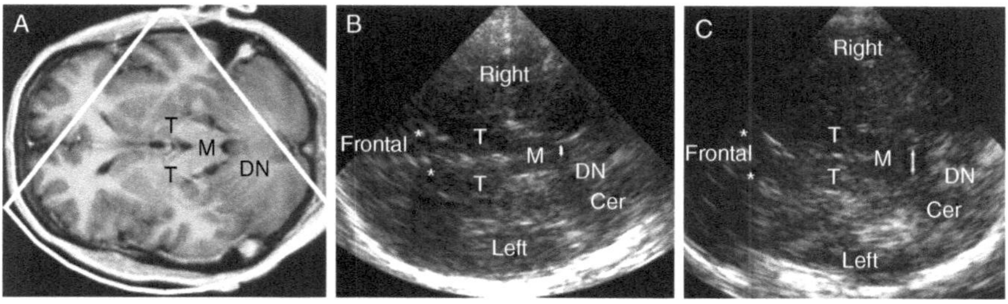

FIG. 6. Transcranial sonography (TCS) findings of cerebellar structures. Cer indicates cerebellum; DN, dentate nucleus; M, midbrain; T, thalamus; *, frontal horn of lateral ventricle. (A) Magnetic resonance image (MRI) corresponding to the TCS images shown in (B) and (C). (B) TCS semi-axial scan through the brain at cerebellar level of a subject without cerebellar atrophy. Note that at this scan, the midbrain and the thalami are displayed within one plane. At this plane, directly dorsal to the midbrain, the fourth ventricle can be displayed and its width can be measured (double arrow). The dentate nucleus is displayed with low echogenicity. (C) TCS semi-axial scan through the brain at cerebellar level of a subject with cerebellar atrophy. The fourth ventricle is enlarged (double arrow), and the dentate nucleus is displayed with increased echogenicity.

the probe should be tilted forward for about 10–30° so that the cerebellum comes more into the center of the display. In patients with advanced spinocerebellar atrophy, the echogenicity of both the cerebellar white matter and the dentate nucleus appears to be brighter compared with healthy controls (Postert *et al.*, 2004). In patients with cerebellar atrophy, a hyperechogenic dentate nucleus can be often identified, bilaterally adjacent to the hypoechogenic fourth ventricle (Postert *et al.*, 2004). Hyperechogenicity of dentate nucleus has also been found in patients with mitochondrial encephalopathy with lactacidosis and stroke-like episodes (MELAS) (Lukas *et al.*, 2009). The fourth ventricle can be displayed usually as a comma-shaped structure if there is no cerebellar atrophy. If a relevant cerebellar atrophy is present, the enlarged fourth ventricle can be identified as a hypo- to anechogenic circular structure in the immediate dorsal vicinity of the mesencephalic brainstem. Here the diameter as well as the area can be measured (Fig. 6).

C. Structures at the Thalamus Level

Starting again at the axial mesencephalic plane (midbrain level), the section through the thalami can be reached by tilting the ultrasound probe for 10–20° in the upward direction. An important landmark of the thalamus level is the usually calcified and therefore highly echogenic pineal gland (Fig. 7). At this plane, the third ventricle, anterior horns of the lateral ventricles, the thalami, and the anatomic site of the basal ganglia are depicted. The thalami are typically displayed as hypoechogenic oval structures; the thalami and the frontal horns help to discern the anatomical site of the caudate nucleus and the lenticular nucleus. At this level, the transverse diameters of the third ventricle and of the frontal horn of the contralateral lateral ventricle can be measured. Furthermore, echogenicities of contralateral thalamus, contralateral lenticular nucleus, and contralateral caudate nucleus should be evaluated semi-quantitatively. Normally, these structures are indistinguishable, i.e., isoechogenic to the surrounding brain parenchyma. Sometimes, the borders of the ipsilateral internal capsule can be detected, allowing a separation of the thalamus from the lenticular nucleus. An increased echogenicity ("hyperechogenicity") of thalamus, lenticular nucleus, or caudate nucleus compared with the surrounding white matter is considered to be abnormal. For further quantification, the size of echogenic area of a structure, especially the lenticular nucleus, can be measured similarly to measurement of SN echogenic size. Table IV summarizes the diagnostically relevant echogenicity changes of basal ganglia.

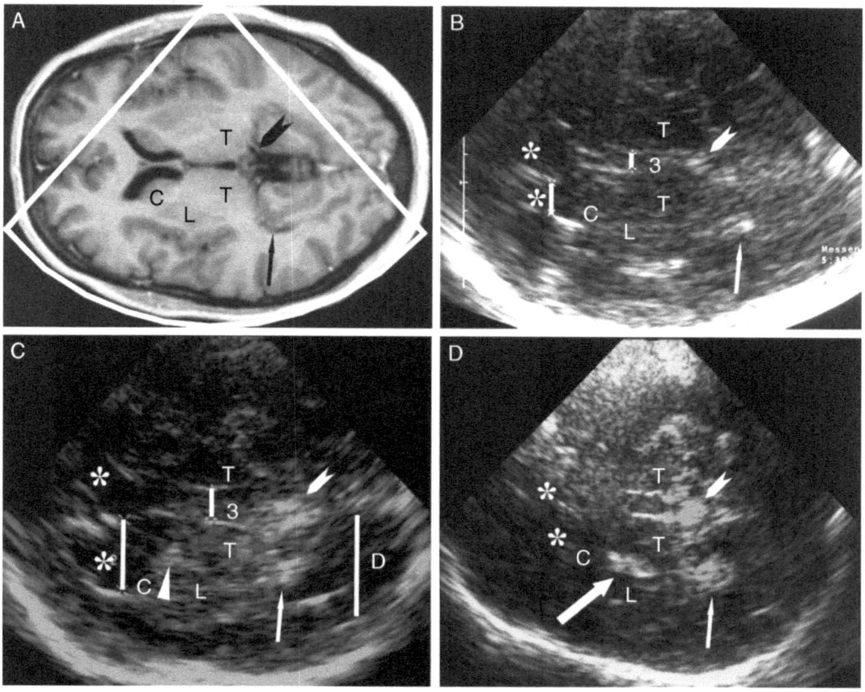

FIG. 7. Transcranial sonography (TCS) findings of ventricles and basal ganglia. C indicates caudate nucleus; D, dorsal horn of lateral ventricle, L, lenticular nucleus; T, thalamus; 3, third ventricle; *, frontal horn of lateral ventricle; arrow, chorioid plexus of posterior part of lateral ventricle; arrow head: pineal gland. (A) Magnetic resonance image (MRI) corresponding to the TCS images shown in (B), (C), and (D). (B) TCS axial scan through the brain at thalamus level of a subject with normal findings. The increased echogenicity of pineal gland (arrow head) and chorioid plexus are normal since these structures usually contain chalk. The thalami and lenticular and caudate nuclei are displayed with low (normal) echogenicity. The widths of ventricles are normal. (C) TCS axial scan through the brain at thalamus level of a patient with advanced Parkinson's disease. Dilatation of the ventricles indicates increased risk of dementia, while increased echogenicity ("hyperechogenicity") of the caudate nucleus (triangle) indicates increased risk of psychosis. (D) TCS axial scan through the brain at thalamus level of a patient with multi-system atrophy. The increased echogenicity ("hyperechogenicity") of the lenticular nucleus (arrow) is a characteristic finding.

1. *Ventricles*

Frontal (anterior) horn, middle part, and posterior horn of lateral ventricles with chorioidal plexus, third and fourth ventricles can be visualized on TCS, adapting the probe position accordingly. At the thalamus plane, the widths of third ventricle and of contralateral frontal horn as parts of the ventricle system can usually be measured. Measurements are performed if the frozen image was zoomed two to threefold. To ensure an accurate and reproducible measurement

TABLE IV
CHARACTERISTIC ECHOGENICITY CHANGES OF DEEP BRAIN STRUCTURES IN MOVEMENT DISORDERS

TCS abnormality	Related condition	Discriminated condition
SN hyperechogenicity	Parkinson's disease	Atypical parkinsonian syndromes
	Corticobasal degeneration	Progressive supranuclear palsy
	Dementia with Lewy bodies	Alzheimer dementia?
SN hypoechogenicity	Restless legs syndrome	
BR hypoechogenicity	Unipolar depression	Bipolar affective disorders
LN hyperechogenicity	Idiopathic dystonia	Psychogenic dystonia
	Atypical parkinsonian syndrome	Parkinson's disease
	Wilson's disease	Parkinson's disease
CN hyperechogenicity	Huntington's disease	Other choreatic disorders?

of ventricle widths, measurements are performed from the ipsilateral to the contralateral inner layers of the hyperechogenic ependyma. We prefer the measurement of the minimal width of third ventricle, and measurement of contralateral frontal-horn widths at the site of fusion of both frontal horns. Other groups perform the measurement of the largest detectable width of the third ventricle (Seidel *et al.*, 1995) whereas the exact site of frontal horn measurement has not been described in detail before. If the local standard of ventricle measurement sites has been defined in an ultrasound lab, TCS measures of ventricle widths correlate well with CT and MRI measures. The correlation is higher for the third ventricle ($r = 0.83 - 0.85$) than for the anterior horns of the lateral ventricles ($r = 0.73$), which may be attributed to the stronger influence of the angulation of the probe (Hernández *et al.*, 2007; Seidel *et al.*, 1995). Age-related normal values of the third ventricle are summarized in Table V. At the thalamus plane, the posterior horn of the lateral ventricle and the embedded highly echogenic choroid plexus can also be visualized (Fig. 7).

TABLE V
AGE-RELATED NORMAL VALUES OF WIDTHS OF THIRD VENTRICLE

	Age	
	<60 years (mean age 40 years)	>60 years (mean age 70 years)
Transcranial sonography	3.14 ± 1.92 mm[a]	6.25 ± 2.91 mm[a]
	4.8 ± 1.9 mm[b]	7.6 ± 2.1 mm[b]
Magnetic resonance	3.16 ± 1.97 mm[a]	5.2 ± 2.51 mm[a]

[a] Published by Hernández *et al.* (2007)
[b] Published by Seidel *et al.* (1995)

2. *Thalamus*

Usually, only the echogenicity of the contralateral thalamus is assessed semi-quantitatively. Normally, the thalami are of low echogenicity. A focal or diffuse increase of echogenicity ("hyperechogenicity") may be considered abnormal. However, thalamus hyperechogenicity has not been described as a diagnostic relevant finding. Diffuse hyperechogenicity is seen sometimes in patients with multiple sclerosis (Walter *et al.*, 2009).

3. *Lenticular Nucleus*

Increased echogenicity ("hyperechogenicity") of the medial parts of the lenticular nucleus as compared to the surrounding brain parenchyma has been first described by Naumann *et al.* (1996) in patients with idiopathic dystonia. Subsequently, this TCS abnormality was found to be frequently present also in atypical parkinsonian syndromes (Behnke *et al.*, 2005; Walter *et al.*, 2003, 2004), Wilson's disease (Walter *et al.*, 2005), welding-related parkinsonism (Walter *et al.*, 2008a), Creutzfeld–Jacob disease (Müller *et al.*, 2008), and multiple sclerosis (Walter *et al.*, 2009). This TCS abnormality can be assessed either visually, simply describing the finding as dot-like or diffuse lenticular nucleus hyperechogenicity (Fig. 7D), or by planimetric measurement of the extent of the hyperechogenic signals of the lenticular nucleus.

4. *Caudate Nucleus*

Caudate nucleus hyperechogenicity, i.e., an increase of caudate nucleus echogenicity compared to the surrounding brain parenchyma (Fig. 7C), was frequently seen in Huntington's disease (Postert *et al.*, 1999) and in advanced stages of Parkinson's disease (Walter *et al.*, 2007c). In Parkinson's disease, this finding was related to an increased risk of psychosis independent from disease duration (Walter *et al.*, 2007c).

D. Cella-Media Level

The cella-media level is approached from the thalamus level if the probe is tilted more cranially with a scanning angle of about 25–30°. At this level, the middle part of the contralateral lateral ventricle (cella media) between the inner layers of the ependyma can be visualized and its width can be measured.

VII. Transfrontal Investigation

In 20–30% of Caucasian adults, a sufficient transfrontal acoustic bone window for diagnostic insonation can be found (Fig. 8). In children, this bone window is sufficient in 30–80%, depending on age. Using this approach, the lateral ventricles, thalami, basal ganglia, corpus callosum, as well as the subcortical white matter of frontal and, partly, of temporal and occipital lobes can be visualized. This approach is the most useful for sonographic measurement of the thickness of the anterior part of corpus callosum (Walter *et al.*, unpublished). Figure 8 shows the typical appearance of the corpus callosum on TCS.

VIII. Causes of Hyperechogenicity of Deep Brain Structures

Generally, the ultrasound echogenicity of any structure is dependent upon the acoustic impedance and the difference of this acoustic impedance between neighboring tissues since the ultrasound waves are reflected at tissue interfaces with diverse acoustic impedance. Enhanced echogenicity of a distinct deep brain structure such as the brainstem midline raphe may result from a specific composition of neuronal, glial, and fiber components. Also, the accumulation of heavy metals such as iron, copper, or manganese may cause increased echogenicity of a certain brain

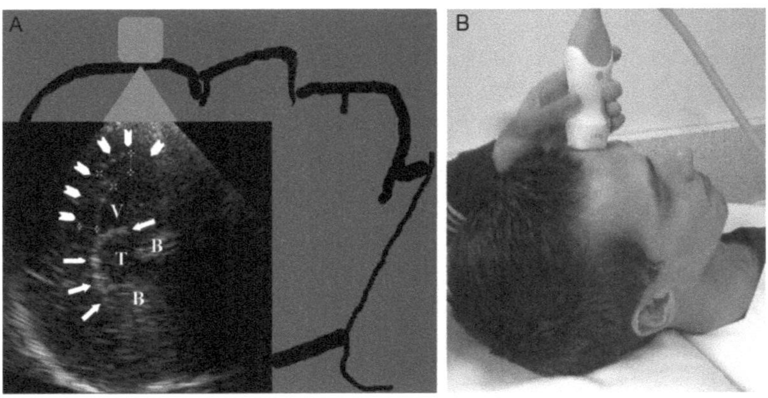

FIG. 8. Transcranial sonography (TCS) of corpus callosum (CC) through the frontal bone. (A) TCS image embedded in a schematic illustration of the anatomic relationship of the brain structures visualized. B indicates basal cisterns surrounding the midbrain; T, thalamus; V, lateral ventricle; arrows, chorioid plexus; arrow heads, CC. The thickness of the CC is measured at three different localisations (measure points). (B) Photograph showing the probe position for TCS of the CC.

structure (Berg *et al.*, 2008). Changes of heavy metal content seem to be at least partly responsible for changes of echogenicity at the SN and the red nucleus and for hyperechogenicity of the lenticular nucleus (dystonia, Wilson's disease).

The pathophysiologic conditions underlying the changes of SN echogenicity are still not fully explained. It is not known why the TCS finding of SN hyperechogenicity appears to be stable during the course of Parkinson's disease even though there is progressive degeneration of SN neurons, which is accompanied by gliosis and further iron accumulation (Berg *et al.*, 2005).

SN hyperechogenicity is a typical sonographic feature of idiopathic PD (Becker *et al.*, 1995; Berg *et al.*, 2001; Huang *et al.*, 2007; Kim *et al.*, 2007; Kolevski *et al.*, 2007; Okawa *et al.*, 2007; Ressner *et al.*, 2007; Schmidauer *et al.*, 2005; Walter *et al.*, 2002). On the other hand, a hypoechogenic SN has primarily been demonstrated in restless legs syndrome (Godau *et al.*, 2007, 2008; Schmidauer *et al.*, 2005). Both disorders have been associated with opposed alterations of iron content of SN. The idea that SN echogenicity is significantly dependent on its iron content was supported by the results of several animal and postmortem studies, which demonstrated a correlation of increased tissue iron concentration and SN hyperechogenicity (Berg *et al.*, 1999a,c, 2002, 2006; Zecca *et al.*, 2005). This, together with an apparently autosomal dominant inheritance of this echofeature in relatives of patients with idiopathic PD (Ruprecht-Dörfler *et al.*, 2003), supports the idea of a primary role of disturbed iron metabolism in PD. Subsequent mutation analyses of genes encoding for iron-transport and iron-regulatory proteins known to be associated with parkinsonism led to the discovery of specific mutations in the *ferritin-H*, the *iron-regulatory protein 2*, and the *hemochromatosis* gene, respectively, in single PD patients with SN hyperechogenicity (Akbas *et al.*, 2006; Deplazes *et al.*, 2003; Felletschin *et al.*, 2003). Moreover, it could be shown that mutations in the *ceruloplasmin* gene were clearly associated with the TCS finding of SN hyperechogenicity in PD patients and healthy control subjects (Hochstrasser *et al.*, 2004). Nevertheless, it is still possible that other factors contribute to SN hyperechogenicity such as abnormal iron–protein bindings, gliosis, and structural changes of neurons or glial cells (atrophy, morphological changes of cells) in the SN.

IX. Limitations of TCS in Movement Disorders

A. Insufficient Bone Window

The main limitation of TCS is its dependency on the quality of the temporal acoustic bone window, accounting for only partial or even missing assessability of brainstem structures in 5–10% of Caucasian subjects (Behnke *et al.*, 2007;

Gaenslen *et al.*, 2008; Godau *et al.*, 2007; Školoudík *et al.*, 2003; Walter *et al.*, 2007a). The number of patients with insufficient bone windows is influenced by age, gender, and ethnicity (Behnke *et al.*, 2007; Gaenslen *et al.*, 2008; Godau *et al.*, 2007; Huang *et al.*, 2007; Kim *et al.*, 2007; Okawa *et al.*, 2007; Tsai *et al.*, 2007; Walter *et al.*, 2007a). It has been demonstrated that especially in elderly women bone windows may often be insufficient to allow full assessment of brainstem structures probably due to osteoporosis (Školoudík *et al.*, 2003; Walter *et al.*, 2007a).

In Caucasian subjects, the sufficient temporal bone window for a minimum partial assessment of brainstem structures is available in 5–10% and of basal ganglia in 10–20% of patients (Walter *et al.*, 2007a). On the other hand, in Asian subjects, an insufficiency of temporal bone windows occurs in 15–60% of patients (Huang *et al.*, 2007; Kim *et al.*, 2007; Okawa *et al.*, 2007; Tsai *et al.*, 2007). No data on the quality of temporal bone windows for TCS in African subjects are available yet.

B. Quality of Ultrasound System

Another limitation of TCS is the dependency on the standard of the ultrasound system as in other sonographic methods. In general, the present-day high-end ultrasound systems of many manufacturers provide a high quality that allows reliable assessments of small echogenic brain structures. Still, as outlined earlier in this chapter, the users are required to generate their own reference values, which account only for their own laboratory and system (Berg *et al.*, 2008). Several studies demonstrated considerable differences of mean values of measured SN echogenic areas. The measurements may vary between two laboratories even when the same ultrasound systems with identical system settings measurements are used (Behnke *et al.*, 2005). These minor differences can be important not only for multicenter scientific investigations, but may also affect the daily routine examinations.

C. Investigator Qualification and Experience

A third limitation of all sonographic methods including TCS is its dependency on the experience and skills of the investigator (Berg *et al.*, 2008; Bogdahn *et al.*, 1998; Kollar *et al.*, 2004; Školoudík *et al.*, 2003, 2007; Walter *et al.*, 2007a).

To ensure adequate reliability of TCS results, the investigator needs to be well trained. In the experience of the groups of Berg and co-workers (2008) and of Walter and co-workers, an examiner who is experienced in other fields of B-mode or cerebrovascular ultrasound can be trained in TCS within 1–3 weeks. A completely inexperienced beginner may need up to 8 weeks to achieve sufficient skill. To allow broader application in the future, current research focuses on technological advances that may solve the problem of investigator dependency.

In spite of dependency on sufficient imaging conditions and adequate qualification of the sonographer, a number of studies have demonstrated high inter-rater reliability of sonographic assessment of midbrain structures and ventricular widths (Becker and Berg, 2001; Berg *et al.*, 1999a; Puls *et al.*, 2000; Seidel *et al.*, 1995; Školoudík *et al.*, 2007; Walter *et al.*, 2007a).

X. Validity and Reproducibility of TCS in Movement Disorders

Several studies showed that TCS is a well-reproducible method and has a high specificity and sensitivity for the diagnosis of PD (Berg *et al.*, 1999a, 2002; Gaenslen *et al.*, 2008; Prestel *et al.*, 2006; Ressner *et al.*, 2007; Walter *et al.*, 2007b, 2007d). Correlations of SN evaluation were statistically significant ($r = 0.8$, Cohen's kappa coefficient 0.83) (Becker and Berg, 2001; Berg *et al.*, 1999a; Walter *et al.*, 2007d).

The study of Školoudík *et al.* (2007) investigated the intra- and inter-investigator as well as the intra- and inter-reader reproducibility of TCS evaluation of SN echogenicity and area measurement. In this study, 22 movement disorder patients were examined twice within 2 weeks by an experienced physician-sonographer. A third examination was performed by a technical assistant experienced in Doppler sonography of brain-supplying vessels who was blinded to the results of the physician-sonographer. All images were stored and analyzed again offline in a blinded setting by the physician-sonographer, technical assistant and a physician without sonographic experience. All intra-investigator and inter-investigator correlations of planimetric SN area measurements ($r = 0.69$–0.88 and $r = 0.5$–0.61) and visual scoring of SN echogenicity ($r = 0.64$–0.92 and 0.51–0.69) were statistically significant ($P < 0.05$). However, intra-reader correlations were significant only for the experienced physician-sonographer ($r = 0.85$–0.96 for SN echogenicity and 0.51–0.69 for SN area, $P < 0.001$). The study confirmed that semi-quantitative TCS evaluation of SN echogenicity and measurement of SN area are highly dependent on the sonographer's experience. In a recent study testing the intra- and inter-rater reliability of SN echogenicity measurements with four highly experienced investigators, a very high diagnostic accuracy and excellent intra (ICC = 0.96–0.93) and interindividual (ICC 0.84–0.89) reproducibility for planimetric SN area measurements was demonstrated (van de Loo *et al.*, in press).

XI. Conclusions

TCS has proven to be reliable and sensitive in detecting abnormalities of very small, echogenic deep brain structures that were not seen before with other neuroimaging methods. Especially with present-generation high-end ultrasound systems, a very high image resolution on TCS of structures located near the midline is achieved, which is today even fairly higher than with MRI under clinical conditions. The specific advantages of TCS, mobility, low interference from patients' movements, short investigation times, and unlimited repeatability support its application especially in patients with movement disorders. Adequate qualification of the investigators is an indispensable condition, which needs a structured training program and the continuing attention of the medical societies. The ongoing technological progress will further enhance the clinical use of TCS for the early diagnosis and monitoring the progression of movement disorders.

References

Aaslid, R., Eden, A., and Ringelstein, E. B. eds. (1986). Transcranial Doppler Sonography. Springer, Wien.

Aaslid, R., Markwalder, T. M., and Nornes, H. (1982). Noninvasive transcranial Doppler ultrasound recording of flow velocity in basal cerebral arteries. *J. Neurosurg.* **57**, 769–774.

Akbas, N., Hochstrasser, H., Deplazes, J., Tomiuk, J., Bauer, P., Walter, U., Behnke, S., Riess, O., and Berg, D. (2006). Screening for mutations of the HFE gene in Parkinson's disease patients with hyperechogenicity of the substantia nigra. *Neurosci. Lett.* **407**, 16–19.

Bartels, E., ed. (1999). Color–Coded Duplex Sonography of the Cerebral Vessels: Atlas and Manual. Schattauer, Stuttgart–New York.

Bártová, P., Skoloudík, D., Ressner, P., Langová, K., Herzig, R., and Kanovsky, P. (2010). Correlation between substantia nigra features detected by sonography and Parkinson disease symptoms. *J. Ultrasound Med.* **29**, 37–42.

Becker, T., Becker, G., Seufert, J., Hofmann, E., Lange, K. W., Naumann, M., Lindner, A., Reichmann, H., Riederer, P., Beckmann, H., and Reiners, K. (1997). Parkinson's disease and depression: Evidence for an alteration of the basal limbic system detected by transcranial sonography. *J. Neurol. Neurosurg. Psychiatr.* **63**, 590–596.

Becker, G., and Berg, D. (2001). Neuroimaging in basal ganglia disorders: Perspectives for transcranial ultrasound. *Mov. Disord.* **16**, 23–32.

Becker, G., Bogdahn, U., Strassburg, H. M., Lindner, A., Hassel, W., Meixensberger, J., and Hofmann, E. (1994a). Identification of ventricular enlargement and estimation of intracranial pressure by transcranial color-coded real-time sonography. *J. Neuroimmunol.* **4**, 17–22.

Becker, G., Perez, J., Krone, A., Demuth, K., Lindner, A., Hofmann, E., Winkler, J., and Bogdahn, U. (1992). Transcranial, color-coded, real-time sonography in the evaluation of intracranial neoplasmas and arteriovenous malformations. *Neurosurgery* **31**, 420–428.

Becker, G., Seufert, J., Bogdahn, U., Reichmann, H., and Reiners, K. (1995). Degeneration of substantia nigra in chronic Parkinson's disease visualized by transcranial color-coded real-time sonography. *Neurology* **45**, 182–184.

Becker, G., Struck, M., Bogdahn, U., and Becker, T. (1994b). Echogenicity of the brainstem raphe in patients with major depression. *Psychiatr. Res.* **55**, 75–84.

Behnke, S., Berg, D., Naumann, M., and Becker, G. (2005). Differentiation of Parkinson's disease and atypical parkinsonian syndromes by transcranial ultrasound. *J. Neurol. Neurosurg. Psychiatr.* **76**, 423–425.

Behnke, S., Double, K. L., Duma, S., Broe, G. A., Guenther, V., Becker, G., and Halliday, G. M. (2007). Substantia nigra echomorphology in the healthy very old: Correlation with motor slowing. *NeuroImage* **34**, 1054–1059.

Berg, D., Becker, G., Zeiler, B., Tucha, O., Hofmann, E., Preier, M., Benz, P., Jost, W., Reiners, K., and Lange, K. W. (1999a). Vulnerability of the nigrostriatal system as detected by transcranial ultrasound. *Neurology* **53**, 1026–1031.

Berg, D., Godau, J., and Walter, U. (2008). Transcranial sonography in movement disorders. *Lancet Neurol.* **7**, 1044–1055.

Berg, D., Grote, C., Rausch, W. D., Mäurer, M., Wesemann, W., Riederer, P., and Becker, G. (1999c). Iron accumulation of the substantia nigra in rats visualized by ultrasound. *Ultrasound Med. Biol.* **25**, 901–904.

Berg, D., Hochstrasser, H., Schweitzer, K. J., and Riess, O. (2006). Disturbance of iron metabolism in Parkinson's disease—Ultrasonography as a biomarker. *Neurotox. Res.* **9**, 1–13.

Berg, D., Merz, B., Reiners, K., Naumann, M., and Becker, G. (2005). Five-year follow-up study of hyperechogenicity of the substantia nigra in Parkinson's disease. *Mov. Disord.* **20**, 383–385.

Berg, D., Roggendorf, W., Schröder, U., Klein, R., Tatschner, T., Benz, P., Tucha, O., Preier, M., Lange, K. W., Reiners, K., Gerlach, M., and Becker, G. (2002). Echogenicity of the substantia nigra: Association with increased iron content and marker for susceptibility to nigrostriatal injury. *Arch. Neurol.* **59**, 999–1005.

Berg, D., Siefker, C., Ruprecht-Dörfler, P., and Becker, G. (2001). Relationship of substantia nigra echogenicity and motor function in elderly subjects. *Neurology* **56**, 13–17.

Berg, D., Supprian, T., Hofmann, E., Zeiler, B., Jäger, A., Lange, K. W., Reiners, K., Becker, T., and Becker. G. (1999b). Depression in Parkinson's disease: Brainstem midline alteration on transcranial sonography and magnetic resonance imaging. *J. Neurol. Neurosurg. Psychiatr.* **246**, 1186–1193.

Berland, L. L., Bryan, C. R., Sekar, B. C., and Moss, C. N. (1988). Sonographic examination of the adult brain. *J. Clin. Ultrasound* **16**, 337–345.

Bogdahn, U., Becker, G., and Schlachetzki, F. eds. (1998). Echoenhancers and Transcranial Color Duplex Sonography. Blackwell Science, Berlin, Vienna.

Deplazes, J., Schöbel, K., Hochstrasser, H., Bauer, P., Walter, U., Behnke, S., Spiegel, J., Becker, G., Riess, O., and Berg, D. (2003). Screening for mutations of the IRP2 gene in Parkinson's disease patients with hyperechogenicity of the substantia nigra. *J. Neuro Oncol.* **111**, 515–521.

Felletschin, B., Bauer, P., Walter, U., Behnke, S., Spiegel, J., Csoti, I., Sommer, U., Zeiler, B., Becker, G., Riess, O., and Berg, D. (2003). Screening for mutations of the ferritin light and heavy genes in Parkinson's disease patients with hyperechogenicity of the substantia nigra. *Neurosci. Lett.* **352**, 53–56.

Furuhata, H. (1989). New evolution of transcranial tomography (TCT) and transcranial color Doppler tomography (TCDT). *Neurosonology* **1**, 8–15.

Gaenslen, A., Unmuth, B., Godau, J., Liepelt, I., Di Santo, A., Schweitzer, K. J., Gasser, T., Machulla, H. J., Reimold, M., Marek, K., and Berg, D. (2008). The specificity and sensitivity of transcranial ultrasound in the differential diagnosis of Parkinson's disease: A prospective blinded study. *Lancet Neurol.* **7**, 417–424.

Glaser, M., Weber, U., Hinrichs, H., and Niehaus, L. (2006). Neuroimaging of midbrain structures with different phased-array ultrasound systems. *Klin. Neurophysiol.* **37**, 165–168.

Godau, J., Schweitzer, K. J., Liepelt, I., Gerloff, C., and Berg, D. (2007). Substantia nigra hypoechogenicity: Definition and findings in restless legs syndrome. *Mov. Disord.* **22**, 187–192.

Godau, J., Wevers, A. K., Gaenslen, A., Di Santo, A., Liepelt, I., Gasser, T., and Berg, D. (2008). Sonographic abnormalities of brainstem structures in restless legs syndrome. *Sleep Med.* **9**, 782–789.

Hernández, N. L., Escrivá, A. G., and Jordà, J. M. (2007). Study of the diameter of the third ventricle with transcranial sonography. *Neurologia* **22**, 507–510.

Hochstrasser, H., Bauer, P., Walter, U., Behnke, S., Spiegel, J., Csoti, I., Zeiler, B., Bornemann, A., Pahnke, J., Becker, G., Riess, O., and Berg, D. (2004). Ceruloplasmin gene variations and substantia nigra hyperechogenicity in Parkinson disease. *Neurology* **63**, 1912–1917.

Huang, Y. W., Jeng, J. S., Tsai, C. F., Chen, L. L., and Wu, R. M. (2007). Transcranial imaging of substantia nigra hyperechogenicity in a Taiwanese cohort of Parkinson's disease. *Mov. Disord.* **22**, 550–555.

Kamphuisen, H. A., Somer, J. C., and Oosterbaan, W. A. (1972). Two-dimensional echoencephalography with electronic sector scanning. Clinical experiences with a new method. *J. Neurol. Neurosurg. Psychiatr.* **35**, 912–918.

Kim, J. Y., Kim, S. T., Jeon, S. H., and Lee, W. Y. (2007). Midbrain transcranial sonography in Korean patients with Parkinson's disease. *Mov. Disord.* **22**, 1922–1926.

Kolevski, G., Petrov, I., and Petrova, V. (2007). Transcranial sonography in the evaluation of Parkinson disease. *J. Ultrasound Med.* **26**, 509–512.

Kollar, J., Schulte-Altedorneburg, G., Sikula, J., Fulesdi, B., Ringelstein, E. B., Mehta, V., Csiba, L., and Droste, D. W. (2004). Image quality of the temporal bone window examined by transcranial Doppler sonography and correlation with postmortem computed tomography measurements. *Cerebrovasc. Dis.* **17**, 61–65.

Lukas, Z., Slonkova, J., Soucek, O., and Bar, M. (2009). Histopathological diagnostics of mitochondrial myopathies: Encephalopathy MELAS with sarcoplasma inclusions. Book of Abstracts of 2nd Neuromuscular Congress, Bratislava, Slovak Republic, 33 [abstract].

Mehnert, S., Reuter, I., Schepp, K., Maaser, P., Stolz, E., and Kaps, M. (2010). Transcranial sonography for diagnosis of Parkinson's disease. *BMC Neurol.* **10**, 9.

Müller, M., Treumann, T., Hewer, E., and Stoeck, K. (2008). Transcranial ultrasound of the basal ganglia in sporadic Creutzfeldt–Jakob disease. *J. Neuroimmunol.* **18**, 154–157.

Naumann, M., Becker, G., Toyka, K. V., Supprian, T., and Reiners, K. (1996). Lenticular nucleus lesion in idiopathic dystonia detected by transcranial sonography. *Neurology* **47**, 1284–1290.

Okawa, M., Miwa, H., Kajimoto, Y., Hama, K., Morita, S., Nakanishi, I., and Kondo, T. (2007). Transcranial sonography of the substantia nigra in Japanese patients with Parkinson's disease or atypical parkinsonism: Clinical potential and limitations. *Intern. Med.* **46**, 1527–1531.

Postert, T., Eyding, J., Berg, D., Przuntek, H., Becker, G., Finger, M., and Schöls, L. (2004). Transcranial sonography in spinocerebellar ataxia type 3. *J. Neuro Oncol.* **68** (Suppl.), 123–133.

Postert, T., Lack, B., Kuhn, W., Jergas, M., Andrich, J., Braun, B., Przuntek, H., Sprengelmeyer, R., Agelink, M., and Büttner, T. (1999). Basal ganglia alterations and brain atrophy in Huntington's disease depicted by transcranial real-time sonography. *J. Neurol. Neurosurg. Psychiatr.* **67**, 457–462.

Prestel, J., Schweitzer, K. J., Hofer, A., Gasser, T., and Berg, D. (2006). Predictive value of transcranial sonography in the diagnosis of Parkinson's disease. *Mov. Disord.* **21**, 1763–1765.

Puls, I., Berg, D., Mäurer, M., Schliesser, M., Hetzel, G., and Becker, G. (2000). Transcranial sonography of the brain parenchyma: Comparison of B-mode imaging and tissue harmonic imaging. *Ultrasound Med. Biol.* **26**, 189–194.

Ressner, P., Skoloudik, D., Hlustik, P., and Kanovsky, P. (2007). Echogenicity of substantia nigra in Parkinson's disease—Pilot study. *J. Neuroimmunol.* **17**, 164–167.

Ruprecht-Dörfler, P., Berg, D., Tucha, O., Benz, P., Meier-Meitinger, M., Alders, G. L., Lange, K. W., and Becker, G. (2003). Echogenicity of the substantia nigra in relatives of patients with sporadic Parkinson's disease. *NeuroImage* **18**, 416–422.

Schmidauer, C., Sojer, M., Seppi, K., Stockner, H., Högl, B., Biedermann, B., Brandauer, E., Peralta, C. M., Wenning, G. K., and Poewe, W. (2005). Transcranial ultrasound shows nigral hypoechogenicity in restless legs syndrome. *Ann. Neurol.* **58**, 630–634.

Schreiber, J., Sojka, E., Licev, L., Sknourilova, P., Gaura, J., and Skoloudik, D. (2008). A new method for the detection of brain stem in transcranial ultrasound images. *Proc. Biosignals.* **2**, 478–483.

Seidel, G., Kaps, M., Gerriets, T., and Hutzelmann, A. (1995). Evaluation of the ventricular system in adults by transcranial duplex sonography. *J. Neuroimmunol.* **5**, 105–108.

Skoloudik, D., Skoda, O., Bar, M., Brozman, M., and Vaclavik, D. (2003). Neurosonologie. Galen, Prague.

Skoloudík, D., Fadrná, T., Bártová, P., Langová, K., Ressner, P., Zapletalová, O., Hlustík, P., Herzig, R., and Kannovský, P. (2007). Reproducibility of sonographic measurement of the substantia nigra. *Ultrasound Med. Biol.* **33**, 1347–1352.

Tsai, C. F., Wu, R. M., Huang, Y. W., Chen, L. L., Yip, P. K., and Jeng, J. S. (2007). Transcranial color-coded sonography helps differentiation between idiopathic Parkinson's disease and vascular parkinsonism. *J. Neurol. Neurosurg. Psychiatr.* **254**, 501–507.

Van de Loo, S., Walter, U., Behnke, S., Hagenah, J., Lorenz, M., Sitzer, M., Hilker, R., and Berg, D. (in press). Reproducibility and diagnostic accuracy of substantia nigra sonography for the diagnosis of Parkinson's disease. *J. Neurol. Neurosurg. Psychiatr.*

Walter, U., Behnke, S., Eyding, J., Niehaus, L., Postert, T., Seidel, G., and Berg, D. (2007a). Transcranial brain parenchyma sonography in movement disorders: State of the art. *Ultrasound Med. Biol.* **33**, 15–25.

Walter, U., Dressler, D., Lindemann, C., Slachevsky, A., and Miranda, M. (2008b). Transcranial sonography findings in welding-related Parkinsonism in comparison to Parkinson's disease. *Mov. Disord.* **23**, 141–145.

Walter, U., Dressler, D., Probst, T., Wolters, A., Abu-Mugheisib, M., Wittstock, M., and Benecke, R. (2007b). Transcranial brain sonography findings in discriminating between parkinsonism and idiopathic Parkinson disease. *Arch. Neurol.* **64**, 1635–1640.

Walter, U., Dressler, D., Wolters, A., Probst, T., Grossmann, A., and Benecke, R. (2004). Sonographic discrimination of corticobasal degeneration vs. progressive supranuclear palsy. *Neurology* **63**, 504–509.

Walter, U., Dressler, D., Wolters, A., Wittstock, M., and Benecke, R. (2007c). Transcranial brain sonography findings in clinical subgroups of idiopathic Parkinson's disease. *Mov. Disord.* **22**, 48–54.

Walter, U., Hoeppner, J., Prudente-Morrissey, L., Horowski, S., Herpertz, S. C., and Benecke, R. (2007d). Parkinson's disease-like midbrain sonography abnormalities are frequent in depressive disorders. *Brain* **130**, 1799–1807.

Walter, U., Kanowski, M., Kaufmann, J., Grossmann, A., Benecke, R., and Niehaus, L. (2008a). Contemporary ultrasound systems allow high-resolution transcranial imaging of small echogenic deep intracranial structures similarly as MRI: A phantom study. *NeuroImage* **40**, 551–558.

Walter, U., Krolikowski, K., Tarnacka, B., Benecke, R., Czlonkowska, A., and Dressler, D. (2005). Sonographic detection of basal ganglia lesions in asymptomatic and symptomatic Wilson disease. *Neurology* **64**, 1726–1732.

Walter, U., Niehaus, L., Probst, T., Benecke, R., Meyer, B. U., and Dressler, D. (2003). Brain parenchyma sonography discriminates Parkinson's disease and atypical parkinsonian syndromes. *Neurology* **60**, 74–77.

Walter, U., Wagner, S., Horowski, S., Benecke, R., and Zettl, U. K. (2009). Transcranial brain sonography findings predict disease progression in multiple sclerosis. *Neurology* **73**, 1010–1017.

Walter, U., Wittstock, M., Benecke, R., and Dressler, D. (2002). Substantia nigra echogenicity is normal in non-extrapyramidal cerebral disorders but increased in Parkinson's disease. *J. Neuro Oncol.* **109**, 191–196.

Zecca, L., Berg, D., Arzberger, T., Ruprecht, P., Rausch, W. D., Musicco, M., Tampellini, D., Riederer, P., Gerlach, M., and Becker, G. (2005). *In vivo* detection of iron and neuromelanin by transcranial sonography: A new approach for early detection of substantia nigra damage. *Mov. Disord.* **20**, 1278–1285.

TRANSCRANIAL SONOGRAPHY—ANATOMY

Heiko Huber

Department of Neurodegeneration, Hertie Institute of Clinical Brain Research and German Center of Neurodegnerative Diseases (DZNE), Tübingen, Germany

Transcranial B-Mode sonography (TCS) allows quick, reliable, and inexpensive depiction of a number of brain structures, which may help in the diagnosis and differentiation of various movement disorders. In the following sections the anatomical structures in three standardized TCS planes (Section I) and the methods of measurement and typical pathological findings in specific structures (Section II) will be described. For better orientation, TCS images are complemented by compatible MRI images.

I. Three Standardized Planes

For the investigation of patients with movement disorders, it is important to know three standardized planes of B-Mode sonography.

The investigation is started by placing the ultrasound probe at the middle or the posterior temporal acoustic bone window (Fig. 1) in parallel to the imagined orbito-meatal line (Fig. 1, OM). By doing so, one usually visualizes the plane of the mesencephalic brainstem (Fig. 1, A1). By tilting the probe slightly upward for about 10–20° the ventricular plane is reached (Fig. 1, A2). These are the two standard planes for routine TCS in a movement disorder patient. In previous papers, the cella media plane has been mentioned as well. However, this does not

INTERNATIONAL REVIEW OF NEUROBIOLOGY, VOL. 90
DOI: 10.1016/S0074-7742(10)90003-2

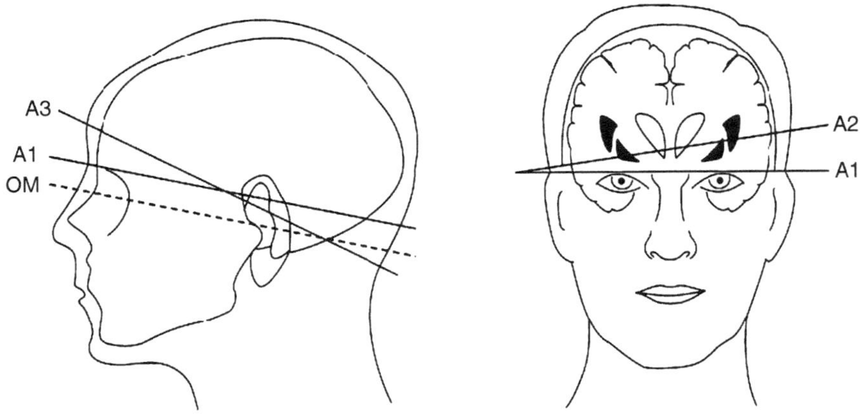

FIG. 1. Standardized planes of B-Mode sonography. OM: orbitomeatal line, A1: mesencephalic plane, A2: third ventricular plane, A3: cerebellar plane.

contribute to the evaluation of movement disorders—therefore, this plane will not be commented upon in this chapter. However, in addition to the two axial planes, the plane of the cerebellum (Fig. 1, A3) can be assessed in a standardized fashion by rotating the ultrasound probe for about 30–45° to visualize the posterior fossa.

A. The Plane of the Mesencephalic Brainstem

The mesencephalic brainstem plane constitutes a nearly axial section through the midbrain. As shown in Fig. 2, the key structure within this plane is the butterfly-shaped midbrain, which is hypoechogenic contrasting the strongly echogenic basal cisterns. In comparison to the standard MRI images, TCS allows to differentiate a number of different structures within the midbrain, i.e., substantia nigra (SN), red nucleus (RN), brainstem midline raphe (BR), and aqueduct. Abnormalities in these midbrain structures have shown to be valuable markers for a variety of disorders, including idiopathic Parkinson's disease, restless legs syndrome, and major depression.

In addition to the midbrain structures, also other structures, including vessels in the basal cisterns, the hippocampus, the lower frontal lobe, parts of the temporal and occipital lobes as well as the upper and medial parts of the cerebellum, can be clearly distinguished within this plane.

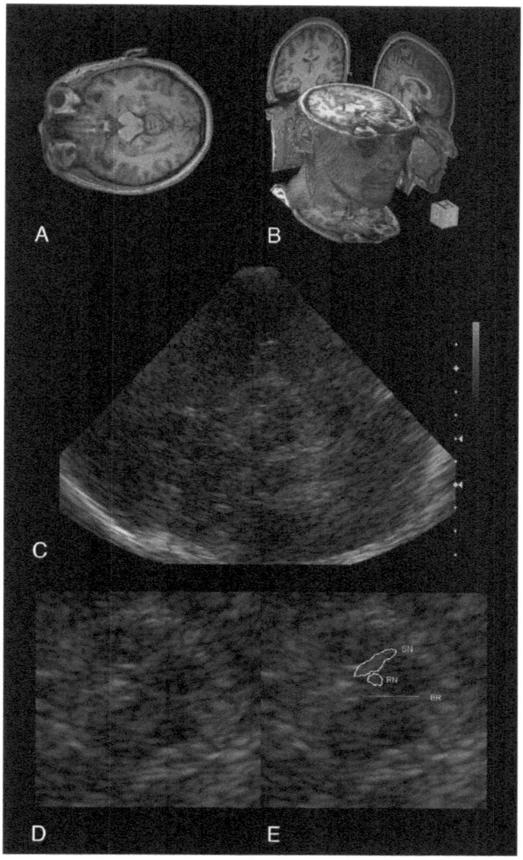

FIG. 2. Plane of the mesencephalic brainstem. (A) MRI section, (B) MRI overview, (C) complementary B-Mode, (D) enlarged midbrain, (E) enlarged labeled midbrain. SN: substantia nigra, RN: red nucleus, BR: brainstem raphe.

B. The Plane of the Third Ventricle

The plane of the third ventricle is reached by tilting the ultrasound probe slightly 10–20° upward starting from the midbrain plane. As shown in Fig. 3, this mild angulation of the probe results in a cross section through the brain, which is not compatible to the axial MRI slices and does not allow direct comparison of the left and right sides. Because of the wavelengths used to penetrate the skull, resolution close to the probe is only limited. Therefore, only structures close to the midline (e.g., the third ventricle) are evaluated at

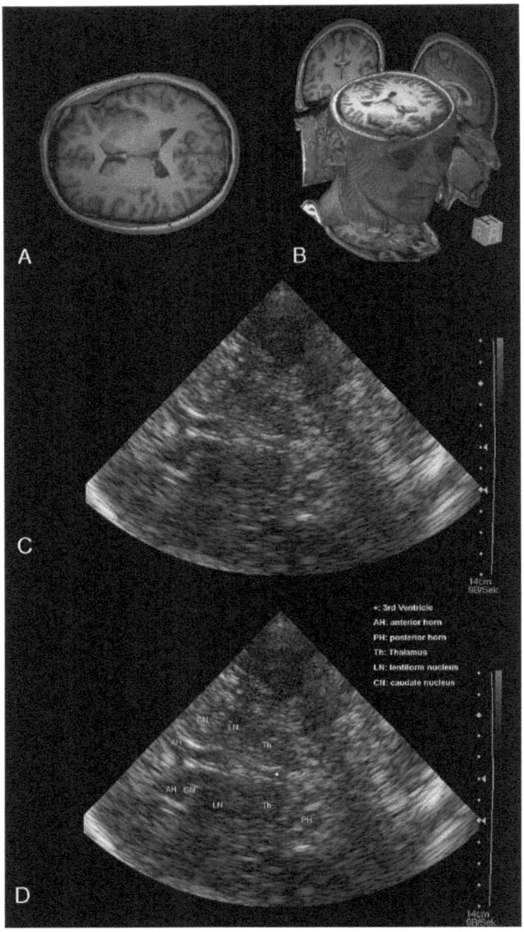

FIG. 3. Plane of the third ventricle. (A) MRI section, (B) MRI overview, (C) complementary B-Mode, (D) complementary labeled B-Mode.

the side ispilateral to the insonating probe while structures more distant from the midline (like the basal ganglia) are generally evaluated from the contralateral side (see below). Landmark structure for orientation in this plane is the pineal gland, which normally appears strongly hyperechogenic due to calcification and is located adjacent to the posterior surface of the third ventricle. This plane allows assessment of the ventricular system including not only the third ventricle, but also the anterior and posterior horns of the lateral ventricles. Measurement of the third ventricle has been shown to be valid

TABLE I
NORMAL VALUES OF THE VENTRICULAR SYSTEM

Age	2nd ventricle	3rd ventricle
<60 years	<17 mm	<7 mm
>60 years	<20 mm	<10 mm

Values from 49 healthy volunteers as sum scores ± 1 standard deviation (See also Berg et al., 2008).

and reliable and correlates well with the findings of other structural neuroimaging methods (Berg et al., 2000). Normal values for the width of the ventricular system have been published (Seidel et al., 1995). From this and other studies, age-related normal values have been derived, which should be referred to when the diagnosis of an enlargement of the ventricular system due to brain atrophy, hydrocephalus, etc. is made (see Table I). Moreover, the thalamus and the basal ganglia including lentiform and caudate nuclei can be assessed. Normally, these structures appear isoechogenic compared to the surrounding tissue. In some cases, however, hyperechogenic areas may occur at the anatomical site of these structures, which is abnormal and related either to a pathological condition or to an abnormality without clear clinical implications, such as slight calcification of the basal ganglia.

C. Plane of the Posterior Fossa

The posterior fossa can be visualized by rotating the anterior pole of the probe upward for about 45° starting from the imagined orbitomeatal line (compare Fig. 1). This plane also does not correspond to the common axial or coronal MRI slices (Fig. 4). It allows assessment of the fourth ventricle as well as of the cerebellar hemispheres, the vermis, and especially the dentate nucleus, which can sometimes be found to be slightly hyperechogenic compared to the surrounding tissue. Other structures that can be pictured in this plane, such as the brainstem, the hippocampus, or the cerebral lobes, are better assessed in one of the other planes and have not been evaluated so far in this plane. The plane of the posterior fossa does not belong to the routine workup so far, and proper depiction of structures in this plane is more dependent on the quality of the temporal acoustic bone windows. Therefore, data on the pathophysiological findings in this plane are limited and standardization of assessments in this plane is still in progress.

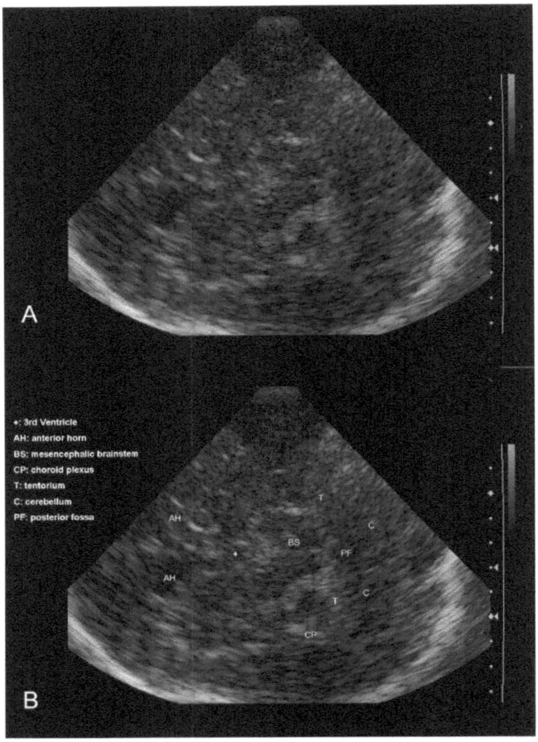

FIG. 4. Plane of the posterior fossa. (A) Complementary B-Mode, (B) complementary labeled B-Mode.

II. Measurements and typical pathological findings

Well examined and of special interest to the evaluation of patients with movement disorders are the ventricles, the lentiform nucleus, the substantia nigra, and the mesencephalic midline raphe. The methods for quantitative assessment and examples of typical pathological findings for these specific structures are shown below.

A. Ventricular System

The width of the third ventricle and the anterior horn of the contralateral lateral ventricle of the ventricular system must be measured in a regular exam.

Especially, the width of the third ventricle has been shown to correlate well with the measurements on CT scans. Both structures are best visualized in the third ventricular plane (compare IB). For assessment of the third ventricle, the largest distance between the inner surfaces of both ependymal linings is measured perpendicularly in the frozen and about a twofold magnified image. In the same plane, the contralateral anterior horn is evaluated by using the midline as orientation. The distance between the medial (nearest to the midline) to the lateral ependymal lining (farthest from the midline)—again perpendicularly from the midline—is used for measurement of the anterior horn. Figure 5B shows a case of a pathologically enlarged ventricular system.

The posterior horn is in general not measured (there exist no standard values so far); however, the impression of an enlarged posterior horn should be documented.

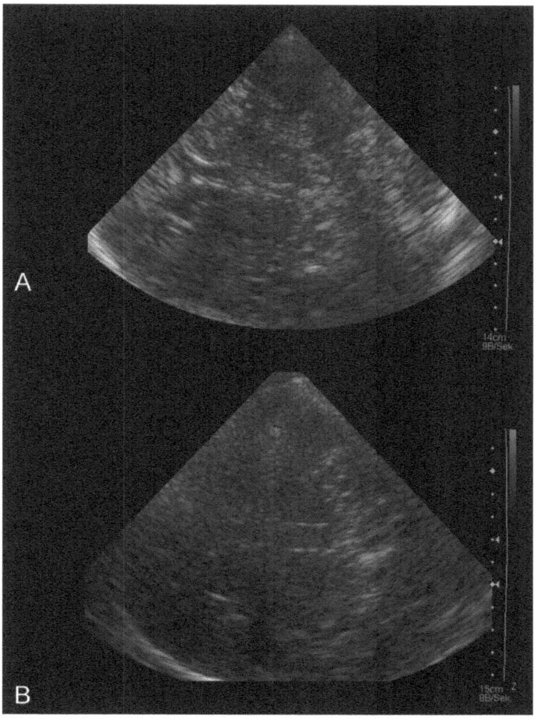

Fig. 5. Measurement of the ventricular system. (A) Normal ventricle system, (B) enlarged ventricular system.

B. Lentiform Nucleus

The lentiform nucleus (LN) is also visualized in the third ventricular plane. Regularly, it is isoechogenic to the surrounding tissue and therefore cannot be quantitatively assessed. Hence, usually a semi-quantitative rating applying a three-item scale is used, rating "normal LN echogenicity," "moderate LN hyperechogenicity," and "marked LN hyperechogenicity." Moderate LN hyperechogenicity has been described in various pathological conditions as in atypical parkinsonian syndromes (MSA, PSP), Wilson's disease, or dystonia, whereas marked LN hyperechogenicity has been found to be mainly correlated to calcifications. Figure 6 shows an example of LN echogenicity.

Recently, approaches to a quantitative assessment of pathological LN echogenicity have also been applied, assessing not only the manually encircled area of LN echogenicity but also the location of the echogenic area in relation to the third ventricle and the anterior horn of the lateral ventricle. To determine the location, the orthogonal distance between the most medial part of the hyperechogenic area

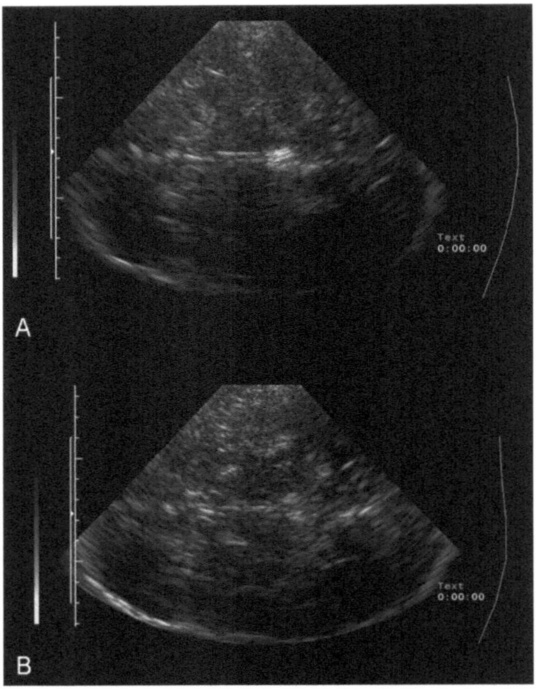

Fig. 6. Lentiform nucleus (LN). (A) Normal LN echogenicity, (B) LN hyperechogenicity (e.g. MSA, PSP, Wilson's disease, dystonia—see respective chapters).

and the external surface of the ependymal lining of the third ventricle in about two to threefold magnified image is measured. Complementarily, also the orthogonal distance between the most anterior part of the hyperechogenic area and the external surface of the ependymal lining of the anterior horn of the lateral ventricle is measured. These measurements are supposed to describe in more detail the location of the hyperechogenic area within the basal ganglia area, and should thus allow allocation to specific parts of the LN. However, as current as January 2010, no validation of these values and their relevance is available, therefore quantitative assessment of the LN is not a standard, yet.

C. Substantia Nigra

The echogenicity of the substantia nigra (SN) is assessed in the midbrain plane ipsilateral to the insonating probe. For proper assessment, the image is frozen and magnified three- to fourfold. Prior to measurements, the SN should be clearly separated from the adjacent red nucleus by visual inspection. The echogenic area of the SN is assessed at the site of its greatest expansion within the butterfly-shaped part of the mesencephalic brainstem ipsilateral to the insonating ultrasound probe. The echogenic area at the anatomical site of the SN is manually encircled with the cursor and then measured planimetrically (Fig. 7). Both, increased and decreased areas of SN echogenicity, have been shown to play a valuable role in the diagnosis and differential diagnosis of various movement disorders, such as Parkinson's disease and restless legs syndrome. Examples of

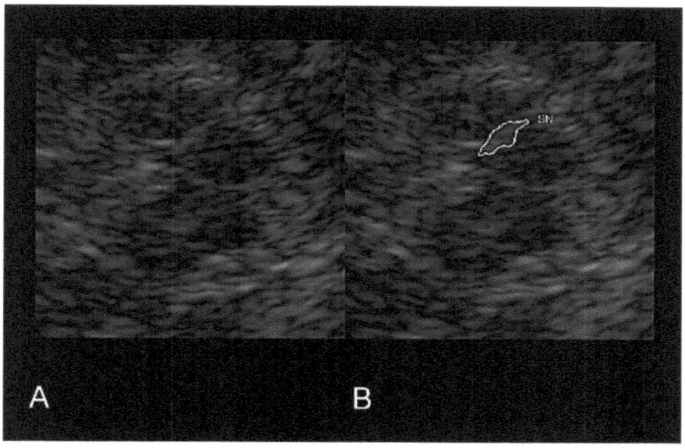

Fig. 7. Substantia nigra (SN). (A) Normal SN, (B) normal SN encircled.

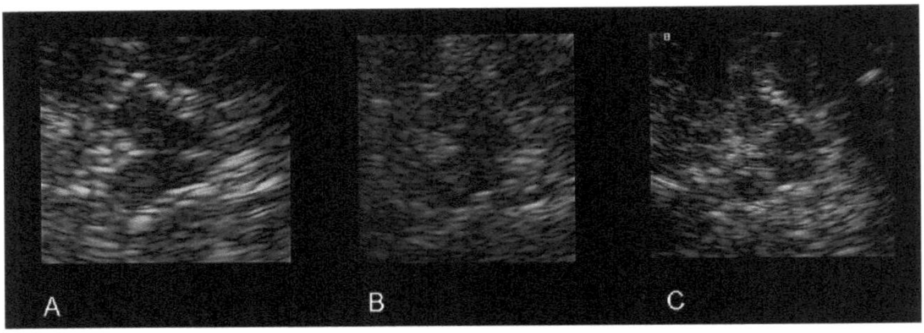

FIG. 8. Abnormalities of SN echogenicity. (A) Decreased (RLS) SN echogenicity, (B) normal SN echogenicity, (C) SN hyperechogenicity (e.g., iPD, DLB).

moderately and markedly increased SN echogenicity as well as of decreased SN echogenicity are given in Fig. 8. More detailed information on pathological alterations of SN echogenicity and its clinical and pathophysiological significance is given in the chapters on Parkinson's disease and RLS.

D. Raphe

The echogenicity of the brainstem midline raphe containing, among others, the serotonergic raphe nuclei is assessed in the lower part of the mesencephalic brainstem plane, often located a little lower compared to the site of SN assessment, close to the intersection with the pons. Still, the butterfly shape of the midbrain needs to be depicted for assessment of its midline. In normal anatomy, the raphe forms a "continuous" white line right in the midline of the lower midbrain. Deviations from this normal finding are rated semi-quantitatively as either "interrupted" or "absent" midline echogenicity. Examples are given in Fig. 9. Decreased echogenicity of the raphe has been attributed to depressive disorders (compare Chapter on depression).

E. Other Structures

For structures not named in the above sections, no standardized assessment is available so far. However, for the evaluation of some structures such as the red nucleus in restless legs syndrome, the dentate nucleus and the fourth ventricle in

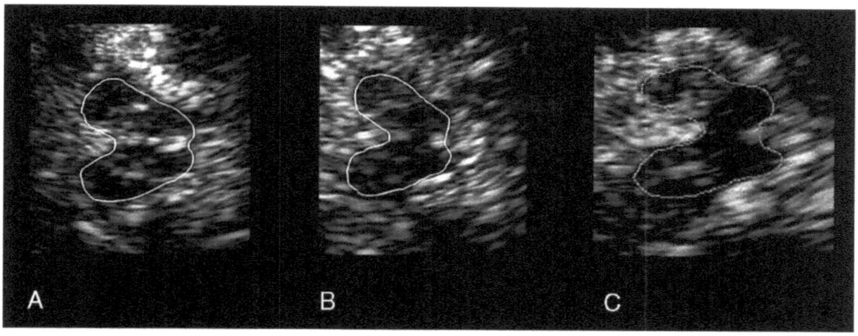

FIG. 9. Brainstem midline raphe (A) Continuous raphe, (B) interrupted raphe, (C) absent raphe.

ataxia, and the caudate nucleus in Huntington's disease, examples and further descriptions are given in the respective chapters.

References

Berg, D., Maurer, M., Warmuth-Metz, M., Rieckmann, P., and Becker, G. (2000). The correlation between ventricular diameter measured by transcranial sonography and clinical disability and cognitive dysfunction in patients with multiple sclerosis. *Arch. Neurol.* **57**(9), 1289–1292.

Berg, D., Godau, J., and Walter, U. (2008). Transcranial sonography in movement disorders. *Lancet Neurol.* (7), 1044–1055.

Seidel, G., Kaps, M., Gerriets, T., and Hutzelmann, A. (1995). Evaluation of the ventricular system in adults by transcranial duplex sonography. *J. Neuroimaging* **5**(2), 105–108.

PART II

TRANSCRANIAL SONOGRAPHY IN PARKINSON'S DISEASE

TRANSCRANIAL SONOGRAPHY IN RELATION TO SPECT AND MIBG

Yoshinori Kajimoto, Hideto Miwa, and Tomoyoshi Kondo

Department of Neurology, Wakayama Medical University, Kimiidera, Wakayama, Japan

Neuroimaging techniques have substantially progressed over the past several years. These developments have provided new information about the degenerative processes involved in Parkinson's disease (PD). Functional imaging approaches such as positron emission tomography (PET) and single photon emission computerized tomography (SPECT) have been successfully employed to detect dopaminergic dysfunction in PD, even in preclinical stages. Myocardial scintigraphy with ^{123}I-metaiodobenzylguanidine (^{123}I-MIBG) has been used to assess cardiac sympathetic function, and recent studies have revealed that the cardiac uptake of MIBG is significantly reduced in patients with PD and Lewy body dementia. In addition, substantia nigra (SN) hyperechogenicity, detected with transcranial sonography (TCS), might provide a marker of susceptibility to PD. We briefly review data regarding the clinical significance of these functional imaging studies in the diagnosis of PD, and discuss the diagnostic potential of the combining TCS of the SN with other functional imaging tools.

INTERNATIONAL REVIEW OF
NEUROBIOLOGY, VOL. 90
DOI: 10.1016/S0074-7742(10)90004-4

I. Introduction

Parkinson's disease (PD) is one of the most common neurodegenerative disorders. However, some evidence has suggested that a substantial proportion of patients clinically diagnosed with PD may be misdiagnosed. For example, among 800 patients diagnosed with mild PD in a study of deprenyl and tocopherol antioxidative therapy, approximately 9% were later given an alternative diagnosis (Jankovic *et al.*, 2000). Another recent study suggested that approximately 5% of patients taking medication for PD are misdiagnosed (Newman *et al.*, 2009). In addition, a study reporting that the error rate evaluated by clinicopathological correlates was as high as 25%, even though most patients were treated by movement-disorder specialists (Hughes *et al.*, 1992). Thus, it is clear that accurate diagnosis of PD still remains a challenge for clinical neurologists.

The use of appropriate diagnostic biomarkers or tests is essential for improving the speed and accuracy of PD diagnosis. Transcranial sonography (TCS) of the substantia nigra (SN) has been proposed as one likely candidate (Becker and Berg, 2001). In addition, a number of ancillary tests are already used to improve the diagnostic accuracy of parkinsonism. Single photon emission computed tomography (SPECT) and positron emission tomography (PET) are scintigraphic techniques that can be used to study neurotransmitter systems. Currently, dopamine transporter SPECT (DAT-SPECT) is widely used for improving PD diagnosis. These techniques examine the uptake or re-uptake of ligands into the dopaminergic nerve terminals in the striatum, enabling the functional visualization of the nigrostriatal dopamine system. Iodine-123-labeled (S)-2-hydroxy-6-methoxy-*N*-((1-ethyl-2-pyrrolidinyl)methyl) benzamide (^{123}I-IBZM) is a highly specific dopamine D2 receptor ligand suitable for measuring striatal receptor activity by SPECT. This agent demonstrates specific binding to dopamine D2 receptors. The dopamine D2 receptor binding potential in the striatum is presumed to be normal or slightly elevated in PD and to be decreased in both MSA and PSP. Therefore, SPECT with ^{123}I-IBZM may be a useful alternative to PET for the discrimination of diseased versus healthy individuals. In addition to functional imaging of the central nervous system, imaging of the peripheral catecholaminergic system has also recently been identified as helpful in the diagnosis of PD. Myocardial scintigraphy with ^{123}I-metaiodobenzylguanidine (^{123}I-MIBG) has been used to assess cardiac sympathetic function. Recent studies have revealed that the cardiac uptake of MIBG is significantly reduced in patients with PD or dementia with Lewy bodies (DLB).

In this chapter, we briefly review data examining the clinical significance of these functional imaging studies in the diagnosis of PD, and discuss the functional correlation between functional imaging studies and TCS of the SN. In addition, we discuss the diagnostic potential of combining TCS of SN with other biomarkers.

II. Dopamine Transporter-SPECT in Parkinson's Disease

The development of radiolabeled ligands to detect striatal dopaminergic dysfunction using SPECT has typically focused on proteins, such as the dopamine transporter (DAT) located in the plasma membrane of presynaptic striatal dopaminergic terminals. Two radiopharmaceuticals can be used for DAT-SPECT: [^{123}I]-CIT: [^{123}I]2-carboxymethoxy-3-(4-iodophenyl) tropane; and [^{123}I]FP-CIT: [^{123}I]N-w-fluoropropyl-2-carbomethoxy-3-(4-iodophenyl) nortropane. DAT-SPECT has been used as a biomarker for disease progression in PD. A progressive decline of striatal DAT binding has been demonstrated with the progression of PD. Increasing disease severity has also been found to have a relationship with the results of [^{18}F]dopa PET studies (Morrish *et al.*, 1996). DAT-SPECT is a highly sensitive (≈100%) method for detecting presynaptic dopaminergic deficits in PD. An abnormal DAT-SPECT result has been found to support the diagnosis of PD or other neurodegenerative parkinsonism in early disease stages or uncertain or incomplete parkinsonian syndromes. On the contrary, normal DAT-SPECT results in a symptomatic patient provide very strong evidence that the condition is not neurodegenerative parkinsonism. While it has been shown that the DAT-SPECT is sensitive in detecting presynaptic nigrostriatal degeneration in PD and related disorders caused by neurodegeneration, it has not been found useful in the differential diagnosis of PD and atypical parkinsonism, such as multi-system atrophy (MSA), progressive supranuclear palsy (PSP), or corticobasal degeneration (CBD) (Benamer *et al.*, 2000; Pirker *et al.*, 2000; Parkinson Study Group, 2000; Varrone *et al.*, 2001).

To date, several studies have examined the diagnostic value of DAT-SPECT for the differentiation between PD and atypical parkinsonism, but failed to show its utility. The amount and pattern of reduced striatal DAT binding in MSA are in the range of PD, with a more pronounced loss of DAT binding exhibited in the posterior putamen (Kim *et al.*, 2002). On the other hand, asymmetry of DAT-binding loss may tend to be more pronounced in PD (Pirker *et al.*, 2000; Varrone *et al.*, 2001). In addition, it has been reported that even clinically pure forms of MSA-C are accompanied by some decrease in [^{18}F]dopa PET, although this decrease is smaller than in MSA-P or PD (Rinne *et al.*, 1995). Similarly, DAT-binding loss is observed in PSP (Pirker *et al.*, 2000). However, with PSP there appears to be a more symmetrical and profound DAT loss in the whole striatum, whereas in PD the decrement is greater in the putamen than in the caudate nucleus (Antonini *et al.*, 2003). In addition, it has been reported that DAT loss in CBD is in the same range as in PD and atypical parkinsonism (Klaffke *et al.*, 2006; Pirker *et al.*, 2000), although DAT loss in CBD is much more asymmetrical and less pronounced than that seen in MSA and PSP (Pirker *et al.*, 2000). Therefore, it should be stressed that

DAT-SPECT is a sensitive method for detecting functional deficits in nigrostriatal dopaminergic terminals in parkinsonism, but is of limited value in discriminating between PD and atypical parkinsonism.

III. Transcranial Sonography and DAT-SPECT in Parkinson's Disease

To understand the pathological substrate leading to abnormal enlargement of SN echogenicity, it would be helpful to determine whether signals of DAT-SPECT are correlated with SN hyperechogenicity. If SN hyperechogenicity appears as a result of neurodegeneration in SN, the size of SN hyperechogenicity would be expected to correlate inversely with DAT signals. To address this question, three studies (Doepp *et al.*, 2008; Spiegel *et al.*, 2006; Weise *et al.*, 2009) examined the use of combined TCS of the SN and DAT-SPECT in PD and/or related disorders (Table I). In two studies (Doepp *et al.*, 2008; Spiegel *et al.*, 2006), no correlation was demonstrated between the results of DAT-SPECT imaging and the degree of SN hyperechogenicity in PD. This pattern of results is compatible with the idea that SN hyperechogenicity does not result from neurodegeneration, but instead represents an iron-related risk marker indicating an increased vulnerability of dopaminergic neurons, as suggested by the results of Berg *et al.* (2006) showing that increased iron content is the most likely cause of SN hyperechogenicity. Contrary to this notion, in a recent study, SN echogenicity was inversely correlated with the DAT signal, suggesting that SN echogenicity might rather be a direct structural marker of degeneration of SN neurons (Weise *et al.*, 2009). The reason for this contradiction is currently unclear. It was reported in one study that SN hyperechogenicity is observable in about 5–10% of normal control subjects regardless of age. In addition, based on our own clinical experience, it is not uncommon to encounter healthy, young control subjects showing marked enlargement of SN hyperechogenicity. We propose that

TABLE I

DOPAMINE TRANSPORTER-SPECT VERSUS TRANSCRANIAL SONOGRAPHY IN PD

Ligand	Relation of echogenic SN-area and measures of dopaminergic function	N	Reference
^{123}I-FP-CIT	Not correlated	53	Spiegel *et al.*, 2006
^{123}I-FP-CIT	Not correlated	46	Doepp *et al.*, 2008
^{123}I-β-CIT	Correlated inversely ($P = 0.003$)	50	Weise *et al.*, 2009

SPECT, single photon emission computerized tomography; PD, Parkinson's disease; SN, substantia nigra; N, number it is important to recognize that SN hyperechogenicity is not a simple result of SN neurodegeneration. However, since the structural substrate determining SN echogenicity has not been determined in detail, the possibility that SN echogenicity has pathogenic significance should not be discounted. It remains conceivable that this index differs in healthy or preclinical subjects compared with patients with PD. Further experiments (particularly prospective studies) addressing this issue are required to understand the relationship between DAT-SPECT and SN hyperechogenicity in more detail.

IV. [^{123}I]MIBG Myocardial Scintigraphy in Parkinson's Disease

[^{123}I]MIBG is a radio-iodinated analog of noradrenalin (norepinephrine) and is transported actively into noradrenalin granules of sympathetic nerve terminals by the noradrenalin transporter (Wieland *et al.*, 1980). Clinically, MIBG myocardial scintigraphy has been used to evaluate regional denervation of the heart in acute coronary syndromes and cardiomyopathies (Glowniak *et al.*, 1989; Hattori and Schwaiger, 2000; Yamashina and Yamazaki, 2007). MIBG studies have been conducted in the field of neurology since the mid-1990s (Braune *et al.*, 1999; Hakusui *et al.*, 1994; Iwasa *et al.*, 1998; Orimo *et al.*, 1999), and the clinical value of these experiments on the diagnosis of PD has attracted much attention (Chung *et al.*, 2009; Hamada *et al.*, 2003; Kashihara *et al.*, 2006; Oka *et al.*, 2007; Orimo *et al.*, 2003; Reinhardt *et al.*, 2000; Suzuki *et al.*, 2007; Takatsu *et al.*, 2000; Taki *et al.*, 2000). Surprisingly, myocardial MIBG uptake has been found to decrease markedly and differentially in patients with PD and DLB (Oka *et al.*, 2007). This finding opened the door to a concept that is pathologically supported but not well recognized clinically, that PD is not specific to the brain but instead is a generalized neuronal disorder. The decrease in myocardial MIBG uptake suggests that functional alterations are not restricted to the central nervous system, and that PD pathology affects sympathetic neuronal function. Neuropathologically, Lewy bodies and α-synuclein-positive neurites have been identified in the hearts of patients with PD and Lewy body disease and have been found in both tyrosine hydroxylase-positive and hydroxylase-negative nerve processes. These findings suggest that postganglionic sympathetic and intrinsic neurons in the heart are involved in PD pathology (Iwanaga *et al.*, 1999). In addition, cardiac plexus is reported to be more heavily involved than the sympathetic ganglia in PD (Mitsui *et al.*, 2006).

In Japan, since 1999, the decrease in cardiac MIBG uptake in patients with PD has become widely accepted as a useful diagnostic marker (Iwasa *et al.*, 1998; Orimo *et al.*, 1999; Yoshita, 1998); meanwhile, this acceptance has gradually been extended to other countries where MIBG is clinically available, including France, Germany, Spain, Great Britain, Canada, and the USA. Previously reported data on myocardial MIBG uptake in PD and related disorders are listed in Table II. The heart-to-mediastinum (H/M) ratios in controls were found to be about 2.0–3.0. Although the exact cutoff value of the H/M ratio should be calculated specifically for each institute, the early H/M ratios ranged from 1.3 to 1.7. The delayed H/M ratios ranged from 1.1 to 1.5, significantly lower than that the values of the controls (Table II).

Reduced MIBG uptake is currently regarded as a characteristic of PD, DLB, and pure autonomic failure, and all are known to exhibit Lewy body pathology. However, approximately 10% of PD patients show normal MIBG uptake, as

TABLE II

^{123}I-MIBG Myocardial Scintigraphy in PD and Control

Number of patients	Early H/M ratio	Delayed H/M ratio	Reference
PD			
7 OH and AF, 22 w/o AF	No uptake in 5 with PD	N/A	Hakusui *et al.*, 1994
12	1.55 ± 0.17	1.37 ± 0.15	Iwasa *et al.*,1998
25	1.36 ± 0.16	1.19 ± 0.15	Yoshita, 1998
46	1.71 ± 0.36	1.53 ± 0.36	Orimo *et al.*, 1999
15	N/A	1.08 ± 0.13	Braune *et al.*, 1999
21	N/A	Median 1.05 (range 0.9–1.15)	Reinhardt *et al.*, 2000
32	1.58 ± 0.37	1.33 ± 0.28	Takatsu *et al.*, 2000
38	1.61 ± 0.29	1.47 ± 0.34	Taki *et al.*, 2000
88	1.51 ± 0.32	1.39 ± 0.33	Hamada *et al.*, 2003
90	1.72 ± 0.33	1.54 ± 0.35	Orimo *et al.*, 2003
130	1.69 ± 0.29	1.37 ± 0.27	Kashihara *et al.*, 2006
143	1.48 ± 0.28	1.40 ± 0.30	Suzuki *et al.*, 2007
27	1.53 ± 0.23	1.36 ± 0.24	Chung *et al.*, 2009
Control			
10	2.08 ± 0.21	2.02 ± 0.24	Iwasa *et al.*, 1998
20	2.03 ± 0.16	2.21 ± 0.23	Yoshita, 1998
10	2.26 ± 0.16	2.3 ± 0.22	Orimo *et al.*, 1999
7	2.42 ± 0.27	2.6 ± 0.15	Takatsu *et al.*, 2000
10	2.24 ± 0.14	2.37 ± 0.14	Taki *et al.*, 2000
25	2.26 ± 0.19	2.19 ± 0.2	Hamada *et al.*, 2003
10	2.20 ± 0.16	2.16 ± 0.22	Orimo *et al.*, 2003
16	2.26 ± 0.21	2.48 ± 0.35	Kashihara *et al.*, 2006
19	20.02 ± 0.12	2.05 ± 0.15	Suzuki *et al.*, 2007
12	2.81 ± 1.2	3.04 ± 1.52	Chung *et al.*, 2009

MIBG, metaiodobenzylguanidine; PD, Parkinson's disease; H/M, heart-to-mediastinum; OH, orthostatic hypotension; AF, autonomic failure; w/o, without reported in several studies (Hamada *et al.*, 2003; Orimo *et al.*, 1999; Taki *et al.*, 2000). In early PD, MIBG uptake does not always decrease, perhaps because the extent of cardiac denervation may differ between subjects and could occasionally be less prominent in early phases of the disease (Orimo *et al.*, 1999). MIBG–myocardial scintigraphy has been found useful for differential diagnosis of parkinsonism. A number of studies have reported that MIBG uptake in MSA is significantly higher than that in PD (Braune *et al.*, 1999; Kashihara *et al.*, 2006; Orimo *et al.*, 1999; Taki *et al.*, 2000; Takatsu *et al.*, 2000; Yoshita, 1998). One exception is a report that MIBG uptake can occasionally decrease in MSA patients (Druschky *et al.*, 2000). This may be due to mild degeneration of the cardiac sympathetic nerve in MSA (Orimo *et al.*, 2007). A meta-analysis of studies of a total of 246 cases of PD and 45 cases of MSA reported that the overall sensitivity for positively identifying patients with PD was 89.7%, and the specificity in discriminating them from patients with MSA was 94.6% (Braune, 2001). In addition, no reduction in MIBG uptake has been reported in most cases of PSP and CBD (Kashihara *et al.*, 2006; Orimo *et al.*, 2003; Taki *et al.*, 2000; Yoshita, 1998), although there are some exceptions (e.g., Nagayama *et al.*, 2005). MIBG uptake is also reported to remain unchanged in essential tremor (Lee *et al.*, 2006; Orimo *et al.*, 1999).

V. Transcranial Sonography and [^{123}I]MIBG Cardiac Scintigraphy

The reduction of the H/M ratio may not always decrease in patients with early-stage PD. As such, we predict that combining TCS and MIBG myocardial scintigraphy would improve the accuracy of the clinical diagnosis of PD. However, we have conducted the only study addressing the correlation of SN hyperechogenicity and the H/M ratio (Kajimoto *et al.*, 2009). In this study, a total of 30 patients with PD were assessed, with cutoff values set at 0.16 cm^2 (mean + 1 S.D.) for TCS, and 1.66 cm^2 (mean - 2 S.D.) for MIBG myocardial scintigraphy. No significant correlation was found between SN echogenicity and H/M ratio ($p = 0.71$, Spearman's correlation; Fig. 1). Eighty percent of patients (24/30) were identified to have abnormally high SN hyperechogenicity on TCS, and 73% patients (22/30) were identified as abnormal using MIBG myocardial scintigraphy. Among eight patients who showed normal MIBG myocardial scintigraphy results, seven (88%) were identified as abnormal by TCS. Only one of the 30 patients exhibited normal TCS and MIBG findings. In total, 29 of the 30 patients were identified as abnormal by combining TCS and MIBG myocardial scintigraphy, resulting in a very high sensitivity (97%). These results suggest that both TCS of SN and MIBG myocardial scintigraphy are useful markers for the diagnosis of

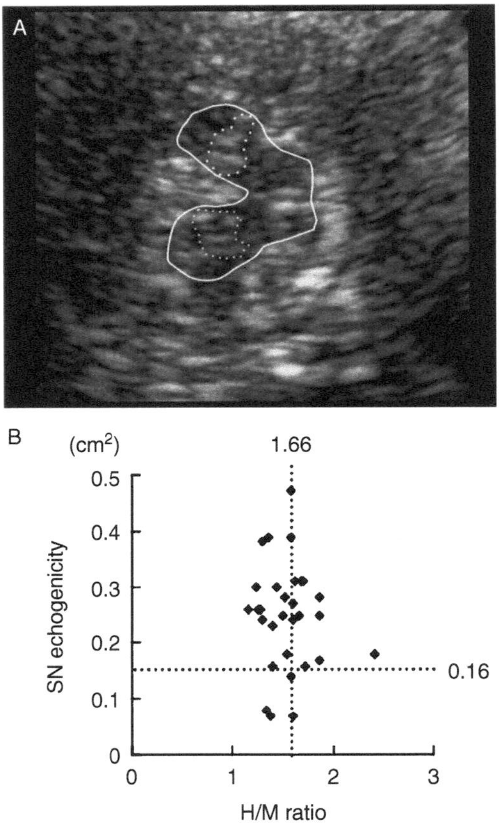

FIG. 1. (A) Substantia nigra (SN) hyperechogenicity in a patient with Parkinson's disease. The midbrain is encircled with continuous lines, and the SN is encircled by dotted lines. (B) Correlation between the heart-to-mediastium (H/M) ratio of MIBG myocardial scintigraphy and the area of larger side of SN echogenicity (cm^2) in patients with early-stage Parkinson's disease. Each patient is represented by one point. There is no correlation between the H/M ratio and the SN hyperechogenicity. The vertical and horizontal lines indicate the cutoff values of SN echogenicity and H/M ratio, respectively.

PD, and that the combination of these two tests can contribute to diagnostic sensitivity in PD. It is thus expected that both TCS of SN and MIBG can make complementary contributions to the early diagnosis of PD. It is not surprising that there is no correlation between MIBG uptake and SN hyperechogenicity in PD. Similarly to the case of DAT-SPECT, SN echogenicity and the reduction of MIBG uptake result from distinct pathological processes. SN hyperechogenicity is not a marker of disease progression but rather a risk marker indicating an increased vulnerability of dopaminergic neurons (Behnke *et al.*, 2003).

VI. Transcranial Sonography of Substantia Nigra: Toward the Future Development of Early or Preclinical Diagnosis of Parkinson's Disease

DAT-SPECT is a sensitive method for detecting presynaptic dopaminergic deficits in parkinsonism, but has low specificity. DAT loss in the striatum has been reported to be observable not only in PD, but also in other forms of parkinsonism, including MSA, PSP, and CBD. MIBG myocardial scintigraphy is a relatively specific method for detecting Lewy body disease, but the reduction of the H/M ratio may not always decrease in patients with early-stage PD. In addition, in practical clinical setting for screening very early PD, there are common disadvantages involved in both tests. First, they both require equipment for handling radioisotopes. Second, from the point of view of medical economics, both techniques are relatively expensive as screening tests. Thus, a combination of other complementary information will likely be necessary for making the diagnosis of early PD more accurate and convenient. Clinical information regarding symptomatic features (such as poor response to levodopa, early falls, limitation of gaze, prominent autonomic failure, asymmetrical dystonic limb, among others) is still clearly useful for indicating atypical parkinsonism. In addition, neuroradiological findings suggesting atypical parkinsonism are also of value, including abnormal density of the striatum, atrophy of the brainstem and cerebellum, asymmetrical widening of cortical sulci, hydrocephalus, and so on. However, these atypical clinical and neuroradiological features are not always observable in the early phases of the illness. Thus, other markers that may be unrelated to disease progression or those that are present before motor symptoms appear would also be useful. As such, olfactory testing, information about REM-sleep behavioral disorders (RBD), and TCS of SN are likely to be helpful markers for diagnosis of early PD. Sommer *et al.* (2004) produced an interesting report relating to this question. In this study, patients with idiopathic olfactory dysfunction, but without PD, underwent TCS of the SN and DAT-SPECT. Interestingly, alterations of the nigrostriatal dopaminergic system were found in about 70% of cases, suggesting that the combination of olfactory testing and TCS of the SN was potentially useful for diagnosis of early PD, and screening of preclinical PD. It has also been shown that TCS supports the discrimination of PD from atypical parkinsonian syndromes, especially MSA and PSP (Gaenslen *et al.*, 2008; Walter *et al.*, 2007).

Finally, we have recently gathered preliminary data aimed at prospectively evaluating the potential usefulness of diagnostic markers (including olfactory testing, TCS of the SN, and MIBG myocardial scintigraphy) for early and accurate diagnosis of PD. The subjects in this study were outpatients who were clinically suspected of having PD ($N = 33$) or atypical parkinsonian syndrome (APS = 6), respectively, from their first visit to our neurological clinic at the

University Hospital. All participants were *de novo* patients who were not receiving any anti-parkinsonian medication. The APS group included patients diagnosed with PSP ($N = 4$) and MSA ($N = 2$). After a tentative clinical diagnosis was made, olfactory testing using the odor stick identification test for Japanese (OSIT-J), TCS, and MIBG myocardial scintigraphy were all performed. In the PD group, significant olfactory loss, abnormal SN hyperechogenicity, and reduced myocardial MIBG uptake were present in 85, 68.2, and 57.6% of patients, respectively. In the APS group, olfactory loss, abnormal SN hyperechogenicity, and reduced myocardial MIBG uptake were present in 33.3, 25, and 0% of patients, respectively. Our results demonstrated that the rate of abnormality in each test was significantly higher in the PD than the APS group. Olfactory loss exhibited the highest positive rate of the three diagnostic tools, and patients showing reduced myocardial MIBG uptake constituted almost half of all PD subjects. This may be because the subjects in this study were in the early stages, and had short disease duration. It is thus expected that olfactory testing, TCS of the SN, and MIBG myocardial scintigraphy used along, or in combination, may increase diagnostic accuracy in very early stages of PD (Fig. 2). Since all subjects in this study, regardless of whether they were diagnosed as PD or APS, were in the early stages of disease and exhibited symptoms of mild severity, their initial diagnosis is not unlikely to change during disease progression. As such, we plan to continue prospective observations for at least 3 years, and to re-evaluate our data to reconfirm the clinical diagnoses.

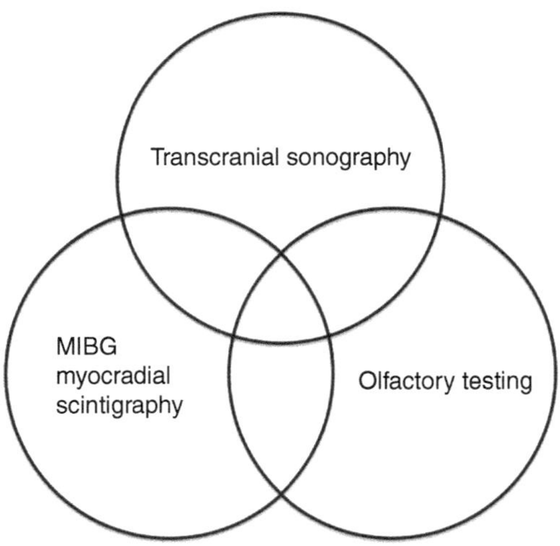

FIG. 2. The schematic representation of diagnostic tools for Parkinson's disease: olfactory testing, transcranial sonography of the substantia nigra, and MIBG myocardial scintigraphy.

VII. Conclusion

Both DAT-SPECT and MIBG myocardial scintigraphy were found to be useful for diagnosing PD. The specificity of DAT-SPECT, however, was found to be relatively low, and the sensitivity of MIBG myocardial scintigraphy was not especially high. As such, it should be noted that the diagnostic power of each method alone is limited in the early phases of illness. To compensate for these limitations, it is useful to combine these techniques with other diagnostic biomarkers. One potential candidate is TCS of the SN. We propose that combining TCS of the SN with functional neuroimaging methods can make complementary contributions to the detection of PD, enabling earlier and more accurate diagnosis.

References

Antonini, A., Benti, R., De Notaris, R., Tesei, S., Zecchinelli, A., Sacilotto, G., Meucci, N., Canesi, M., Mariani, C., Pezzoli, G., and Gerundini, P. (2003). ^{123}I-Ioflupane/SPECT binding to striatal dopamine transporter (DAT) uptake in patients with Parkinson's disease, multiple system atrophy, and progressive supranuclear palsy. *Neurol. Sci.* **24**, 149–150.

Becker, G., and Berg, D. (2001). Neuroimaging in basal ganglia disorders: perspectives for transcranial ultrasound. *Mov. Disord.* **16**, 23–32.

Behnke, S., Berg, D., and Becker, G. (2003). Does ultrasound disclose a vulnerability factor for Parkinson's disease? *J. Neurol.* **250**(Suppl. 1), I24–I27.

Benamer, T. S., Patterson, J., Grosset, D. G., Booij, J., de Bruin, K., van Royen, E., Speelman, J. D., Horstink, M. H., Sips, H. J., Dierckx, R. A., Versijpt, J., and Decoo, D. (2000). Accurate differentiation of parkinsonism and essential tremor using visual assessment of [^{123}I]-FP-CIT SPECT imaging: the [^{123}I]-FP-CIT study group. *Mov. Disord.* **15**, 503–510.

Berg, D., Hochstrasser, H., Schweitzer, K. J., and Riess, O. (2006). Disturbance of iron metabolism in Parkinson's disease—Ultrasonography as a biomarker. *Neurotox. Res.* **9**, 1–13.

Braune, S., Reinhardt, M., Schnitzer, R., Riedel, A., and Lücking, C. H. (1999). Cardiac uptake of [^{123}I]MIBG separates Parkinson's disease from multiple system atrophy. *Neurology* **53**, 1020–1025.

Braune, S. (2001). The role of cardiac metaiodobenzylguanidine uptake in the differential diagnosis of parkinsonian syndromes. *Clin. Auton. Res.* **11**, 351–355.

Chung, E. J., Lee, W. Y., Yoon, W. T., Kim, B. J., and Lee, G. H. (2009). MIBG scintigraphy for differentiating Parkinson's disease with autonomic dysfunction from Parkinsonism-predominant multiple system atrophy. *Mov. Disord.* **24**, 1650–1655.

Doepp, F., Plotkin, M., Siegel, L., Kivi, A., Gruber, D., Lobsien, E, Kupsch, A., and Schreiber, S. J. (2008). Brain parenchyma sonography and ^{123}I-FP-CIT SPECT in Parkinson's disease and essential tremor. *Mov. Disord.* **23**, 405–410.

Druschky, A., Hilz, M. J., Platsch, G., Radespiel-Tröger, M., Druschky, K., Kuwert, T., and Neundörfer, B. (2000). Differentiation of Parkinson's disease and multiple system atrophy in early disease stages by means of I-123-MIBG-SPECT. *J. Neurol. Sci.* **175**, 3–12.

Gaenslen, A., Unmuth, B., Godau, J., Liepelt, I., Di Santo, A., Schweitzer, K. J., Gasser, T., Machulla, H. J., Reimold, M., Marek, K., and Berg, D. (2008). The specificity and sensitivity of transcranial ultrasound in the differential diagnosis of Parkinson's disease: a prospective blinded study. *Lancet Neurol.* **7**, 417–24.

Glowniak, J. V., Turner, F. E., Gray, L. L., Palac, R. T., Lagunas-Solar, M. C., and Woodward, W. R. (1989). Iodine-123 metaiodobenzylguanidine imaging of the heart in idiopathic congestive cardiomyopathy and cardiac transplants. *J. Nucl. Med.* **30**, 1182–1191.

Hakusui, S., Yasuda, T., Yanagi, T., Tohyama, J., Hasegawa, Y., Koike, Y, Hirayama, M., and Takahashi, A. (1994). A radiological analysis of heart sympathetic functions with meta-[^{123}I] iodobenzylguanidine in neurological patients with autonomic failure. *J. Auton. Nerv. Syst.* **49**, 81–84.

Hamada, K., Hirayama, M., Watanabe, H., Kobayashi, R., Ito, H., Ieda, T, Koike, Y., and Sobue, G. (2003). Onset age and severity of motor impairment are associated with reduction of myocardial 123I-MIBG uptake in Parkinson's disease. *J. Neurol. Neurosurg. Psychiatr.* **74**, 423–426.

Hattori, N., and Schwaiger, M. (2000). Metaiodobenzylguanidine scintigraphy of the heart: what have we learnt clinically? *Eur. J. Nucl. Med.* **27**, 1–6.

Hughes, A. J., Daniel, S. E., Kilford, L., and Lees, A. J. (1992). Accuracy of clinical diagnosis of idiopathic Parkinson's disease: a clinico-pathological study of 100 cases. *J. Neurol. Neurosurg. Psychiatr..* **55**, 181–184.

Iwanaga, K., Wakabayashi, K., Yoshimoto, M., Tomita, I., Satoh, H., Takashima, H., Satoh, A., Seto, M., Tsujihata, M., and Takahashi, H. (1999). Lewy body-type degeneration in cardiac plexus in Parkinson's and incidental Lewy body diseases. *Neurology* **52**, 1269–1271.

Iwasa, K., Nakajima, K., Yoshikawa, H., Tada, A., Taki, J., and Takamori, M. (1998). Decreased myocardial 123I-MIBG uptake in Parkinson's disease. *Acta Neurol. Scand.* **97**, 303–306.

Jankovic, J., Rajput, A. H., McDermott, M. P., Perl, D. P., for Parkinson Study Group. (2000). The evolution of diagnosis in early Parkinson disease. *Arch. Neurol.* **57**, 369–372.

Kajimoto, Y., Miwa, H., Okawa-Izawa, M., Hironishi, M., and Kondo, T. (2009). Transcranial sonography of the substantia nigra and MIBG myocardial scintigraphy: complementary role in the diagnosis of Parkinson's disease. *Parkinsonism Relat. Disord.* **15**, 270–272

Kashihara, K., Ohno, M., Kawada, S., and Okumura, Y. (2006). Reduced cardiac uptake and enhanced washout of ^{123}I-MIBG in pure autonomic failure occurs conjointly with Parkinson's disease and dementia with Lewy bodies. *J. Nucl. Med.* **47**, 1099–1101.

Kim, Y. J., Ichise, M., Ballinger, J. R., Vines, D., Erami, S. S., Tatschida, T., and Lang, A. E. (2002). Combination of dopamine transporter and D2 receptor SPECT in the diagnostic evaluation of PD, MSA, and PSP. *Mov. Disord.* **17**, 303–312.

Klaffke, S., Kuhn, A. A., Plotkin, M., Amthauer, H., Harnack, D., Felix, R., and Kupsch, A. (2006). Dopamine transporters, D2 receptors, and glucose metabolism in corticobasal degeneration. *Mov. Disord.* **21**, 1724–1727.

Lee, P. H., Kim, J. W., Bang, O. Y., Joo, I. S., Yoon, S. N., and Huh, K. (2006). Cardiac ^{123}I-MIBG scintigraphy in patients with essential tremor. *Mov. Disord.* **21**, 1235–1238.

Merlet, P., Valette, H., Dubois-Randé, J. L., Moyse, D., Duboc, D., Dove, P., Bourguignon, M. H., Benvenuti, C., Duval, A. M., Agostini, D., Loisance, D., and Castaigne, A. (1992). Prognostic value of cardiac metaiodobenzylguanidine imaging in patients with heart failure. *J. Nucl. Med.* **33**, 471–477.

Mitsui, J., Saito, Y., Momose, T., Shimizu, J., Arai, N., Shibahara, J., Ugawa, Y., Kanazawa, I., Tsuji, S., and Murayama, S. (2006). Pathology of the sympathetic nervous system corresponding to the decreased cardiac uptake in 123I-metaiodobenzylguanidine (MIBG) scintigraphy in a patient with Parkinson's disease. *J. Neurol. Sci.* **243**, 101–104.

Morrish, P. K., Sawle, G. V., and Brooks, D. J. (1996). An [18F]dopa-PET and clinical study of the rate of progression in Parkinson's disease. *Brain* **119**, 585–591.

Nagayama, H., Hamamoto, M., Ueda, M., Nagashima, J., and Katayama, Y. (2005). Reliability of MIBG myocardial scintigraphy in the diagnosis of Parkinson's disease. *J. Neurol. Neurosurg. Psychiatr.* **76**, 249–251.

Newman, E. J., Breen, K., Patterson, J., Hadley, D. M., Grosset, K. A., and Grosset, D. G. (2009). Accuracy of Parkinson's disease diagnosis in 610 general practice patients in the West of Scotland. *Mov. Disord.* **24**, 2379–2385.

Oka, H., Yoshioka, M., Morita, M., Onouchi, K., Suzuki, M., Ito, Y., Hirai, T., Mochio, S., and Inoue K. (2007). Reduced cardiac 123I-MIBG uptake reflects cardiac sympathetic dysfunction in Lewy body disease. *Neurology* **69**, 1460–1465.

Orimo, S., Kanazawa, T., Nakamura, A., Uchihara, T., Mori, F., Kakita, A., Wakabayashi, K., and Takahashi H. (2007). Degeneration of cardiac sympathetic nerve can occur in multiple system atrophy. *Acta Neuropathol.* **113**, 81–86

Orimo, S., Ozawa, E., Nakade, S., Hattori, H., Tsuchiya, K., Taki, K., and Takahashi, A. (2003). [^{123}I] meta-iodobenzylguanidine myocardial scintigraphy differentiates corticobasal degeneration from Parkinson's disease. *Intern. Med.* **42**, 127–128.

Orimo, S., Ozawa, E., Nakade, S., Sugimoto, T., and Mizusawa, H. (1999). (123)I-metaiodobenzylguanidine myocardial scintigraphy in Parkinson's disease. *J. Neurol. Neurosurg. Psychiatr.* **67**, 189–194.

Parkinson Study Group. (2000). A multicenter assessment of dopamine transporter imaging with DOPASCAN/SPECT in parkinsonism. *Neurology* **55**, 1540–1547.

Pirker, W., Asenbaum, S., Bencsits, G., Prayer, D., Gerschlager, W., Deecke, L., and Brücke, T. (2000). [^{123}I]beta-CIT SPECT in multiple system atrophy, progressive supranuclear palsy, and corticobasal degeneration. *Mov. Disord.* **15**, 1158–1167.

Reinhardt, M. J., Jüngling, F. D., Krause, T. M., and Braune, S. (2000). Scintigraphic differentiation between two forms of primary dysautonomia early after onset of autonomic dysfunction: value of cardiac and pulmonary iodine-123 MIBG uptake. *Eur. J. Nucl. Med.* **27**, 595–600.

Rinne, J. O., Burn, D. J., Mathias, C. J., Quinn, N. P., Marsden, C. D., and Brooks, D. J. (1995). Positron emission tomography studies on the dopaminergic system and striatal opioid binding in the olivopontocerebellar atrophy variant of multiple system atrophy. *Ann. Neurol.* **37**, 568–573

Sommer, U., Hummel, T., Cormann, K., Mueller, A., Frasnelli, J., Kropp, J., and Reichmann, H. (2004). Detection of presymptomatic Parkinson's disease: combining smell tests, transcranial sonography, and SPECT. *Mov. Disord.* **19**, 1196–1202.

Spiegel, J., Hellwig, D., Möllers, M. O., Behnke, S., Jost, W., Fassbender, K, Samnick, S., Dillmann, U., Becker, G., and Kirsch, C. M. (2006). Transcranial sonography and [^{123}I]FP-CIT SPECT disclose complementary aspects of Parkinson's disease. *Brain* **129**, 1188–1193.

Suzuki, M., Urashima, M., Oka, H., Hashimoto, M., and Taira, K. (2007). Cardiac sympathetic denervation in bradykinesia-dominant Parkinson's disease. *Neuroreport* **18**, 1867–1870.

Takatsu, H., Nishida, H., Matsuo, H., Watanabe, S., Nagashima, K., Wada, H, Noda, T., Nishigaki, K., and Fujiwara, H. (2000). Cardiac sympathetic denervation from the early stage of Parkinson's disease: clinical and experimental studies with radiolabeled MIBG. *J. Nucl. Med.* **41**, 71–77.

Taki, J., Nakajima, K., Hwang, E. H., Matsunari, I., Komai, K., Yoshita, M., Sakajiri, K., and Tonami, N. (2000). Peripheral sympathetic dysfunction in patients with Parkinson's disease without autonomic failure is heart selective and disease specific. *Eur. J. Nucl. Med.* **27**, 566–573.

Varrone, A., Marek, K. L., Jennings, D., Innis, R. B., and Seibyl, J. P. (2001). [(123)I]beta-CIT SPECT imaging demonstrates reduced density of striatal dopamine transporters in Parkinson's disease and multiple system atrophy. *Mov. Disord.* **16**, 1023–1032.

Walter, U., Dressler, D., Probst, T., Wolters, A., Abu-Mugheisib, M., Wittstock, M., and Benecke, R. (2007). Transcranial brain sonography findings in discriminating between parkinsonism and idiopathic Parkinson disease. *Arch. Neurol.* **64**, 1635–1640.

Weise, D., Lorenz, R., Schliesser, M., Schirbel, A., Reiners, K., and Classen, J. (2009). Substantia nigra echogenicity: A structural correlate of functional impairment of the dopaminergic striatal projection in Parkinson's disease. *Mov. Disord.* **24**, 1669–1675.
Wieland, D. M., Wu, J., Brown, L. E., Mangner, T. J., Swanson, D. P., and Beierwaltes, W. H. (1980). Radiolabeled adrenergi neuron-blocking agents: adrenomedullary imaging with [131I]iodobenzylguanidine. *J. Nucl. Med.* **21**, 349–353.
Yamashina, S., and Yamazaki, J. (2007). Neuronal imaging using SPECT. *Eur. J. Nucl. Med. Mol. Imaging.* **34**, 939–950.
Yoshita, M. (1998). Differentiation of idiopathic Parkinson's disease from striatonigral degeneration and progressive supranuclear palsy using iodine-123 meta-iodobenzylguanidine myocardial scintigraphy. *J. Neurol. Sci.* **155**, 60–67.

DIAGNOSIS OF PARKINSON'S DISEASE—TRANSCRANIAL SONOGRAPHY IN RELATION TO MRI

Ludwig Niehaus,[*,†] and Kai Boelmans[‡]

[*]Department of Neurology, Otto-von-Guericke University, Magdeburg, Germany
[†]Department of Neurology, Castle Hospital Winnenden, Winnenden, Germany
[‡]Department of Neurology, University Hospital Hamburg-Eppendorf, Hamburg, Germany

Displaying the echo pattern (echogenicity) of brain tissue transcranial sonography (TCS) may provide new and complementary information to other neuroimaging methods. In contrast to conventional magnetic resonance imaging (MRI), TCS is able to detect highly characteristic changes in signal brightness of the substantia nigra (SN) in patients with idiopathic Parkinson's disease. In this review, TCS findings are related to conventional and advanced high-field brain MRI findings. On the basis of the MRI findings, especially with T2-relaxometry, the possible role of trace metals in the genesis of altered echogenicity on TCS of brain parenchyma, especially of the SN, are discussed.

I. Introduction

Transcranial B-mode sonography (TCS) has been increasingly used as a diagnostic tool in Parkinson's disease (PD) and related disorders (Berg *et al.*, 2008). TCS through the preauricular bone window allows the depiction of characteristic abnormalities in the echogenicity of substantia nigra (SN) and

INTERNATIONAL REVIEW OF
NEUROBIOLOGY, VOL. 90
DOI: 10.1016/S0074-7742(10)90005-6

0074-7742/10 $35.00

basal ganglia. Increased echogenicity ("hyperechogenicity") of the SN could be detected in more than 90% of PD patients (Berg *et al.*, 2001a, 2008; Walter *et al.*, 2003, 2007a).

Various efforts have been made with magnetic resonance imaging (MRI) to measure structural alterations in the brain related to PD. However, conventional field-strength MRI is usually normal in PD patients (Seppi and Schocke, 2005). In contrast to TCS, structural MRI usually failed to depict signal abnormalities in the SN in parkinsonian patients.

This review focuses on the similarities and differences in imaging of deep brain structures using TCS and conventional structural MRI. Optimized MRI techniques allowing sophisticated investigation of midbrain nuclei are demonstrated. Applying advanced 3-T MRI, anatomical localization and configuration of the SN are exemplarily illustrated. Furthermore, we discuss the potential impact of MRI to elucidate the nature of increased echogenicity in deep brain structures. To address this topic, studies correlating ultrasonographic and MRI findings are presented.

II. Neuroimaging with TCS and Conventional MRI Techniques

Ultrasound techniques have developed to an extent allowing two-dimensional real-time imaging of intracranial structures with high spatial image resolution of about 0.8 mm in the axial direction of insonation (Berg and Becker, 2002; Kallmann *et al.*, 2004). Walter *et al.* (2008) recently demonstrated that contemporary TCS systems achieve a higher image resolution of echogenic deep intracranial structures than conventional field-strength MRI under clinical conditions.

It may be surprising that the ability of TCS to visualize brain parenchyma and to depict subtle changes in deep brain structures has been widely neglected. A possible reason for this might be the divergent and unusual appearance of cerebral structures on transcranial sonography compared to MRI. Although shape and volume of intracranial structures appear similar in both TCS and MRI, signal brightness in TCS and signal intensity in MRI significantly differ (Fig. 1).

The inconsistency of imaging findings may be exemplarily illustrated by the appearance of cerebral spinal fluid in TCS and MRI. Basal cisterns and ventricles are shown with high signal intensity in T2-weighted MR images. TCS demonstrates low signal brightness within the ventricles, whereas the perimesencephalic cisterns are depicted with a highly echogenic signal. TCS and MRI also outline basal ganglia with appreciable differences. Although TCS presents both putamen

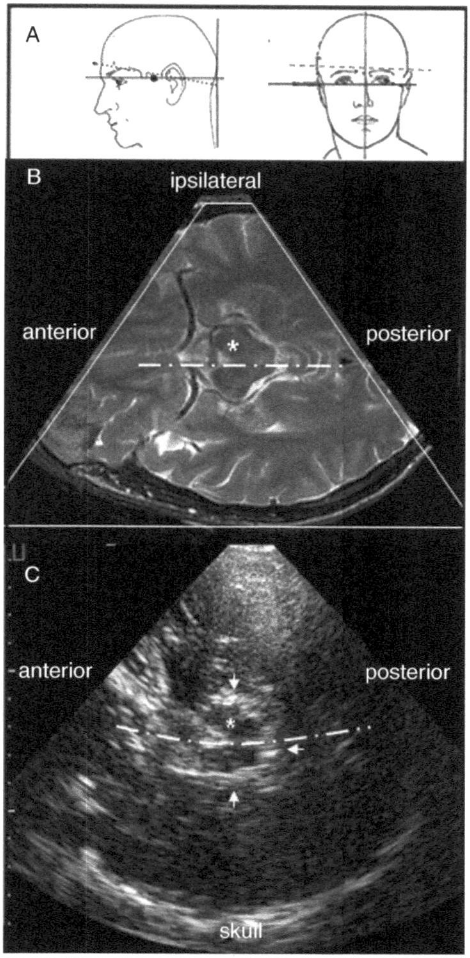

FIG. 1. Brain magnetic resonance imaging (MRI) and transcranial sonography (TCS): (**A**) Axial scanning plane in a–p direction and lateral view. (**B**) T2-weighted axial MRI depicting the midbrain and the substantia nigra (*). (**C**) TCS showing axial transection at midbrain level corresponding to the T2-weighted MRI shown in (B). Note the image distortion due to the phased-array scanning through the preauricular skull.

and globus pallidus with low echogenicity, T2-weighted MRI delineates the globus pallidus with strongly hypointense and the putamen with isointense signals (Table I).

If both neuroimaging methods are compared, the implication of the physical image generation needs to be considered. In contrast to MRI, which is

TABLE I

SIGNAL INTENSITY OF INTRACRANIAL STRUCTURES ON TRANSCRANIAL SONOGRAPHY (TCS) AND CONVENTIONAL MRI (T1-, PD-, AND T2-WEIGHTED SE-SEQUENCES)

	Substantia nigra	Red nucleus	Globus pallidus	Putamen	Ventricles	Cisterna ambiens	Skull
TCS	O	(+)	O	O	–	+	+
MR-T1w	O	O	O	O	–	–	–
MR-PD	+	(–)	(+)	+	–	–	–
MR-T2w	(–)	–	–	O	+	+	–

Signal intensity is given in relation to that of surrounding white matter (hypointense, –; isointense, O; hyperintense, +).

based on magnetic tissue properties, TCS is based on reflection and scattering of ultrasound waves at interfaces with diverse acoustic impedance.

III. MRI Studies of the Substantia Nigra in Parkinson's Disease

The basal ganglia volume and biochemical composition normally change as a function of age, but can be abnormally altered in neurodegenerative disorders. MRI can visualize differences in brain volume and tissue composition, including the content of water and trace metals.

Various efforts have been made with conventional field-strength MRI to detect structural alterations in the brain related to PD (Adachi *et al.*, 1999; Schrag *et al.*, 2000). However, T1-, proton density-, and T2-weighted sequences are usually normal in PD. MRI planimetric and volumetric techniques evaluating the SN were not able to reliably differentiate PD patients from controls (Adachi *et al.*, 1999; Geng *et al.*, 2006). At a first glance it seems paradoxical that MRI failed to demonstrate disease-specific changes in the SN of PD patients. This might at least in part be caused by the fact that no optimal MRI sequence has been established for evaluating the SN. Before turning to the various MRI techniques in detail and how these may be used to assess nigral changes, it is useful to briefly consider the configuration and the MR characteristics of the SN area. The volume of SN is small and accounts only for 0.5% of the total brain volume. The nigral nucleus is located between the crural fibers and the red nucleus. The SN runs in the anterosuperolateral aspect to the posteroinferomedial aspect through the entire midbrain (Oikawa *et al.*, 2002). It can be best identified as a hyperintense gray matter structure on proton density-weighted images (Adachi *et al.*, 1999; Geng *et al.*, 2006; Oikawa *et al.*, 2002). The anatomical boundaries separating the SN pars compacta (SNc) and pars reticulata (SNr) are poorly defined on MRI scans.

IV. Structural Changes of the Substantia Nigra Revealed by MRI

It is essential to apply MR sequences with high spatial resolution to evaluate the SN boundaries. SN is located as a tilted, band-like structure in the midbrain. Thus, MR scanning of SN with axial slice orientation is strongly impaired by partial volume effects (Oikawa *et al.*, 2002). MRI of the midbrain using 45° angulations to commissural line (AC-PC) or perpendicular to the midbrain axis (Fig. 2) may overcome this limitation. Using 3-T MRI, we found that the anatomic margin of the SN can be viewed clearly by using proton density- and T2-weighted sequences with optimized angulation (Fig. 3).

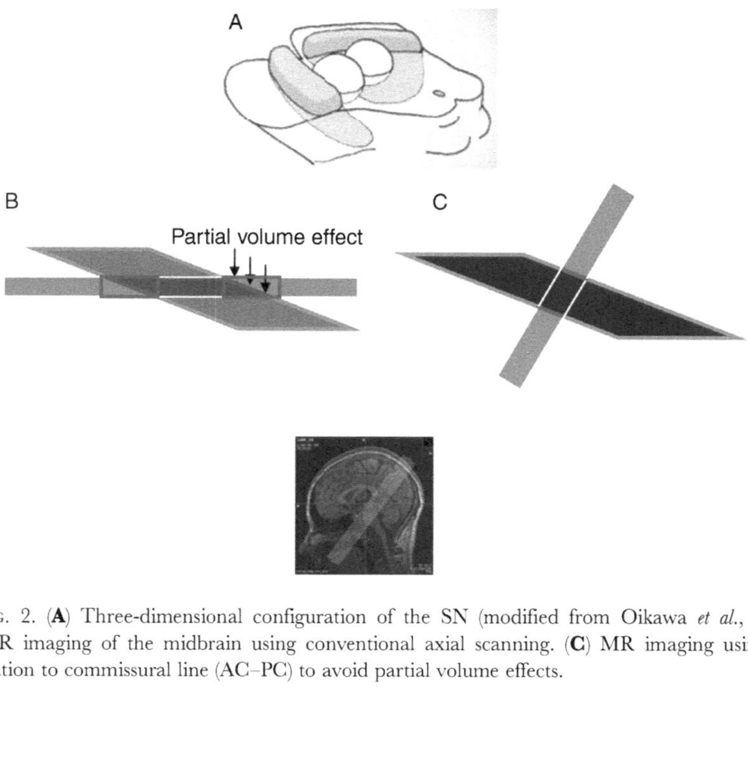

FIG. 2. (**A**) Three-dimensional configuration of the SN (modified from Oikawa *et al.*, 2002). (**B**) MR imaging of the midbrain using conventional axial scanning. (**C**) MR imaging using 45° angulation to commissural line (AC–PC) to avoid partial volume effects.

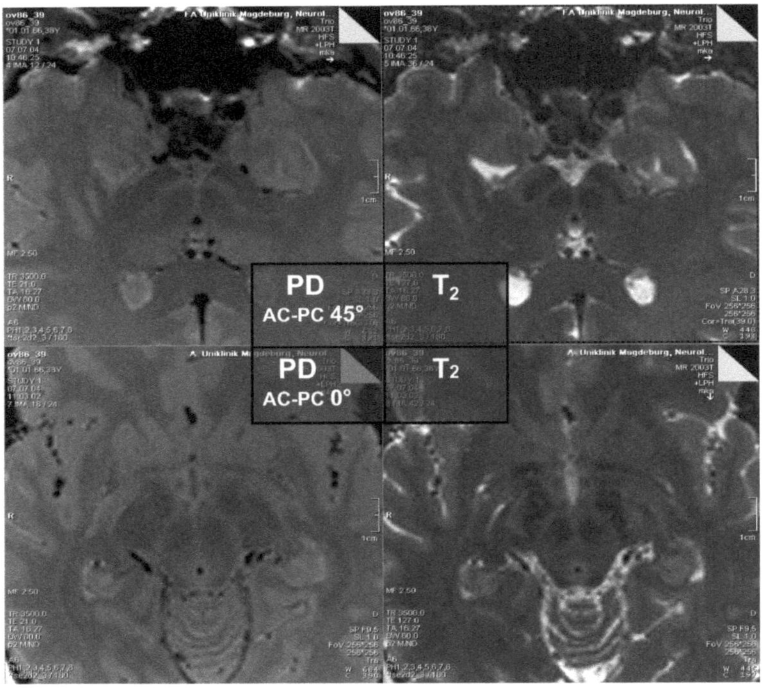

FIG. 3. High-resolution MR imaging of the midbrain with a 3-T scanner (TRIO, Siemens, Erlangen, Germany). Proton density- (left side) and T2-weighted (right side) spinecho sequences (TR = 3000 ms, TE = 20 ms, 118 ms; voxel size 1 × 1 × 1 mm^2) were applied angulated 45° to the commissural line as determined on the sagittal section. Axial images were taken parallel to commissural line (0° to AC–PC). Optimized angulation significantly reduces partial volume effects, which are notable on axial images.

A. MRI-Based Morphometric Analysis of the Substantia Nigra

The SN has been divided into SNc and SNr. It was difficult to define the SN boundary precisely on current MR images, since the ventral parts of the SNr extend and intermix with crural fibers. Many researchers believed that SNr and red nucleus are both hypointense on T2-weighted images because of their high iron concentration. Regarding the signal appearance of the SN on T2-weighted images, the darker hypointense region was defined as SNr, and the region of signal hyperintensity between this and the red nucleus as SNc (Antonini *et al.*, 1993; Duguid *et al.*, 1986; Sohmiya *et al.*, 2004). However, there has been no MRI-pathoanatomical investigation published yet supporting this view.

Based on T2-weighted images, there were a few reports measuring the size of the SNc and SNr in PD patients. Some studies reported a narrowing of the band of hyperintensity (SNc) between the SNr and the red nucleus in PD, whereas others could not replicate this finding (Duguid *et al.*, 1986; Pujol *et al.*, 1992; Sohmiya *et al.*, 2004). The small size and irregular appearance make the width of the SNc difficult to assess, and its measurement is currently beyond the sensitivity of volumetric analysis.

Oikawa *et al.* (2002) correlated findings of the SN on T2-weighted images with findings on proton density-weighted SE imaging and fast STIR imaging. The hypointense area on T2-weighted images, proposed to represent the SNr (Duguid *et al.*, 1986), is located in the medial part of the crural fibers and did not correspond to the SN. Moreover, the hypointense area is not evident at the level of lower midbrain, where SNr does exist in anatomic specimens.

Using conventional field-strength MRI, it appears to be impossible to distinguish reliably both SN regions on T2-weighted images (Geng *et al.*, 2006; Oikawa *et al.*, 2002). This has to be kept in mind if differences in volume and signal intensity of SNc and SNr should be evaluated.

Magnetization transfer (MT) imaging was also employed to reveal pathological changes of the SN (Eckert *et al.*, 2004; Tambasco *et al.*, 2003). MT imaging is based on the interactions between bound protons within structures as myelin or cell membranes and the mobile protons of free water. In patients with PD, the magnetization transfer ratio (MTR) in the SN was significantly smaller than in control subjects. However, there may be a considerable overlap between the patient and the control groups. In contrast to the SN echogenic size in TCS, there was no cutoff value reported which could be used to discriminate between the healthy subjects and the patients at an individual level.

B. T2-Weighted MRI of the Substantia Nigra in Parkinson's Disease

The pathoanatomical hallmark of PD is a progressive degeneration and atrophy of nigrostriatal neurons within the midbrain (Ma *et al.*, 1996). Neurodegenerative processes characterized by cell loss, by astrogliosis or microglial proliferation or accumulation of iron or other paramagnetic substances, should lead to signal changes in affected brain areas on MRI (Gerlach *et al.*, 2006). However, signal intensity and contrast on MR images are based on a variety of factors, including proton density, relaxation time, diffusion, magnetic susceptibility, and field strength. For example, gliosis may increase the signal intensity on T2-weighted MR images. On the contrary, accumulation of iron leads to signal loss and hypointensity on T2-weighted images with minimal intensity changes on T1-weighted images.

The most consistently reported MRI finding in PD is the detection of signal attenuation of the SN on T2-weighted images (Antonini *et al.*, 1993; Bartzokis *et al.*, 1999; Graham *et al.*, 2000; Ryvlin *et al.*, 1995; Vymazal *et al.*, 1999). It has been suggested that increased iron deposition in SN may lead to signal attenuation via a mechanism involving diffusion of water protons through local magnetic field inhomogeneities. T2-relaxation time should be shortened in approximate proportion to regional iron content, but a weak correlation was found in some studies (Brooks *et al.*, 1989; Bartazokis *et al.*, 1993). It is known that iron is present in different concentrations in the basal ganglia nuclei and that increased iron in the SN is a feature of PD. In accordance to histochemical findings in PD, many, but not all, MRI studies showed that T2-relaxation time is decreased in the SN than in age-matched controls (Antonini *et al.*, 1993; Bartzokis *et al.*, 1999; Behnke *et al.*, 2009; Gorell *et al.*, 1995, 2000; Mondino *et al.*, 2002; Ryvlin *et al.*, 1995; Vymazal *et al.*, 1999).

As illustrated above for the delineation of midbrain nuclei (Oikawa *et al.*, 2002), possible imperfections of ROI placement may contribute to the inconsistent findings. Furthermore, the observed decrease in the T2-weighted image-based measurements was small and there was a substantial overlap between patients and controls.

The sensitivity for signal changes can be increased by using $T2^*$-weighted gradient echo sequences (Ordidge *et al.*, 1994). $T2^*$ is a combination of transverse relaxation and magnetic field inhomogeneity, with tissue iron being a major determinant for local field inhomogeneity.

Attempts have been made to monitor changes in iron content in PD using MR relaxometry, a combination of T2- and $T2^*$-weighted images (Ordidge *et al.*, 1994). However, previous studies applying this MRI approach to examine iron deposition in PD yielded inconsistent MRI findings compared with postmortem histoanatomical investigations. T2-relaxation time is also found to be decreased in the putamen (Antonini *et al.*, 1993; Bartzokis *et al.*, 1999) and the globus pallidus (Bartozokis *et al.*, 1999), but these regions did not show an alteration of iron content on autopsy (Dexter *et al.*, 1991).

V. Morphological and Biochemical Alterations of SN Assessed by TCS and MRI

In PD, TCS studies demonstrated an increase of echogenicity ("hyperechogenicity") of the SN in the above 90% of patients (Becker and Berg, 2001; Berg *et al.*, 2008; Niehaus *et al.*, 2004; Walter *et al.*, 2003, 2007a). In relatives of patients with sporadic PD, SN hyperechogenicity was found to be associated with a subclinical impairment of the nigrostriatal system detected by ^{18}F-Dopa positron emission tomography (PET) examination (Ruprecht-Dörfler *et al.*, 2003). Marked hyperechogenicity of the SN was also seen in correlation to abnormal nigrostriatal PET findings in asymptomatic *parkin* mutation carriers who had a significant risk to develop PD during their lifetime (Walter *et al.*, 2004).

Noteworthy, the same echo pattern (Fig. 4) can also be detected in 8–14% of healthy subjects at ages ranging from 16 to 83 years (Berg *et al.*, 1999a, 2001b; Walter *et al.*, 2007a) and with some increase of incidence in higher age groups (Behnke *et al.*, 2007). SN hyperechogenicity is related to motor slowing and to neuropsychological impairment in the elderly population (Berg *et al.*, 2001b; Behnke *et al.*, 2007; Liepelt *et al.*, 2008; Todd *et al.*, 2010). In young and middle-aged subjects, SN hyperechogenicity was found to be associated with decreased striatal ^{18}F-Dopa uptake as measured by PET (Berg *et al.*, 2002; Behnke *et al.*, 2009).

On the basis of the above-mentioned findings, it has been suggested that SN hyperechogenicity may be a marker for a subclinical nigrostriatal alteration and an increased susceptibility for PD (Behnke *et al.*, 2009; Berg *et al.*, 1999a, 2002; Ruprecht-Dörfler *et al.*, 2003; Sommer *et al.*, 2004; Walter *et al.*, 2004, 2007b). Given the similarity to the echo feature of SN hyperechogenicity in PD, it has been speculated that SN hyperechogenicity in healthy subjects reflects histopathological changes or displays an alteration in tissue impedance due to an abnormal iron accumulation, a process associated with neurodegeneration in PD (Behnke *et al.*, 2009; Berg *et al.*, 2001b; Double *et al.*, 2000; Jellinger *et al.*, 1990; Sommer *et al.*, 2004; Walter *et al.*, 2007b). On the other hand, changes in tissue echogenicity may be an epiphenomenon and are not linked to a disease-associated impairment. Even if this applies, SN hyperechogenicity might be a phenotype of a genetically determined variant associated with an alteration

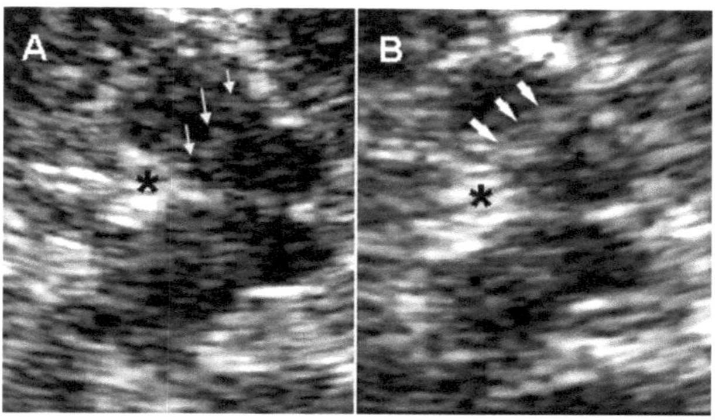

FIG. 4. Transcranial sonography (TCS) of the butterfly-shaped hypoechogenic mesencephalic midbrain in two healthy subjects. (**A**) TCS of midbrain with normal echogenic substantia nigra (SN) (arrows) characterized by low-to-moderate echogenicity. (**B**) TCS of midbrain with a distinctly hyperechogenic SN (ipsilateral to the transducer). Arrow heads indicate the SN. The interpeduncular cistern marked by asterisks shows high echogenicity, which is similar in both TCS images.

within the dopaminergic nigrostriatal system. The present knowledge about the mechanism leading to the changes in SN echomorphology is limited. To clarify the underlying cause, it would be helpful to get new insights and pathophysiological concepts of the process of neurodegeneration.

Hyperechogenicity of brain structures has been attributed to structural changes as gliosis and an increased deposition of trace metals such as copper or iron (Berg *et al.*, 1999a, 1999b; Walter *et al.*, 2005, 2009). For example, TCS may detect increased copper concentration in the lenticular nucleus with higher sensitivity than MRI (Walter *et al.*, 2005). Although copper can potentially affect MRI contrast if present in sufficient quantities, concentrations of these paramagnetic ions are normally too small to produce detectable MR signal changes.

There is some evidence from animal experiments and postmortem studies in humans that SN hyperechogenicity is associated with higher iron tissue level (Berg *et al.*, 1999b, 2002; Berg, 2006). A significant positive correlation was found between the echogenic area of the SN and the concentration of iron, H- and L-ferritins (Berg, 2006; Zecca *et al.*, 2005). In the human brain, iron accumulates during adolescence and early adulthood. The highest levels of iron are found in the globus pallidus and the SN. It is not known how iron enters the brain, as Fe^{2+} and Fe^{3+} ions are not able to cross the blood–brain barrier. Transferrin-binding sites are the most likely means of entry. Nevertheless, comparison of transferrin-binding sites and iron concentration has shown a mismatch (Hill and Switzer, 1984). The cellular and subcellular distribution of iron is also uncertain. In the midbrain, most iron is sited in the oligodendrozytes of the SN pars reticulata. In PD, neurodegeneration is associated with increased iron deposition in the SN, particularly in the pars compacta (Riederer *et al.*, 1989; Sofic *et al.*, 1991). Increased iron deposits in the SN in PD may have several reasons that may reside in disturbances of iron uptake, storage, and transport as neurodegeneration progresses. Elevated brain iron may be an important component of neurodegenerative cell damage through oxidative stress mechanisms (Riederer *et al.*, 1989). Since iron levels in the SN were found to be normal in preclinical cases of PD (Dexter *et al.*, 1991), such mechanisms are likely to contribute to the pathology of PD rather than being of direct etiological relevance.

It is thought that most of the brain iron is stored intracellular as ferritin (Dexter *et al.*, 1991; Griffiths *et al.*, 1999; Hallgren and Sourander, 1958). In PD, ferritin level has been found to be raised in the SN in most, but not all, studies (Dexter *et al.*, 1991; Griffiths *et al.*, 1999; Jellinger *et al.*, 1990; Riederer *et al.*, 1989). Griffiths *et al.* (1999) demonstrated an increased density of ferritin cores within the SN, and that ferritin molecules are more heavily loaded with iron. This environment might encourage free radical generation and hence neuronal damage. There is increasing evidence that SN in PD may be particularly susceptible to oxidative stress mechanisms. However, the role of iron and ferritin in the pathogenesis of PD is not fully understood.

Iron-related changes of the SN may be an important factor contributing to its increased echogenicity in healthy subjects and patients with PD (Berg *et al.*, 2002; Berg, 2006). With respect to the distribution and time course of iron accumulation in human brain (Griffiths *et al.*, 1999), one may query whether elevated iron concentration is the main reason for the increasing signal brightness of the SN in healthy subjects. On the one hand, iron is very low in the brain at birth and accumulates gradually a little in very young children (Hallgreen and Sourander, 1958). On the other hand, in one TCS study, SN hyperechogenicity was frequently seen in newborns (Iova *et al.*, 2004). In contrast to the slow accumulation of brain iron, this echofeature was found to decrease substantially during the first years of life. Furthermore, there is no close correlation of the amount of iron as well as ferritin and signal brightness of subcortical nuclei in healthy adults. Both globus pallidus and red nucleus showing the highest iron concentration usually do not exhibit hyperechogenicity on TCS (Morris *et al.*, 1992; Walter *et al.*, 2007a).

In PD, the disparity in signal appearance of iron-rich nuclei is thought to reflect increased amounts of iron bound to proteins other than ferritin in the SN. However, SN hyperechogenicity is also present at preclinical and early stages of PD, at which iron levels in the SN were found to be normal (Dexter *et al.*, 1993). Furthermore, SN signal brightness did not change with PD duration or ongoing clinical impairment (Berg *et al.*, 2005). As shown in radiotracer imaging studies in PD, SN hyperechogenicity and degeneration of dopaminergic neuron exist independently from each other (Behnke *et al.*, 2009; Niehaus *et al.*, 2006; Spiegel *et al.*, 2006). Behnke *et al.* (2009) hypothesized that the echofeature of SN hyperechogenicity in PD patients and in healthy subjects with reduced striatal ^{18}F-Dopa uptake is based on another pathomechanism than neurodegeneration.

VI. Correlation of MRI and TCS Findings of Substantia Nigra

Using TCS and MRI, Walter *et al.* (2009) investigated SN echogenicity and signal intensity of the SN on T2-weighted MR images in patients with multiple sclerosis. SN echogenic area correlated with T2-signal attenuation within in the SNc (Fig. 5), but not within SNr. Based on a close correlation of iron tissue concentration and signal loss on T2-weighted MRI, the authors suggested iron accumulation as a possible factor causing hyperechogenicity of SN.

Behnke *et al.* (2009) intended to elucidate the reasons for an increase of SN echogenicity by correlating TCS and MRI findings. Additionally, functional imaging of the dopaminergic nigrostriatal system was performed by ^{18}F-Dopa PET. They examined patients with PD and healthy subjects with SN

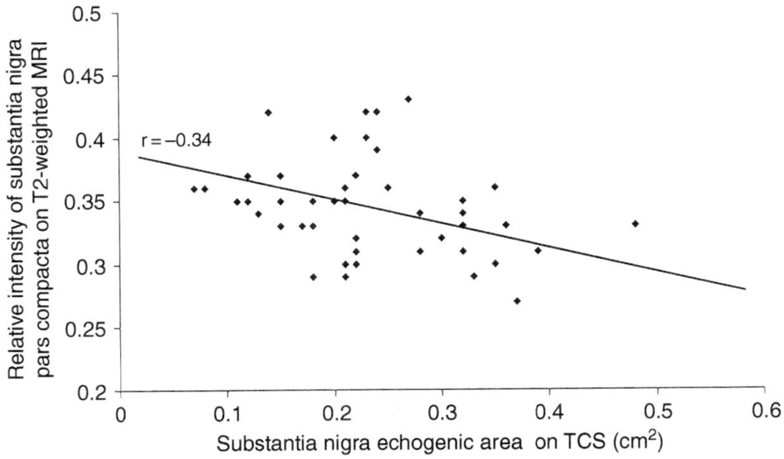

FIG. 5. Diagram showing the relation between transcranial sonography and MRI measures of the substantia nigra (SN) pars compacta in patients with multiple sclerosis (Provided by Professor U. Walter, Department of Neurology, University of Rostock, Rostock, Germany. Corresponding data published in: Walter *et al.*, 2009). SN echogenic size correlated inversely to T2-signal intensity, which is presented as a ratio to ventricular CSF.

hyperechogenicity and healthy controls without SN hyperechogenicity using 1.5-T MRI. Assuming that sophisticated MRI techniques like T2-relaxometry are highly sensitive in detecting parenchymal iron aggregations (Graham *et al.*, 2000), they applied a multiecho sequence and calculated T2 times within the SN area. Interestingly, healthy subjects with SN hyperechogenity demonstrated significantly shorter SN T2-relaxation times than probands without this echofeature.

More recently, our group investigated tissue integrity of basal ganglia in young healthy subjects with and without SN hyperechogenicity by MR relaxometry on a 3-T MR scanner (TRIO, Siemens, Erlangen, Germany) (unpublished). As a preliminary result, we found that the T2-relaxation times within the SN and red nucleus were lower in subjects with SN hyperechogenicity than without the echomarker. Transversal relaxation assessed in both groups of subjects did not differ in other basal ganglia or subcortical regions.

On the basis of these findings, conclusions should be drawn with cautions. By physical principle, transversal relaxation is characterized by a time constant T2, which is shortened in approximate proportion to regional iron content (Bartzokis *et al.*, 1993). Microscopic *in vitro* studies indicate that ferritin is the only storage protein that is detectable in both control and parkinsonian brain (Griffiths *et al.*, 1999). The T2 effect of nonheme iron in ferritin is believed to arise from the superparamagnetic moment at the core which induces local field inhomogenities.

The resulting phase incoherence leads to a decrease of the transversal magnetization and appears as signal attenuation on conventional field-strength T2-weighted MRI. This additional dephasing due to local field inhomogenities can be detected by MRI because radiofrequency cannot compensate for it.

However, the theory that iron is directly related to T2 signal loss has been challenged by some authors and appears to be far more complicated than originally thought. Knowledge of iron storage form and its interaction with cellular compartments is essential to interpreting the MRI data (Chen *et al.*, 1993). Although MRI studies demonstrated a strong proton spin dephasing effect of ferritin in aqueous solution, the signal changes are much less prominent in tissue. Moreover, it has been shown that regional iron content is only one of several determinants of transverse relaxation in tissue. For example, water density, which may vary independently of brain iron, has a major impact on relaxation time measurements.

Keeping these limitations in mind, it might be argued that additional factors other than iron accumulation contribute to signal attenuation in hyperechogenic SN. Different T2 relaxation may also result from differences in macromolecular structure of the tissue associated with a shift of the hydrogen pool bound to myelin or cell membranes. Acoustic properties of the anatomical SN region are probably based on such pathophysiological changes at inter- and intracellular compartments in PD patients and in susceptible asymptomatic subjects. However, the results of several studies suggest that echogenicity of the SN does not change in parallel with ongoing neurodegeneration (Berg *et al.*, 1999a, 2005; Spiegel *et al.*, 2006) even though this concept has been challenged recently (Behnke *et al.*, 2007; Weise *et al.*, 2009).

To further investigate the mechanism of the signal intensity shift in the SN, high-resolution MRI studies with susceptibility-weighted and magnetization transfer imaging providing additional insight into the structural integrity are necessary.

References

Adachi, M., Hosoya, T., Haku, T., Yamaguchi, K., and Kawanami, T. (1999). Evaluation of the substantia nigra in patients with parkinsonian syndrome accomplished using multishot diffusion-weighted MR imaging. *AJNR Am. J. Neuroradiol.* **20**, 1500–1506.

Antonini, A., Leenders, K. L., Meier, D., Oertel, W. H., Boesiger, P., and Anliker, M. (1993). T2 relaxation time in patients with Parkinson's disease. *Neurology* **43**, 697–700.

Bartzokis, G., Aravagiri, M., Oldendorf, W. H., Mintz, J., and Marder, S. R. (1993). Field dependent transverse relaxation rate increase may be a specific measure of tissue iron stores. *Magn. Reson. Med.* **29**, 459–464.

Bartzokis, G., Cummings, J. L., Markham, C. H., Marmarelis, P. Z., Treciokas, L. J., Tishler, T. A., Marder, S. R., and Mintz, J. (1999). MRI evaluation of brain iron in earlier- and later-onset Parkinson's disease and normal subjects. *Magn. Reson. Imaging.* **17**, 213–222.

Becker, G., and Berg, D. (2001). Neuroimaging in basal ganglia disorders, perspectives for transcranial ultrasound. *Mov. Disord.* **16**, 23–32.

Behnke, S., Double, K. L., Duma, S., Broe, G. A., Guenther, V., Becker, G., and Halliday, G. M. (2007). Substantia nigra echomorphology in the healthy very old: Correlation with motor slowing. *Neuroimage* **34**, 1054–1059.

Behnke, S., Schroeder, U., Dillmann, U., Buchholz, H. G., Schreckenberger, M., Fuss, G., Reith, W., Berg, D., and Krick, C. M. (2009). Hyperechogenicity of the substantia nigra in healthy controls is related to MRI changes and to neuronal loss as determined by F-Dopa PET. *Neuroimage* **47**, 1237–1243.

Berg, D., Becker, G., Zeiler, B., Tucha, O., Hofmann, E., Preier, M., Benz, P., Jost, W., Reiners, K., and Lange, K. W. (1999a). Vulnerability of the nigrostriatal system as detected by transcranial ultrasound. *Neurology* **53**, 1026–1031.

Berg, D., Grote, C., Rausch, W. D., Mäurer, M., Wesemann, W., Riederer, P., and Becker, G. (1999). Iron accumulation of the substantia nigra in rats visualized by ultrasound. *Ultrasound Med. Biol.* **25**, 901–904.

Berg, D., Siefker, C., and Becker, G. (2001a). Echogenicity of the substantia nigra in Parkinson's disease and its relation to clinical findings. *J. Neurol.* **248**, 684–689.

Berg, D., Siefker, C., Ruprecht-Dörfler, P., and Becker, G. (2001b). Relationship of substantia nigra echogenicity and motor function in elderly subjects. *Neurology* **56**, 13–17.

Berg, D., and Becker, G. (2002). Perspectives of B-mode transcranial ultrasound. *Neuroimage* **15**, 463–473.

Berg, D., Roggendorf, W., Schröder, U., Klein, R., Tatschner, T., Benz, P., Tucha, O., Preier, M., Lange, K. W., Reiners, K., Gerlach, M., and Becker, G. (2002). Echogenicity of the substantia nigra, association with increased iron content and marker for susceptibility to nigrostriatal injury. *Arch. Neurol.* **59**, 999–1005.

Berg, D., Merz, B., Reiners, K., Naumann, M., and Becker, G. (2005). Five-year follow-up study of hyperechogenicity of the substantia nigra in Parkinson's disease. *Mov. Disord.* **20**, 383–385.

Berg, D. (2006). *In vivo* detection of iron and neuromelanin by transcranial sonography - a new approach for early detection of substantia nigra damage. *J. Neural. Transm.* **113**, 775–780.

Berg, D., Godau, J., and Walter, U. (2008). Transcranial sonography in movement disorders. *Lancet Neurol.* **7**, 1044–1055.

Brooks, D. J., Luthert, P., Gadian, D., and Marsden, C. D. (1989). Does signal attenuation on high-field T2-weighted MRI of the brain reflect regional cerebral iron deposition? Observations on the relationship between regional cerebral water proton T2 values and iron levels. *J. Neurol. Neurosurg. Psychiatr.* **52**, 108–111.

Chen, J. C., Hardy, P. A., Kucharczyk, W., Clauberg M, Joshi JG, Vourlas A, Dhar, M., and Henkelman, R. M. (1993). MR of human post-mortem brain tissue, correlative study between T2 and assays of iron and ferritin in Parkinson and Huntington disease. *AJNR Am. J. Neuroradiol.* **14**, 275–281.

Dexter, D. T., Carayon, A., Javoy-Agid, F., Agid, Y., Wells, F. R., Daniel, S. E., Lees, A. J., Jenner, P., and Marsden, C. D. (1991). Alterations in the levels of iron, ferritin and other trace metals in Parkinson's disease and other neurodegenerative diseases affecting the basal ganglia. *Brain* **114**, 1953–1975.

Dexter, D. T., Sian, J., Jenner, P., and Marsden, C. D. (1993). Implications of alterations in trace element levels in brain in Parkinson's disease and other neurological disorders affecting the basal ganglia. *Adv Neurol.* **60**, 273–81.

Double, K. L., Gerlach, M., Youdim, M. B., and Riederer, P. (2000). Impaired iron homeostasis in Parkinson's disease. *J. Neural. Transm.* **60**(Suppl.), 37–58.

Duguid, J. R., De La Paz, R., and DeGroot, J. (1986). Magnetic resonance imaging of the midbrain in Parkinson's disease. *Ann. Neurol.* **20**, 744–747.

Eckert, T., Sailer, M., Kaufmann, J., Schrader, C., Peschel, T., Bodammer, N., Heinze, H. J., and Schoenfeld, M. A. (2004). Differentiation of idiopathic Parkinson's disease, multiple system atrophy, progressive supranuclear palsy, and healthy controls using magnetization transfer imaging. *Neuroimage* **21**, 229–235.

Geng, D. Y., Li, Y. X., and Zee, C. S. (2006). Magnetic resonance imaging-based volumetric analysis of basal ganglia nuclei and substantia nigra in patients with Parkinson's disease. *Neurosurgery* **58**, 256–262.

Gerlach, M., Double, K. L., Youdim, M. B., and Riederer, P. (2006). Potential sources of increased iron in the substantia nigra of parkinsonian patients. *J. Neural. Transm.* **70**(Suppl.), 133–142.

Gorell, J. M., Ordidge, R. J., Brown, G. G., Deniau, J. C., Buderer, N. M., and Helpern, J. A. (1995). Increased iron-related MRI contrast in the substantia nigra in Parkinson's disease. *Neurology* **45**, 1138–1143.

Graham, J. M., Paley, M. N., Grünewald, R. A., Hoggard, N., and Griffiths, P. D. (2000). Brain iron deposition in Parkinson's disease imaged using the PRIME magnetic resonance sequence. *Brain* **123**, 2423–2431.

Griffiths, P. D., Dobson, B. R., Jones, G. R., and Clarke, D. T. (1999). Iron in the basal ganglia in Parkinson's disease. An in vitro study using extended X-ray absorption fine structure and cryo-electron microscopy. *Brain* **122**, 667–673.

Hallgreen, B., and Sourander, P. (1958). The effect of age on the non-haemin iron in the human brain. *J. Neurochem.* **3**, 41–51.

Hill, J. M., and Switzer, R. C. 3rd. (1984). The regional distribution and cellular localization of iron in the rat brain. *Neuroscience* **11**, 595–603.

Hutchinson, M., Raff, U., and Lebedev, S. (2003). MRI correlates of pathology in parkinsonism, segmented inversion recovery ratio imaging (SIRRIM). *Neuroimage* **20**, 1899–1902.

Iova, A., Garmashov, A., Androuchtchenko, N., Kehrer, M., Berg, D., Becker, G., and Garmashov, Y. (2004). Postnatal decrease in substantia nigra echogenicity. Implications for the pathogenesis of Parkinson's disease. *J. Neurol.* **251**, 1451–1454.

Jellinger, K., Paulus, W., Grundke-Iqbal, I., Riederer, P., and Youdim, M. B. (1990). Brain iron and ferritin in Parkinson's and Alzheimer's diseases. *J. Neural Transm. Park. Dis. Dement. Sect.* **2**, 327–340.

Kallmann, B. A., Sauer, J., Schliesser, M., Warmuth-Metz, M., Flachenecker, P., Becker, G., Rieckmann, P., and Maurer, M. (2004). Determination of ventricular diameters in multiple sclerosis patients with transcranial sonography (TCS) - a two year follow-up study. *J. Neurol.* **251**, 30–34.

Liepelt, I., Wendt, A., Schweitzer, K. J., Wolf, B., Godau, J., Gaenslen, A., Bruessel, T., and Berg, D. (2008). Substantia nigra hyperechogenicity assessed by transcranial sonography is related to neuropsychological impairment in the elderly population. *J. Neural Transm.* **115**, 993–999.

Ma, S. Y., Rinne, J. O., Collan, Y., Röyttä, M., and Rinne, U. K. (1996). A quantitative morphometrical study of neuron degeneration in the substantia nigra in Parkinson's disease. *J. Neurol. Sci.* **140**, 40–45.

Martin W R. (2009). Quantitative estimation of regional brain iron with magnetic resonance imaging. *Parkinsonism Relat. Disord.* **15**, S215–S218.

Mondino, F., Filippi, P., Magliola, U., and Duca, S. (2002). Magnetic resonance relaxometry in Parkinson's disease. *Neurol. Sci.* **23**(Suppl. 2), S87–S88.

Morris, C. M., Candy, J. M., Keith, A. B., Oakley, A. E., Taylor, G. A., Pullen, R. G., Bloxham, C. A., Gocht, A., and Edwardson, J. A. (1992). Brain iron homeostasis. *J. Inorg. Biochem.* **47**, 257–265.

Niehaus, L., Savyer, N., Weber, U., Hertel, R., Trottenberg, T., and Kupsch, A. (2004). Brain parenchyma sonography in patients with essential tremor and Parkinson's disease. *Cerebrovasc. Dis.* **17**(Suppl. 4), 3.

Niehaus, L., Steinke, R., Glaser, M., Heinze, H. J., and Schott, B. (2006). Is substantia nigra echogenicity related to the dopaminergic nigrostriatal impairment in Parkinson's disease? *Klin Neurophysiol.* **37**, 72–73.

Oikawa, H., Sasaki, M., Tamakawa, Y., Ehara, S., and Tohyama, K. (2002). The substantia nigra in Parkinson Disease, Proton density-weighted spin-echo and fast short inversion time inversion-recovery MR findings. AJNR *Am. J. Neuroradiol.* **23**, 1747–1756.

Ordidge, R. J., Gorell, J. M., Deniau, J. C., Knight, R. A., and Helpern, J. A. (1994). Assessment of relative brain iron concentrations using T2-weighted and T2*-weighted MRI at 3 Tesla. *Magn. Reson. Med.* **32**, 335–341.

Pujol, J., Junqué, C., Vendrell, P., Grau, J. M., and Capdevila, A. (1992). Reduction of the substantia nigra width and motor decline in aging and Parkinson's disease. *Arch. Neurol.* **49**, 1119–1122.

Riederer, P., Sofic, E., Rausch, W. D., Schmidt, B., Reynolds, G. P., Jellinger, K., and Youdim, M. B. (1989). Transition metals, ferritin, glutathione, and ascorbic acid in parkinsonian brains. *J. Neurochem.* **52**, 515–520.

Ruprecht-Dörfler, P., Berg, D., Tucha, O., Benz, P., Meier-Meitinger, M., Alders, G. L., Lange, K. W., and Becker, G. (2003). Echogenicity of the substantia nigra in relatives of patients with sporadic Parkinson's disease. *NeuroImage* **18**, 416–422.

Ryvlin, P., Broussolle, E., Piollet, H., Viallet, F., Khalfallah, Y., and Chazot, G.(1995). Magnetic resonance imaging evidence of decreased putamenal iron content in idiopathic Parkinson's disease. *Arch. Neurol.* **52**, 583–588.

Schrag, A., Good, C. D., Miszkiel, K., Morris, H. R., Mathias, C. J., Lees, A. J., and Quinn, N. P. (2000). Differentiation of atypical parkinsonian syndromes with routine MRI. *Neurology* **54**, 697–702.

Seppi, K., and Schocke, F. H. (2005). An update on conventional and advanced magnetic resonance imaging techniques in the differential diagnosis of neurodegenerative parkinsonism. *Curr. Opin. Neurol.* **18**, 370–375.

Sofic, E., Riederer, P., Heinsen, H., Beckmann, H., Reynolds, G. P., Hebenstreit, G., and Youdim, M. B. (1988). Increased iron (III) and total iron content in post mortem substantia nigra of parkinsonian brain. *J. Neural. Transm.* **74**, 199–205.

Sofic, E., Paulus, W., Jellinger, K., Riederer, P., and Youdim, M. B. (1991). Selective increase of iron in substantia nigra zona compacta of parkinsonian brains. *J. Neurochem.* **56**, 978–982.

Sohmiya, M., Tanaka, M., Aihara, Y., and Okamoto, K. (2004). Structural changes in the midbrain with aging and Parkinson's disease, an MRI study. *Neurobiol. Aging* **25**, 449–453.

Sommer, U., Hummel, T., Cormann, K., Mueller, A., Frasnelli, J., Kropp, J., and Reichmann, H. (2004). Detection of presymptomatic Parkinson's disease, Combining smell tests, transcranial sonography, and SPECT. *Mov. Disord* **19**, 1196–1202.

Spiegel, J., Hellwig, D., Möllers, M. O., Behnke, S., Jost, W., Fassbender, K., Samnick, S., Dillmann, U., Becker, G., and Kirsch, C. M. (2006). Transcranial sonography and [123I]FP-CIT SPECT disclose complementary aspects of Parkinson's disease. *Brain* **129**, 1188–1193.

Tambasco, N., Pelliccioli, G. P., Chiarini, P., Montanari, G. E., Leone, F., Mancini, M. L., Paciaroni, M., and Gallai, V., 2003. Magnetization transfer changes of grey and white matter in Parkinson's disease. *Neuroradiology* **45**, 224–230.

Todd, G., Taylor, J. L., Baumann, D., Butler, J. E., Duma, S. R., Hayes, M., Carew-Jones, F., Piguet, O., Behnke, S., Ridding, M. C., Berg, D., and Double, K. L. (2010). Substantia nigra echomorphology and motor cortex excitability. *Neuroimage* **50**, 1351–1356.

Vymazal, J., Righini, A., Brooks, R. A., Canesi, M., Mariani, C., Leonardi, M., and Pezzoli, G. (1999). T1 and T2 in the brain of healthy subjects, patients with Parkinson disease, and patients with multiple system atrophy, relation to iron content. *Radiology* **211**, 489–495.

Walter, U., Niehaus, L., Probst, T., Benecke, R., Meyer, B. U., and Dressler, D. (2003). Brain parenchyma sonography discriminates Parkinson's disease from atypical Parkinsonian syndromes. *Neurology* **60**, 74–77.

Walter, U., Klein, C., Hilker, R., Benecke, R., Pramstaller, P., and Dressler, D. (2004). Brain parenchyma sonography detects preclinical parkinsonism. *Mov. Disord.* **19**, 1445–1449.

Walter, U., Krolikowski, K., Tarnacka, B,, Benecke, R., Czlonkowska, A., and Dressler, D. (2005). Sonographic detection of basal ganglia lesions in asymptomatic and symptomatic Wilson disease. *Neurology* **64**, 1726–1732.

Walter, U., Behnke, S., Eyding, J., Niehaus, L., Postert, T., Seidel, G., and Berg, D. (2007a). Transcranial sonography of brain parenchyma in movement disorders: state of the art. *Ultrasound Med. Biol.* **33**, 15–25.

Walter, U., Hoeppner, J., Prudente-Morrissey, L., Horowski, S., Herpertz, S. C., and Benecke, R. (2007b). Parkinson's disease-like midbrain sonography abnormalities are frequent in depressive disorders. *Brain* **130**, 1799–1807.

Walter, U., Kanowski, M., Kaufmann, J., Grossmann, A., Benecke, R., and Niehaus, L. (2008). Contemporary ultrasound systems allow high-resolution transcranial imaging of small echogenic deep intracranial structures similarly as MRI, A phantom study. *Neuroimage* **40**, 551–558.

Walter, U., Wagner, S., Horowski, S., Benecke. R., and Zettl, U. K. (2009). Transcranial brain sonography findings predict disease progression in multiple sclerosis. *Neurology* **73**, 1010–1017.

Weise, D., Lorenz, R., Schliesser, M., Schirbel, A., Reiners, K., and Classen, J. (2009). Substantia nigra echogenicity: A structural correlate of functional impairment of the dopaminergic striatal projection in Parkinson's disease. *Mov. Disord.* **24**, 1669–1675.

Zecca, L., Berg, D., Arzberger, T., Ruprecht, P., Rausch, W. D., Musicco, M., Tampellini, D., Riederer, P., Gerlach, M., and Becker, G. (2005). *In vivo* detection of iron and neuromelanin by transcranial sonography, a new approach for early detection of substantia nigra damage. *Mov. Disord.* **20**, 1278–1285.

EARLY DIAGNOSIS OF PARKINSON'S DISEASE

Alexandra Gaenslen and Daniela Berg

Department of Neurodegeneration, Hertie Institute of Clinical Brain Research and German Center of Neurodegenerative Diseases (DZNE), Tübingen, Germany

The accuracy of the clinical diagnosis of Parkinson's disease (PD) is still limited. Especially in the early stages, when cardinal symptoms are not conclusive, diagnosis can be delayed as structural neuroimaging methods such as CCT or MRI do not provide characteristic features that allow the diagnosis of this chronic neurodegenerative disorder. Functional neuroimaging using PET and SPECT techniques is helpful in patients with first signs of parkinsonism, but expensive and not broadly available. In this scenario, transcranial sonography (TCS) has proven to be helpful. Up to 90% of PD patients show hyperechogenicity of the substantia nigra (SN) on TCS. Already in the early stages of PD this echofeature is visible, allowing the differentiation of very mildly affected patients with idiopathic PD from healthy persons and from patients with atypical parkinsonism with high sensitivity and specificity. Additionally, specific ultrasound features for some forms of secondary parkinsonism can be detected by TCS, helping in the early identification, for example, of patients with Wilson's disease. Therefore, especially in the early diagnosis, TCS can be recommended as a supplementary tool to facilitate the diagnostic classification of patients with first signs of parkinsonism.

INTERNATIONAL REVIEW OF
NEUROBIOLOGY, VOL. 90
DOI: 10.1016/S0074-7742(10)90006-8

I. Difficulties in the Diagnosis of Parkinson's Disease

The accuracy of the clinical diagnosis of Parkinson's disease (PD) is still limited. According to population-based studies in the United Kingdom, there is still a high rate of false diagnoses, when comparing the initial diagnosis with later diagnoses according to the clinical UK Brain Bank criteria (Gelb *et al.*, 1999; Litvan *et al.*, 2003). Only 83% of 202 patients with the initial diagnosis of probable PD could be clinically confirmed as probable or possible PD by movement disorders specialists applying the UK Brain Bank criteria (Schrag *et al.*, 2002). As many as 5.4% of the misdiagnosed patients had atypical parkinsonian syndromes (aPS; progressive supranuclear palsy—PSP or multiple system athrophy—MSA), 4.6% were diagnosed as having vascular parkinsonism (VP), and 3.1% as having non-parkinsonian tremor. When the positive response to different forms of dopaminergic medication is considered, at least 5% of 610 patients on these anti-parkinsonian medications have diagnoses other than PD (Newman *et al.*, 2009). The most frequent misdiagnoses in this study were essential tremor (ET), VP, and tremors of other genes. In this cohort, patients had a mean disease duration of at least 5 years. In another population-based cohort of 402 PD patients re-examined at a movement disorder department, PD could not be verified in 29% of the patients with a mean disease duration of 8 years (Meara *et al.*, 1999).

From the DATATOP (deprenyl and tocopherol antioxidative therapy of Parkinson's disease) study population ($n = 800$), diagnosis other than PD had to be made in 8% of patients because of additional clinical, neuroimaging, or postmortem findings after 7 years' follow-up (Jankovic *et al.*, 2000). This clinical uncertainty becomes even more obvious when postmortem findings are considered. In a study published in 1992, including 100 PD patients with the clinical diagnosis of PD, Lewy bodies could only be found in 76% (Hughes *et al.*, 1992). In another sample of 100 brains of patients with the clinical diagnosis of PD according to the diagnostic criteria for definite PD (Calne *et al.*, 1992), sensitivity for PD according to clinical diagnosis was 90% in the neuropathologically confirmed PD cases (Hughes *et al.*, 2001).

Especially in the very early stages of the disease, when no full spectrum of clinical signs necessary to establish the clinical diagnosis is obvious, diagnosis of PD can be a real challenge. Mixed tremor (including resting and postural/action tremor) could be a sign of both ET or a parkinsonian syndrome; bradykinesia and rigidity may occur not only in PD, but also in metabolic disorders like Wilson's disease or atypical parkinsonian syndromes (aPS); hypokinesia may well be a sign of both depression and PD; and slowness and shuffling of gait may not only occur in PD, but also be associated with hydrocephalus and vascular parkinsonism, to name just a few of the frequent differential diagnoses.

A. Is it Early PD? Confirmation of the Diagnosis of PD in Very Early Affected Subjects

Since the establishment of functional neuroimaging techniques, a new era of verification of an affected nigrostriatal system has begun, allowing the diagnosis of a presynaptic dopaminergic impairment even before all symptoms of PD are evident. Different methods of neuroimaging are applied to help in the diagnosis of early parkinsonism. The conventional computer tomography (CT) and magnetic resonance imaging (MRI) allow the exclusion of atypical or secondary forms of parkinsonism, but in the routine sequences no PD-specific changes can be observed (Seppi and Poewe, 2010; Tolosa *et al.*, 2006). Functional neuroimaging of the brain and heart are more helpful. In the very early stages of PD (Hoehn and Yahr 1) it is possible to see contralaterally pronounced deficits in dopaminergic function—sometimes only contralateral to the clinically affected side, sometimes already bilaterally on [123I] beta-CIT single-photon emission computed tomography (SPECT) (Filippi *et al.*, 2005; Marek *et al.*, 1996). Later in the disease progression, dopamine transporter imaging and ^{18}F-fluorodopa positron emission tomography (^{18}F-dopa-PET), applied for visualizing the uptake and metabolism of dopa allows to differentiate PD from healthy controls with up to 100% accuracy (Brooks, 2000).

Early PD patients with a mean duration of PD symptoms of 1.26 years (SD 0.67) could be separated with ^{18}F-dopa neuroimaging from healthy volunteers with a sensitivity and specificity of 99% (Jokinen *et al.*, 2009). By contrast, routine structural neuroimaging methods such as CCT or MRI do not give any additional diagnostic information in the comparison of healthy subjects and those with first signs of parkinsonism, as there is no routine CCT or MRI parameter that allows the diagnosis of PD.

Here, the structural neuroimaging method of TCS has been proven to be a valuable supplementary tool for investigations. Up to 90% of PD patients show hyperechogenicity of the substantia nigra (SN) on TCS (Becker *et al.*, 1995; Berg *et al.*, 2001; Huang *et al.*, 2007; Ressner *et al.*, 2007; Walter *et al.*, 2002) (Fig. 1A and B). In a 5-year follow-up study, the area of hyperechogenicity did not change with disease progression (Berg *et al.*, 2005). In addition, most of the studies published so far did not find a correlation of the area of hyperechogenicity and disease stages according to Hoehn and Yahr or disease severity according to the UPDRS (Berg *et al.*, 2005; Kim *et al.*, 2007; Okawa *et al.*, 2007; Walter *et al.*, 2007). Therefore, a rather stable signal can be expected during the whole disease course, from the initial symptoms to the stage of most severe disability. Only in some studies a positive correlation with higher Hoehn and Yahr stages was found (Kolevski *et al.*, 2007; Weise *et al.*, 2009; Tsai *et al.*, 2007).

Importantly, SN hyperechogenicity can also be found in healthy subjects. The prevalence of this echofeature is about 10% in the healthy adult population (Berg *et al.*, 1999). Only in the old and especially very old an increased prevalence of

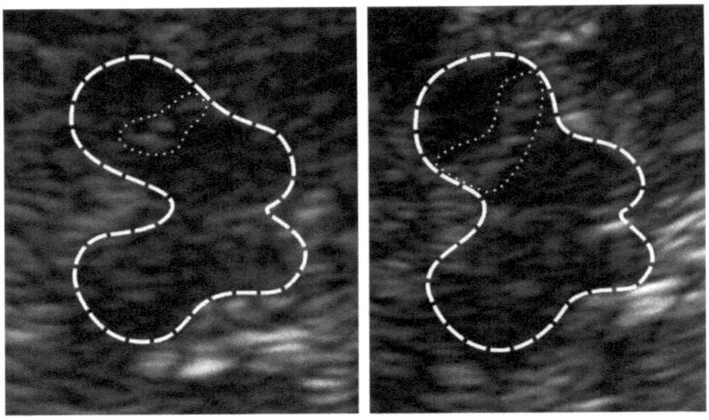

FIG. 1. (A) Zoomed image of the butterfly-shaped hypoechogenic brainstem (encircled with a dashed line) in the mesencephalic scanning plane. A small area of hyperechogenicity is visible at the anatomical area of the SN (encircled with a dotted line) ipsilateral to the insonating ultrasound probe—subject with normal echogenicity of the SN. (B) Subject with hyperechogenicity of the SN. Here a large area of hyperechogenicity is visible at the anatomical area of the SN (encircled with a dotted line) ipsilateral to the insonating ultrasound probe.

hyperechogenicity of SN has been found. The prevalence of hyperechogenicity of the SN in adults older than 60 years has been described to be 13.9% (Berg *et al.*, 2001) and increases up to 25% in very old healthy persons (mean age 89.2 years) (Behnke *et al.*, 2007). However, in these studies with elderly people, no correlation between the area of SN hyperechogenicity and the age of probands could be found, as also described in other studies (Hoeppner *et al.*, 2009). Whether there is a correlation of area of SN echogenicity and age is still a matter of debate, because other studies have seen a positive correlation with age (Huang *et al.*, 2007).

However, it seems that these changes may primarily become obvious in the old and very old population. Here, it could have another cause and meaning than in the younger ones. At least up to 60 years of age the prevalence of SN hyperechogenicity is stable. Here, a distinct meaning in subjects carrying this echofeature may be supposed.

Nevertheless sensitivity of SN hyperechogenicity to separate PD patients from healthy persons in early stages (mean disease duration 2.5 ± 1.5 years) is high, reaching 86% with a positive predictive value of 95% for the diagnosis PD (Doepp *et al.*, 2008).

B. Differentiation of PD from aPS in the Early Phase

In the differentiation of PD and atypical parkinsonism, neuroimaging techniques displaying the functional status of the presynaptic nigrostriatal system (i.e., 18F-dopa PET and DATScan) are less specific. Only 64% of PD patients could be correctly differentiated from MSA patients in a study using 18F-dopa PET, which included 28 PD patients, 25 MSA, and 10 PSP patients (Burn *et al.*, 1994). This uncertainty of diagnosis can at least up to a certain extent be improved by postsynaptic neuroimaging. For studies using SPECT with [123I] iodobenzamide (IBZM) a sensitivity of up to 80% and specificity of up to 71% for the correct diagnosis for MSA has been shown (Seppi *et al.*, 2004). A greater diagnostic accuracy may be obtained by (18)F-fluorodeoxyglucose PET ((18)F-FDG PET). With a sensitivity of 95.8% and a specificity of 100%, 24 patients with MSA of the parkinsonian type could be differentiated from eight PD patients in the first 3 years after symptom onset (Kwon *et al.*, 2008). However, so far, it has not been established how early in the disease course can metabolic alterations typical for atypical parkinsonian syndromes be found.

Using MRI, typical structural alterations can be detected in aPS such as PSP (midbrain atrophy (Mickey and Pinguin Sign) enlarged third ventricle), MSA (hot cross bun sign, putaminal rim), or corticobasal degeneration (CBD) (unilateral cortical atrophy) that allow the differentiation of these aPS from idiopathic PD. However, these features usually occur as signs of advanced atrophy and are therefore usually not helpful for the early differential diagnosis (Seppi and Schocke, 2005).

Here, TCS again provides helpful supplementary diagnostic information. Typically, in patients with aPS echogenicity of the SN is normal, whereas often basal ganglia (BG) are hyperechogenic—see also Chapter IV.6 (Berg *et al.*, 2008). Especially for the very early diagnosis of PD a recently published study yields promising results (Fig. 2) (Gaenslen *et al.*, 2008). This study was designed to test the diagnostic accuracy of TCS for the diagnosis of idiopathic PD and atypical PS in 60 subjects presenting with first parkinsonian symptoms. Patients included were those early in the disease course (mean duration of symptoms, 15.5 months), and at the time of first diagnosis a diagnosis, according to the UK Brain Bank criteria, could be made only in 63%. After 12 months' follow-up the diagnosis of PD could be established in 39 patients, aPS in 10, and parkinsonism in four was excluded (initial symptoms were mainly caused by depressive disorders) whereas in seven the diagnosis was still unclear. In contrast to the clinical uncertainty at baseline, TCS findings at baseline corresponded very well with the diagnosis after 1-year follow-up. Indeed, SN hyperechogenicity at baseline was found to have a sensitivity of 90.7% for the diagnosis of PD, the specificity was 82.4%, with a positive predictive value of 92.9%, and a negative predictive value of 77.8% (Gaenslen *et al.*, 2008).

C. Differentiation of PD from ET in the Early Phase

Differentiation of PD and ET in the early disease phase is often a major challenge because both diseases can mimic each other, especially because mixed forms of tremor can occur in both.

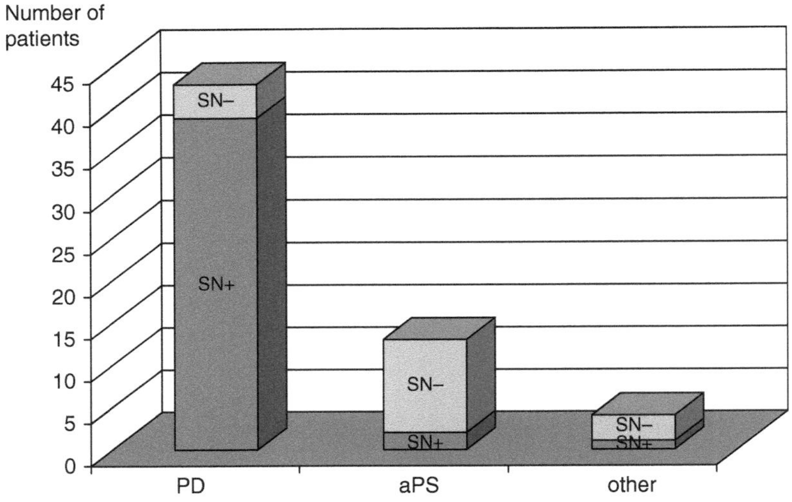

FIG. 2. Results of the study of Gaenslen et al., 2008. Number of patients with a clinical diagnosis established after 1 year of follow-up since presentation of their first symptoms. In the vast majority of patients, who were classified as having idiopathic PD already at baseline, SN hyperechogenicity could be detected. The sensitivity of this ultrasound sign for the diagnosis of PD was 90.7%. Only one of the patients was classified as having atypical form of parkinsonism had SN hyperechogenicity at baseline. In four patients other diagnoses, mainly depression, led to the presentation with parkinsonism at baseline. Of these, only one presented with SN hyperechogenicity. Whether this echofeature indicates an increased risk for PD in this patient needs to be determined in follow-up investigations.

Conventional MRI and CT are normal in these patients. Only advanced MRI techniques seem to reveal disease-specific changes, that is, high-field MRI (3-tesla) using voxel-based morphometry showed structural changes in 19 patients with ET in the midbrain, occipital lobe, frontal lobe, and cerebellum (Benito-Leon *et al.*, 2009). In contrast to these structural neuroimaging techniques, functional presynaptic neuroimaging has proven to be very helpful in the early differential diagnosis of ET and PD. As an example, ([11C]dMP or [11-C]FE-CIT) scans showed in early PD patients in (H & Y 1–2) reduced dopaminergic function and normal imaging in the patients with ET (Breit *et al.*, 2006).

Using TCS, also a high classification accuracy concerning ET and PD can be reached (see also chapter on Transcranial Sonography in Essential Tremor). In general, patients with ET display normal echogenicity of the SN. However, there is a slightly higher percentage of ET subjects with SN hyperechogenicity than found in the normal population (13–16%), which may be due to a higher risk for conversion to PD in ET patients compared with healthy subjects. (Budisic *et al.*, 2009; Stockner *et al.*, 2007). Still, sensitivity and specificity of TCS for the diagnosis of PD are high (78% and 92%, respectively) (Doepp *et al.*, 2008). The additional implementation of 123I-FP-CIT-SPECT could improve it up to 100% (Doepp *et al.*, 2008), indicating that TCS could be used as a first-line imaging tool, which may be complemented by presynaptic functional imaging in still unclear cases.

D. Differentiation of PD from Wilson's Disease in the Early Phase

Although seldom, the diagnosis of Wilson's disease (WD) is extremely important in the early phase of the disease, because causal treatment is essential for the disease course. Young-onset PD is one of the major differential diagnoses of WD. Especially at the onset of first symptoms of WD, which may include slight tremor, slight rigidity, and/or hypokinesia and also slight gait ataxia or dystonia, patients can be mistaken for having PD. Using functional neuroimaging, an affection of the presynaptic dopaminergic system can be seen in both disease entities. MRI does show typical lesions in the BG; however, at very mild stages these can be missed (Kim *et al.*, 2006). Using TCS, hyperechogenicity of the SN can be found in up to 50% of patients with Wilson's disease, implicating that this echofeature is not helpful for the differentiation of PD and Wilson's disease (Walter *et al.*, 2005). However, one study demonstrated that in all of the 21 Wilson's disease patients hyperechogenicity of the lenticular nucleus (LN) could be found. The area of hyperechogenicity seems to correlate with clinical disease severity, and important for the early diagnosis, this alteration can often already be seen before structural abnormalities are evident on MRI (Walter *et al.*, 2005; see also chapter on Transcranial Sonography in Brain Disorders with Trace Metal Accumulation).

E. Differentiation of PD from Motor Slowing Associated with Depression in the Early Disease Stage

Clinically, slight motor slowing in depressed subjects cannot be differentiated from early parkinsonian symptoms. The question whether a patient suffers from motor slowing associated with depression or PD occurs frequently, even more so, as depression frequently antedates PD. Again, functional neuroimaging may help to depict changes of the presynaptic dopaminergic system typical for PD but not for depression. By contrast, routine MRI and CCT do not give additional information to help to differentiate these disease entities.

Two features visible on TCS, however, can alleviate diagnosis. Normally, no hyperechogenicity of the SN is found in depressed patients, although it needs to be considered that more depressed patients show SN hyperechogenicity (up to 40%) than normal subjects (Walter *et al.*, 2007), which is attributed to the fact that depression may antecede PD and that depressed subjects with SN+ have a higher chance to convert to PD. The second sign for differential diagnosis is a reduced echogenicity of the mesencepahlic midbrain raphe, which is either interrupted or completely missing in depressed patients (see also Chapter Transcranial Sonography In Depression; Becker *et al.*, 1997). From the combination of these two signs it may therefore be derived:

- Reduced raphe echogenicity and normal SN—depression
- Hyperechogenicity of the SN and normal raphe—PD
- Reduced raphe and hyperechogenicity of the SN—depression plus PD or depression and a vulnerability of PD, which, however, may never manifest during life time.

F. Is it PD or Vascular Parkinsonism?

Elderly people often suffer from years of hypertension, which have left their tracks in the brain. On MRI, typically small vascular lesions are seen in the white matter surrounding the ventricular system. If a patient with parkinsonism presents such lesions, it needs to be investigated whether the lesions have led to the clinical presentation of the patient suffering from additional PD (see also chapter on Transcranial Sonography in the Discrimination of Parkinson's Disease Versus Vascular Parkinsonism). If functional neuroimaging can be performed and shows a typical presynaptic pattern of degeneration with side difference and stronger affection of the dorsal part of the putamen, diagnosis of (additional) PD is possible. However, sometimes results from functional neuroimaging are not that typical, and vascular lesions may also lead to abnormalities on the scans. Also, using TCS the differentiation between VP and PD may sometimes be difficult (see also chapter on Transcranial Sonography in the Discrimination of Parkinson's Disease Versus Vascular Parkinsonism). The prevalence of hyperechogenicity of the SN in patients with VP is 20% higher than in healthy controls (about 10%) but lower than in PD patients (84–90%), as shown in a first study of Tsai *et al.* (Becker *et al.*, 1995; Tsai *et al.*, 2007). Here, the additional use of color-coded sonography can be of value, as patients with VP often have signs of atherosclerosis, which are generally missing in typical idiopathic PD (Tsai *et al.*, 2007). So far, studies matching clinical diagnosis, neuroimaging including TCS, and postmortem findings are missing. This is important with regard to the fact that some patients with VP do have a beneficial effect from dopaminergic therapy. Therefore, all studies relating neuroimaging finding to clinical diagnosis need to be evaluated with caution.

Clinically, patients with *hydrocephalus* often present with a type of parkinsonism that is very similar to the parkinsonian features observed in patients suffering from VP, that is, primarily gait is affected. Here, visualization of the ventricular system is of utmost importance. Defined normal values for the ventricular system help in the decision whether pressure of the ventricular system may be responsible for the extrapyramidal features, which are often associated with cognitive decline and incontinence. Based on the assessment done by TCS, further diagnostic and therapeutic steps such as lumbar puncture can be initiated. Normal values are given in Chapter III.

G. TCS in the Early Differential Diagnosis of Parkinsonian Syndromes—Summary

Regarding this summary of early differential diagnoses of parkinsonian syndromes, it becomes evident that TCS is an important supplementary tool to alleviate early diagnosis. As both therapeutic strategies and prognoses differ in the different forms of parkinsonism, it is essential to obtain the most secure diagnosis possible as early as possible (Maetzler *et al.*, 2009).

In sum, the prevalence difference of SN hyperechogenicity in PD patients and in the healthy population allows to differentiate these two groups with TCS with high predictable value of 85–91% (Budisic *et al.*, 2009; Prestel *et al.*, 2006).

This diagnostic accuracy could also be demonstrated in the above-mentioned study of 60 *de novo* patients with first parkinsonian symptoms—see above (Gaenslen *et al.*, 2008), demonstrating at baseline a sensitivity of 90.7% for the diagnosis of PD with a specificity as 82.4%, rendering a positive predictive value of 92.9%. This example shows that TCS can be very helpful and elusive for the very early diagnosis of PD, even when the clinical signs are not prominent enough.

II. Perspectives for Early Diagnosis of PD

The fact that SN hyperechogenicity can also be found in healthy subjects who have additional premotor markers for PD and the finding of a conversion to PD in healthy subjects with SN hyperechogenicity implies that in the future early diagnosis may even be shifted to an even earlier period of PD (see also chapter Transcranial Sonography in the Premotor Diagnosis of Parkinson's Disease).

A standardized screening procedure with acquisition of very early motor (such as slowing hand movements or reduced arm swing) and nonmotor symptoms (such as olfactory dysfunction, depression, REM sleep disorders, or constipation) could help to identify persons at high risk for PD, which may be followed by examination with TCS and functional neuroimaging for early detection of dopaminergic dysfunction. This group may benefit the most from the yet-to-develop neuroprotective medication.

References

Becker, G., Seufert, J., Bogdahn, U., Reichmann, H., and Reiners, K. (1995). Degeneration of substantia nigra in chronic Parkinson's disease visualized by transcranial color-coded real-time sonography. *Neurology* **45**(1), 182–184.

Becker, T., Becker, G., Seufert, J., Hofmann, E., Lange, K. W., Naumann, M., Lindner, A., Reichmann, H., Riederer, P., Beckmann, H., and Reiners, K. (1997). Parkinson's disease and depression: evidence for an alteration of the basal limbic system detected by transcranial sonography. *J. Neurol. Neurosurg Psychiatr.* **63**(5), 590–596.

Behnke, S., Double, K. L., Duma, S., Broe, G. A., Guenther, V., Becker, G., and Halliday, G. M. (2007). Substantia nigra echomorphology in the healthy very old: Correlation with motor slowing. *Neuroimage* **34**(3), 1054–1059.

Benito-Leon, J., Alvarez-Linera, J., Hernandez-Tamames, J. A., Alonso-Navarro, H., Jimenez-Jimenez, F. J., and Louis, E. D. (2009). Brain structural changes in essential tremor: Voxel-based morphometry at 3-Tesla. *J. Neurol. Sci.* **287**(1–2), 138–142.

Berg, D., Becker, G., Zeiler, B., Tucha O., Hofmann E., Preier M., Benz, P., Jost, W., Reiners, K., and Lange, K. W. (1999). Vulnerability of the nigrostriatal system as detected by transcranial ultrasound. *Neurology* **53**(5), 1026–1031.

Berg, D., Godau, J., and Walter, U. (2008). Transcranial sonography in movement disorders. *Lancet Neurol.* **7**(11), 1044–1055.

Berg, D., Merz, B., Reiners, K., Naumann, M., and Becker, G. (2005). Five-year follow-up study of hyperechogenicity of the substantia nigra in Parkinson's disease. *Mov. Disord.* **20**(3), 383–385.

Berg, D., Siefker, C., and Becker, G. (2001). Echogenicity of the substantia nigra in Parkinson's disease and its relation to clinical findings. *J. Neurol.* **248**(8), 684–689.

Berg, D., Siefker, C., Ruprecht-Dorfler, P., and Becker, G. (2001). Relationship of substantia nigra echogenicity and motor function in elderly subjects. *Neurology* **56**(1), 13–17.

Breit, S., Reimold, M., Reischl, G., Klockgether, T., and Wullner, U. (2006). [(11)C]d-threo-methylphenidate PET in patients with Parkinson's disease and essential tremor. *J. Neural Transm.* **113**(2), 187–193.

Brooks, D. J. (2000). Morphological and functional imaging studies on the diagnosis and progression of Parkinson's disease. *J. Neurol.* **247**(Suppl. 2), II11–II18.

Budisic, M., Trkanjec, Z., Bosnjak, J., Lovrencic-Huzjan, A., Vukovic, V., and Demarin, V. (2009). Distinguishing Parkinson's disease and essential tremor with transcranial sonography. *Acta Neurol. Scand.* **119**(1), 17–21.

Burn, D. J., Sawle, G. V., and Brooks, D. J. (1994). Differential diagnosis of Parkinson's disease, multiple system atrophy, and Steele–Richardson–Olszewski syndrome: discriminant analysis of striatal 18F-dopa PET data. *J. Neurol. Neurosurg. Psychiatr.* **57**(3), 278–284.

Calne, D. B., Snow, B. J., and Lee, C. (1992). Criteria for diagnosing Parkinson's disease. *Ann. Neurol.* **32**(Suppl), S125–S127.

Doepp, F., Plotkin, M., Siegel, L., Kivi, A., Gruber, D., Lobsien, E., Kupsch, A., and Schreiber, S. J. (2008). Brain parenchyma sonography and 123I-FP-CIT SPECT in Parkinson's disease and essential tremor. *Mov. Disord.* **23**(3), 405–410.

Filippi, L., Manni, C., Pierantozzi, M., Brusa, L., Danieli, R., Stanzione, P., and Schillaci, O. (2005). 123I-FP-CIT semi-quantitative SPECT detects preclinical bilateral dopaminergic deficit in early Parkinson's disease with unilateral symptoms. *Nucl. Med. Commun.* **26**(5), 421–426.

Gaenslen, A., Unmuth, B., Godau, J., Liepelt, I., Di Santo, A., Schweitzer, K. J., Gasser, T., Machulla, H. J., Reimold, M., Marek, K., and Berg, D. (2008). The specificity and sensitivity of transcranial ultrasound in the differential diagnosis of Parkinson's disease: a prospective blinded study. *Lancet Neurol.* **7**(5), 417–424.

Gelb, D. J., Oliver, E., and Gilman, S. (1999). Diagnostic criteria for Parkinson disease. *Arch. Neurol.* **56**(1), 33–39.

Hoeppner, J., Prudente-Morrissey, L., Herpertz, S. C., Benecke, R., and Walter, U., (2009). Substantia nigra hyperechogenicity in depressive subjects relates to motor asymmetry and impaired word fluency. *Eur. Arch. Psychiatr. Clin. Neurosci.* **259**(2), 92–97.

Huang, Y. W., Jeng, J. S., Tsai, C. F., Chen, L. L., and Wu, R. M. (2007). Transcranial imaging of substantia nigra hyperechogenicity in a Taiwanese cohort of Parkinson's disease. *Mov. Disord.* **22**(4), 550–555.

Hughes, A. J., Daniel, S. E., Kilford, L., and Lees, A. J. (1992). Accuracy of clinical diagnosis of idiopathic Parkinson's disease: a clinico-pathological study of 100 cases. *J. Neurol. Neurosurg. Psychiatr.* **55**(3), 181–184.

Hughes, A. J., Daniel, S. E., and Lees, A. J. (2001). Improved accuracy of clinical diagnosis of Lewy body Parkinson's disease. *Neurology* **57**(8), 1497–1499.

Jankovic, J., Rajput, A. H., McDermott, M. P., and Perl, D. P. (2000). The evolution of diagnosis in early Parkinson disease. Parkinson Study Group. *Arch. Neurol.* **57**(3), 369–372.

Jokinen, P., Helenius, H., Rauhala, E., Bruck, A., Eskola, O., and Rinne, J. O. (2009). Simple ratio analysis of 18F-fluorodopa uptake in striatal subregions separates patients with early Parkinson disease from healthy controls. *J. Nucl. Med.* **50**(6), 893–899.

Kim, J. Y., Kim, S. T., Jeon, S. H., and Lee, W. Y. (2007). Midbrain transcranial sonography in Korean patients with Parkinson's disease. *Mov. Disord.* **22**(13), 1922–1926.

Kim, T. J., Kim, I. O., Kim, W. S., Cheon, J. E., Moon, S. G., Kwon, J. W., Seo, J. K., and Yeon, K. M. (2006). MR imaging of the brain in Wilson disease of childhood: findings before and after treatment with clinical correlation. *AJNR Am. J. Neuroradiol.* **27**(6), 1373–1378.

Kolevski, G., Petrov, I., and Petrova, V. (2007). Transcranial sonography in the evaluation of Parkinson disease. *J. Ultrasound Med.* **26**(4), 509–512.

Kwon, K. Y., Choi, C. G., Kim, J. S., Lee, M. C., and Chung, S. J. (2008). Diagnostic value of brain MRI and F-18-FDG PET in the differentiation of parkinsonian type multiple system atrophy from Parkinson's disease. *Eur. J. Neurol.* **15**(10), 1043–1049.

Litvan, I., Bhatia, K. P., Burn, D. J., Goetz, C. G., Lang, A. E., McKeith, I., Sethi K.D, Shults C, Wenning G. K., and Movement Disorders Society Scientific Issues Committee. (2003). Movement Disorders Society Scientific Issues Committee report: SIC Task Force appraisal of clinical diagnostic criteria for Parkinsonian disorders. *Mov. Disord.* **18**(5), 467–486.

Maetzler, W., Liepelt, I., and Berg, D. (2009). Progression of Parkinson's disease in the clinical phase: potential markers. *Lancet Neurol.* **8**(12), 1158–1171.

Marek, K. L., Seibyl, J. P., Zoghbi, S. S., Zea-Ponce, Y., Baldwin, R. M., Fussell, B., Charney, D. S., van Dyck, C., Hoffer, P. B., and Innis, R. P. (1996). [123I] beta-CIT/SPECT imaging demonstrates bilateral loss of dopamine transporters in hemi-Parkinson's disease. *Neurology* **46**(1), 231–237.

Meara, J., Bhowmick, B. K., and Hobson, P. (1999). Accuracy of diagnosis in patients with presumed Parkinson's disease. *Age Ageing.* **28**(2), 99–102.

Newman, E. J., Breen, K., Patterson, J., Hadley, D. M., Grosset, K. A., and Grosset, D. G. (2009). Accuracy of Parkinson's disease diagnosis in 610 general practice patients in the West of Scotland. *Mov. Disord.*

Okawa, M., Miwa, H., Kajimoto, Y., Hama, K., Morita, S., Nakanishi, I., and Kondo T. (2007). Transcranial sonography of the substantia nigra in Japanese patients with Parkinson's disease or atypical parkinsonism: clinical potential and limitations. *Intern. Med.* **46**(18), 1527–1531.

Prestel, J., Schweitzer, K. J., Hofer, A., Gasser, T., and Berg, D. (2006). Predictive value of transcranial sonography in the diagnosis of Parkinson's disease. *Mov. Disord.* **21**(10), 1763–1765.

Ressner, P., Skoloudik, D., Hlustik, P., and Kanovsky, P. (2007). Hyperechogenicity of the substantia nigra in Parkinson's disease. *J. Neuroimaging* **17**(2), 164–167.

Schrag, A., Ben-Shlomo, Y., and Quinn, N. (2002). How valid is the clinical diagnosis of Parkinson's disease in the community? *J. Neurol. Neurosurg. Psychiatr.* **73**(5), 529–534.

Seppi, K., and Poewe, W. (2010). Brain magnetic resonance imaging techniques in the diagnosis of parkinsonian syndromes. *Neuroimaging Clin. N. Am.* **20**(1), 29–55.

Seppi, K., and Schocke, M. F. (2005). An update on conventional and advanced magnetic resonance imaging techniques in the differential diagnosis of neurodegenerative parkinsonism. *Curr. Opin. Neurol.* **18**(4), 370–375.

Seppi, K., Schocke, M. F., Donnemiller, E., Esterhammer, R., Kremser, C., Scherfler, C., Diem, A., Jaschke, W., Wenning, G. K., and Poewe, W. (2004). Comparison of diffusion-weighted imaging and [123I]IBZM-SPECT for the differentiation of patients with the Parkinson variant of multiple system atrophy from those with Parkinson's disease. *Mov. Disord.* **19**(12), 1438–1445.

Stockner, H., Sojer, M.K, K. S., Mueller, J., Wenning, G. K., Schmidauer, C., and Poewe, W. (2007). Midbrain sonography in patients with essential tremor. *Mov. Disord.* **22**(3), 414–417.

Tolosa, E., Wenning, G., and Poewe, W. (2006). The diagnosis of Parkinson's disease. *Lancet Neurol.* **5** (1), 75–86.

Tsai, C. F., Wu, R. M., Huang, Y. W., Chen, L. L., Yip, P. K., and Jeng, J. S. (2007). Transcranial color-coded sonography helps differentiation between idiopathic Parkinson's disease and vascular parkinsonism. *J. Neurol.* **254**(4), 501–507.

Walter, U., Dressler, D., Wolters, A., Wittstock, M., and Benecke, R., (2007). Transcranial brain sonography findings in clinical subgroups of idiopathic Parkinson's disease. *Mov. Disord.* **22**(1), 48–54.

Walter, U., Hoeppner, J., Prudente-Morrissey, L., Horowski, S., Herpertz, S. C., and Benecke, R. (2007). Parkinson's disease-like midbrain sonography abnormalities are frequent in depressive disorders. *Brain* **130**(Pt 7), 1799–1807.

Walter, U., Krolikowski, K., Tarnacka, B., Benecke, R., Czlonkowska, A., and Dressler, D. (2005). Sonographic detection of basal ganglia lesions in asymptomatic and symptomatic Wilson disease. *Neurology* **64**(10), 1726–1732.

Walter, U., Wittstock, M., Benecke, R., and Dressler, D. (2002). Substantia nigra echogenicity is normal in non-extrapyramidal cerebral disorders but increased in Parkinson's disease. J Neural Transm. **109**(2), 191–196.

Weise, D., Lorenz, R., Schliesser, M., Schirbel, A., Reiners, K., and Classen, J. (2009). Substantia nigra echogenicity: A structural correlate of functional impairment of the dopaminergic striatal projection in Parkinson's disease. Mov Disord. **24**(11), 1669–1675.

TRANSCRANIAL SONOGRAPHY IN THE PREMOTOR DIAGNOSIS OF PARKINSON'S DISEASE

Stefanie Behnke,* Ute Schröder,* and Daniela Berg†

*Department of Neurology, University Hospital of the Saarland, Homburg/Saarland, Germany
†Department of Neurodegeneration, Hertie Institute of Clinical Brain Research and German Center of Neurodegenerative Diseases (DZNE), Tübingen, Germany

Substantia nigra hyperechogenicity (SN+) is a common finding in transcranial ultrasound studies of parkinsonian patients. However, this echofeature is also found in 10% of healthy controls and is here associated with PD risk factors such as male gender and positive family history for PD, as well as with putative premotor symptoms such as hyposmia, depression, sleep disturbances, and subtle neuropsychological and motor impairment. SN+ is hypothesized to indicate an increased risk to develop PD during lifetime. PET studies substantiate this hypothesis, given that SN+ controls showed reduced striatal [^{18}Fluoro] Dopa uptake in PET studies. First results of longitudinal follow-up studies show a higher relative risk for future PD in SN+ asymptomatic individuals.

I. Introduction

In 1995, Georg Becker *et al.* (1995a) described for the first time in transcranial ultrasound (TCS) studies an abnormal increase of echogenicity within the mesencephalic brainstem at the anatomical site of the substantia nigra (SN) in patients

INTERNATIONAL REVIEW OF NEUROBIOLOGY, VOL. 90
DOI: 10.1016/S0074-7742(10)90007-X

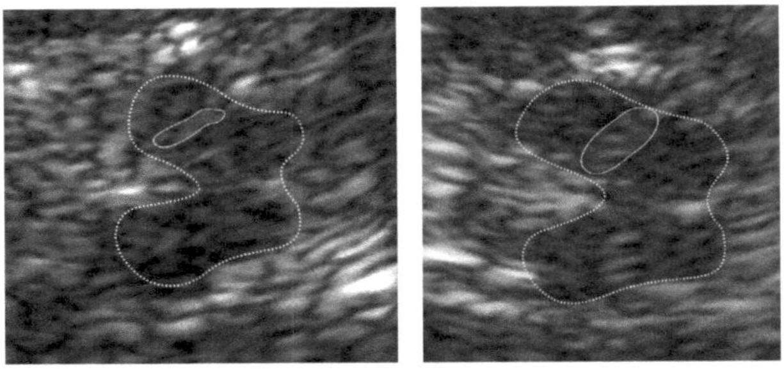

FIG. 1. (**A**) Sonographic image of the mesencephalic brainstem, which is encircled with a dotted line. The normoechogenic SN ipsilateral to the insonating probe is encircled with a normal line. (**B**) Sonographic image of the mesencephalic brainstem with bilateral increase of echogenicity at the anatomical site of the substantia nigra in a patient with Parkinson's disease. The brainstem is encircled with a dotted line, the SN ipsilateral to the insonating probe with normal line.

with Parkinson's disease (Fig. 1). He suggested this hyperechogenicity of the substantia nigra (SN+) to constitute an independent biomarker of the disease. This was revolutionary as other standard structural neuroimaging techniques do not show abnormalities indicative for Parkinson's disease. Interestingly, 2 of the 30 healthy control subjects in this pilot study exhibited SN+ as well. When one of both developed PD in the further course (unpublished data), this was a thrilling finding because it suggested that the newly described disease marker SN+ may be apparent *before* clinical symptoms emerge and may therefore indicate a vulnerability of the nigrostriatal system with the risk to develop PD during lifetime.

II. Prevalence of SN+ in Healthy Controls

In the following years, several studies examined the prevalence of the echomarker within the normal population. In the first large series, 330 healthy volunteers aged 20–80 years were included. Although none of them showed any clinical sign of extrapyramidal disorders, 8.6% were found to exhibit SN+ (Berg *et al.*, 1999a). This prevalence rate was adapted to ~10% in light of other cross-sectional studies (Berg *et al.*, 2002) and with regard to the definition that hyperechogenicity constitutes an area of echogenicity at the anatomical site of the SN larger than the 90th percentile of the normal population. Under certain conditions such as a positive family history of PD or in very advanced age groups, the prevalence of SN+ turned out to be even higher. Ruprecht-Doerfler *et al.* could show that in unaffected first-degree relatives of parkinsonian patients, SN+ is present in up to 40% (2003). In a non-parkinsonian study population with very advanced age (mean age of 86 years) the SN+ proportion reached 30% (Behnke *et al.*, 2007). More recently examined large study populations with more than 1000 healthy individuals aged >50 years showed SN+ in about 16% of cases probably due to both the higher age group and a high proportion of first-degree PD relatives participating in the trial (Schweitzer *et al.*, 2007). Interestingly, in this cohort, it could be shown that the known epidemiological risk factors for PD, male gender and positive family history, are associated with the echostatus SN+, too (Schweitzer *et al.*, 2007), corroborating the hypothesis that SN+ may show an individually increased risk to develop PD later in life.

III. Association of SN+ with Putative Premotor Symptoms of PD

Apart from the typical motor symptoms of PD leading to the diagnosis according to the British Brain Bank criteria—bradykinesia, rigidity, tremor, and postural instability—the presence of a number of nonmotor signs in PD patients is well established. Given that according to the pathological stages proposed by Braak, neurodegeneration in PD ascends through the brainstem, starting far apart from the substantia nigra it does not surprise that nonmotor symptoms may precede motor retardation for years to decades (Tolosa *et al.*, 2007) before the neurodegenerative process reaches the mesencephalon and crosses the threshold toward clinical or better motor manifestation in the late Braak stage 3 (Braak *et al.*, 2003). Among these putative premotor symptoms of PD olfactory deficits, dysautonomia especially regarding the gastrointestinal system, REM sleep behavior disorder, mood disturbances, and mild cognitive deficits are the most intensively examined signs suggested to frequently occur already in disease stages prior to the development of motor impairment (Tolosa *et al.*, 2007). Braak staging describes the PD-associated neurodegeneration as an ascending process of PD-related pathology involving earliest the dorsal vagus nucleus in not only the lower brainstem and the olfactory bulb but also the peripheral nervous system with alpha-synucleinopathy in the enteric plexus in stage 1, then ascending through the pontine brainstem with affection of the locus ceruleus and pedunculopontine as well as raphe nuclei in stage 2 (Braak *et al.*, 2003). Lewy bodies and neurites in the mesencephalon with impairment of the substantia nigra have only been detected in the later stage 3 before the neurodegenerative process proceeds involving first the mesocortex and then neocortical association areas and the whole neocortex in the stages 4–6 (Braak *et al.*, 2003). Not only these pathological findings but also evidence from epidemiological surveys and clinical studies have further substantiated the existence of a premotor phase in PD (de Lau *et al.*, 2006; Abbott *et al.*, 2003; Iranzo *et al.*, 2006; Becker

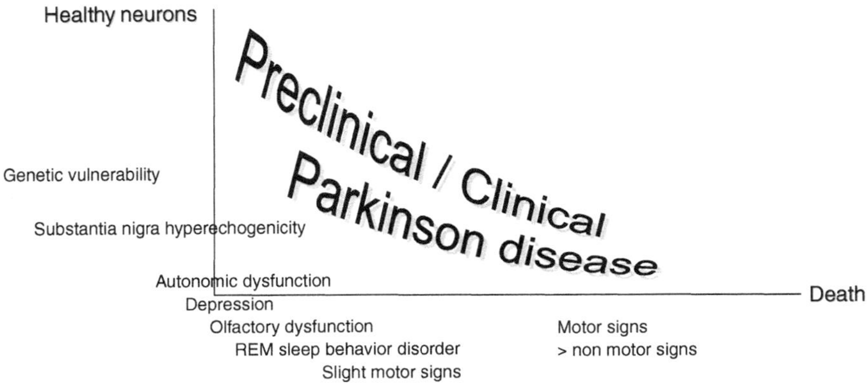

FIG. 2. Putative premotor signs in the development of Parkinson's disease.

et al., 2002). It has been suggested that, at the time point of PD diagnosis according to the established criteria, as many as 60% of nigral neurons in the ventrolateral part of the substantia nigra have degenerated (Fearnley and Lees, 1991) and, consecutively, the dopamine content in the striatum has been reduced by 60–80% (Riederer and Wuketich, 1976; Hornykiewicz and Kish, 1987). Hence, it is conclusive that early clinical symptoms, dopamine depending or not, may be detected if looked for carefully (Fig. 2). With the question whether the vulnerability marker SN+ goes along with clinical signs of the premotor stages of PD, a number of studies examined the association between the SN echostatus in healthy controls and the presence of nonmotor symptoms of PD.

A. OLFACTORY DEFICITS

The pathoanatomic correlate of Lewy bodies and neurites in PD can be detected in the olfactory system in the early Braak stages of neurodegeneration 1 and 2. Accordingly, a high proportion of 80–90% of PD patients exhibits olfactory dysfunction including reduced sense of odor detection, odor discrimination, as well as identification (Berendse and Ponsen, 2006; Doty *et al.*, 1984). Hyposmia may antecede PD as has been shown not only in asymptomatic mutation carriers for monogenetic PD (Markopoulou *et al.*, 1997) but also in the follow-up of healthy subjects with idiopathic hyposmia (Ponsen *et al.*, 2009; Ross *et al.*, 2008). In a large two-center TCS study comprising 1204 healthy controls, hyposmia as assessed in screening tests of odor identification was significantly associated with the echostatus SN+ (Liepelt *et al.*, 2009). In patients diagnosed with idiopathic olfactory loss, TCS studies turned out to show SN+ in more than 30%. Among these

SN+ hyposmic individuals, 70% had signs of nigrostriatal dysfunction in (123)I-FP-CIT-SPECT scans, 50% with a clearly pathological reduction of median striatal uptake ratios (Sommer *et al.*, 2004). In the 4-year follow-up of these subjects, three out of the 11 SN+ patients had developed motor abnormalities in the neurological examination (Haehner *et al.*, 2007).

B. Depression

Depression is another highly frequent nonmotor complaint ofparkinsonian patients occurring in at least one-third of PD patients (Brooks and Doder, 2001), in many of them depressive symptoms even preceded motor symptoms. The fact that depression may therefore constitute a premotor symptom of PD is underscored by the observations that depressive patients have an up to threefold increased risk to develop PD (Leentjens *et al.*, 2003). According to the Braak stages, brainstem structures such as serotonergic raphe nuclei and the locus ceruleus are involved in the preclinical stage 2 of the neurodegenerative process in PD (Braak *et al.*, 2003). Especially the locus ceruleus is supposed to play a key role in the development of mood and anxiety disturbances in PD patients (Chan-Palay and Asan, 1989).

In early TCS studies examining depressive patients, Becker *et al.* could show that the echomorphology of the mesencephalic brainstem is typically altered with an interrupted or absent echogenic midline signal, the so-called brainstem raphe (Becker *et al.*, 1995b; Berg *et al.*, 1999b, see also Chapter on TCS in depression). Walter *et al.* confirmed this abnormality of brainstem raphe echogenicity in depressive patients and showed additionally that depressed non-parkinsonian individuals display SN+ in about 40% of cases when examined by TCS compared to 13% of SN+ in non-parkinsonian controls without affective disorder (2007). He found the highest prevalence of SN+ (87%) in PD patients with concomitant depression, with larger hyperechogenic SN areas in those with depressive symptoms prior to PD onset. This may lead to the conclusion that in depressive patients the presence of SN+ might indicate a higher risk of developing PD over time.

C. REM Sleep Behavior Disorder

According to the Braak stages, alpha-synucleinopathy involves sleep-related structures such as the locus ceruleus, pedunculopontine nuclei, and the dorsal nucleus of the vagus nerve in stages 1 and 2 (Braak *et al.*, 2003). Various sleep anomalies including insomnia, parasomnias, and excessive daytime sleepiness have been described in PD patients (Stocchi *et al.*, 2001). Idiopathic REM sleep behavior disorder (iRBD), which is regarded as another nonmotor symptom of PD, precedes motor symptoms in up to 20% of patients (Iranzo *et al.*, 2006). Moreover, follow-up of patients with iRBD revealed that after 12 years about 30% had developed PD (Schenck *et al.*, 1996). Therefore, iRBD seems to be one of the strongest premotor markers for PD. TCS examination in patients with iRBD could demonstrate that the echogenic SN area in 55 iRBD patients is significantly larger than the one in 165 age- and gender-matched nonaffected controls. SN+ was present in 37% of patients compared to about 10% of SN+ controls without sleep abnormalities, a proportion corresponding to the earlier described prevalence of SN+ within healthy control groups (Stockner *et al.*, 2009). The high prevalence of SN+ within this group of individuals believed to be at higher risk for PD corroborates the hypothesis of SN+ as an independent PD risk marker, although a longitudinal follow-up of SN+ iRBD patients is necessary to prove whether subjects of this special subgroup do indeed convert to manifest PD.

D. Dysautonomia

Several forms of dysfunction of the autonomic nervous system such as constipation and other gastrointestinal motility problems, urinary and sexual disturbances, and orthostatic hypotension may accompany motor symptoms in PD, although most of them are more severe in advanced disease stages (Tolosa *et al.*, 2007). Especially, constipation has been described in a number of epidemiological studies to precede the motor manifestation of PD in some patients and the daily bowel rate has been shown to have an impact on the future risk for PD (Abbott *et al.*, 2001). These epidemiological findings are in line with the neuropathological description of early Lewy bodies and neurites in the dorsal nucleus of the vagus and even in the peripheric nervous system with affecting of enteric nervous plexus in the Braak stage 1 (Braak *et al.*, 2003) suggesting that consecutive gastrointestinal symptoms or other signs of dysautonomia may occur in premotor stages of PD.

Studies on a possible relation of quantitatively assessed autonomic symptoms and SN echostatus are currently under way. Results need to be awaited before evaluation of a possible association of SN+ with evidence for premotor autonomic dysfunction.

E. Cognitive Decline

Whereas relevant cognitive deficits occur in the majority of PD patients in the later course of the disease (Emre *et al.*, 2003), subtle neuropsychological impairment is not uncommon already at the time point of diagnosis when early motor symptoms develop (Muslimovic *et al.*, 2005). Minor cognitive abnormalities have been suggested to be detectable prior to the diagnosis of PD based on the findings of studies in unaffected co-twins or first-degree relatives of PD patients without any motor retardation showing a reduced performance in special neuropsychological tasks compared to control groups (Dujardin *et al.*, 1999; Holthoff *et al.*, 1994). In the cognitively impaired co-twins, [^{18}Fluoro] Dopa-PET studies could prove a nigrostriatal dysfunction as shown by a reduced radiotracer binding (Holthoff *et al.*, 1994) leading to the hypothesis that cognitive changes may indicate preclinical PD in this group of individuals. In SN+ healthy controls, two independent studies using different neuropsychological tasks could demonstrate an association between the echostatus of SN+ and subtle cognitive deficits (Liepelt *et al.*, 2008; Behnke *et al.*, 2009a). In both the studies, changes were mild and perceptible only in defined tests regarding especially memory function, which is known to be affected in PD.

F. Subtle Signs of Motor Slowing

Prior to the conversion toward a clinically manifest parkinsonian syndrome, a number of patients suffer from minor or subjective symptoms of motor slowing such as feelings of stiffness or pain, or from mild forms of tremor, and balance difficulties. These complaints have been shown to precede the diagnosis of PD and to be associated with an increased risk of developing the full clinical picture of the disease (de Lau *et al.*, 2006). The occurrence of these preclinical signs of slight motor impairment can be understood when the process of neurodegeneration at the substantia nigra is considered. Only when about 60% of dopaminergic cells at the ventrolateral part of the substantia nigra are lost, the clinical diagnosis of PD can be made (Fearnley and Lees, 1991). However, even at stages of minor degeneration of dopaminergic neurons the remaining effective neurons may not be sufficient to allow extreme motor tasks (Lees, 1992—Kennedy story) or to enable unimpaired movements under challenging conditions like fatigue or other illnesses. In most of the above-cited TCS studies examining healthy controls, subjects were free of motor symptoms suggestive for an extrapyramidal disorder. Nevertheless, it turned out that several conditions challenging the nigrostriatal system may go along with mostly rather mild but clear symptoms of motor retardation in control groups exhibiting the echofeature SN+. Such challenging conditions are as follows:

(i) Blocking of the dopaminergic system. Interestingly, it could be shown that patients receiving high-potency neuroleptic therapy known to exert antidopaminergic effects experience more often extrapyramidal side effects if SN+ has been described on TCS (Berg *et al.*, 2001a; Jabs *et al.*, 2003).

(ii) Age, known to increase the risk to develop PD. Signs of brady- or hypokinesia become more frequent with advancing age even if they do not suffice to reach the diagnosis of PD. This condition has been suggested to be related to an age-dependent decline in cell numbers in the SN and the striatum modifying nigrostriatal functioning (Prettyman *et al.*, 1998). Indeed, in TCS studies, mild signs of motor slowing or motor asymmetry were found in healthy controls in advanced age groups. They were closely correlated to the echostatus of SN+ with SN+ individuals exhibiting more often and more severe extrapyramidal symptoms (Berg *et al.*, 2001; Behnke *et al.*, 2007).

(iii) Very demanding motor tasks. These can, for example, be achieved during virtuosic playing of instruments or challenging sports activities. In a study investigating a healthy group of tap dancers, some showed motor asymmetry with unilaterally reduced arm motions while tapping (Ruprecht-Doerfler *et al.*, 2007). Compared to their dancing colleagues without arm-swinging abnormalities, these dancers showed significantly more often and more extended SN+ suggesting that SN+ indicates a functional impairment of the nigrostriatal system that becomes evident under challenging conditions. At rest, neurological examination was normal in all dancers.

Since all these thoroughly examined putative nonmotor and very early motor preclinical signs are not specific for PD when regarded alone, they may not be sufficient to identify those healthy individuals being at high risk for the development of PD even if combined with the SN echostatus in TCS. At present, there is a huge body of evidence that SN+ represents indeed a relevant risk marker for PD but the prevalence of about 10% SN+ within the overall population exceeds the prevalence of PD, which is age-dependently 1–2%, by far. In order to identify the subgroup of SN+ individuals experiencing a proceeding neuodegeneration toward the clinically overt parkinsonian syndrome, a large multicenter prospective study is being conducted examining healthy individuals aged >50 years not only by TCS but also by a thorough neurological examination looking for mild parkinsonian signs and reduced arm swing and a *nonmotor test battery* comprising the history of known epidemiological risk factors and autonomic dysfunction, testing of the olfactory system, and a questionnaire for depressive symptoms (Liepelt *et al.*, 2009). In this cross-sectional study, the co-occurrence of different risk factors and premotor markers was significantly more frequent in the SN+ subgroup of the large cohort of PD-wise healthy subjects, leading to the hypothesis that with SN+ as a screening tool the presence of a combination of premotor abnormalities may be more suitable to define a high-risk group for PD (Liepelt *et al.*, 2009).

Follow-up investigations of this cohort are underway. First analysis show that SN+ is indeed a potent risk marker for the development of PD (Berg *et al.*, in preparation).

IV. [^{18}Fluoro] Dopa-PET in SN+ Healthy Controls

All the above-cited findings of an association of SN+ with putative premotor symptoms of PD in healthy controls gain special implications in light of results of functional imaging of the nigrostriatal system using [^{18}Fluoro] Dopa-PET in SN+ healthy individuals. As many as 60% of 20 SN+ subjects out of the first large control population showed a significant reduction of striatal [^{18}Fluoro] Dopa uptake in PET scans compared to the SN– control group, although at this time point, there was no clinical symptom indicative for an extrapyramidal syndrome (Berg *et al.*, 1999a, 2002). These findings give further evidence that the echostatus SN+ indeed goes along with a premotor impairment of the nigrostriatal system. Another study could show that the quantitative tracer uptake in [^{18}Fluoro] Dopa-PET scans lies between the uptake ratios of the SN-control group and a group of patients with manifest PD (Fig. 3; Behnke *et al.*, 2009b). Tracer binding was again significantly reduced both in the p‘utamen and in the caudate nucleus in SN+ subjects compared to SN– controls. Though these findings suggest that a relevant proportion of SN+ clinically healthy controls

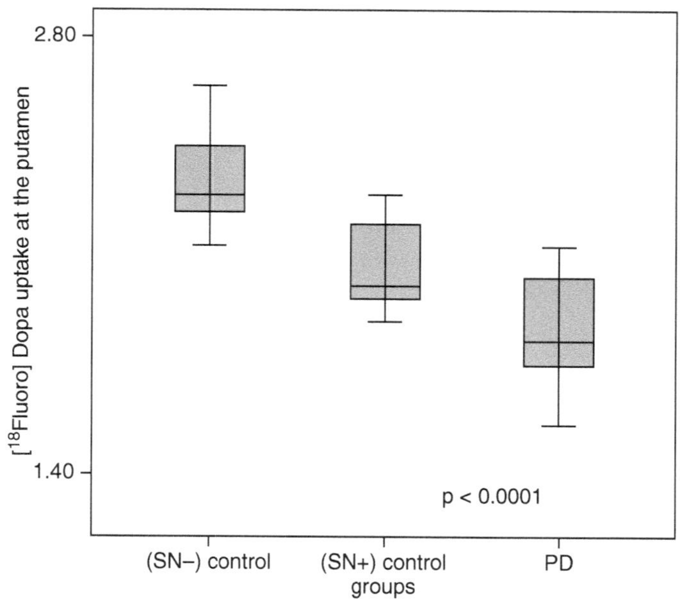

FIG. 3. Putaminal [^{18}Fluoro] Dopa uptake in a PET study examining healthy controls with or without hyperechogenicity of the substantia nigra (SN+ and SN-, respectively) and patients with Parkinson's disease (PD). The intergroup difference was statistically significant (Kruskal–Wallis test).

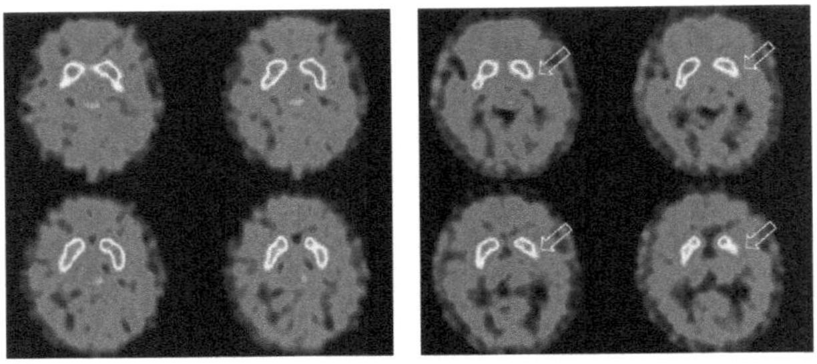

FIG. 4. (**A**) Baseline [^{18}Fluoro] Dopa-PET of a 40-years-old healthy control with hyperechogenicity of the substantia nigra. (**B**) Follow-up [^{18}Fluoro] Dopa-PET in the same individual 9 years later showing a significant reduction of [^{18}Fluoro] Dopa uptake in the left putamen (arrow). At the time point of follow-up examination the man was still free of motor symptoms allowing the diagnosis of PD but converted to a mild right-sided tremor dominant parkinsonian syndrome after another year.

suffer from a PET-wise demonstrable nigrostriatal dysfunction, it has been unclear whether there are SN+ controls proceeding toward overt parkinsonism. A most recent follow-up study comprised 10 individuals who took part in the initial PET studies. These still asymptomatic SN+ individuals underwent a second PET scan after a mean time of 7.25 years as well as a thorough neurological examination and a follow-up via telephone interviews by a neurologist experienced in movement disorders. [^{18}Fluoro] Dopa uptake remained unchanged in 8 of 10 study participants compared to the baseline examination. In the other two, a marked unilateral reduction of [^{18}Fluoro] Dopa uptake ratios in the putamen appeared over the time followed by a clinically manifest PD in one of these individuals after another year (Fig. 4A and B; Schroeder *et al.*, submitted). These results with 20% of progressively pathological PET findings and 10% of PD converters matches the expected proportion of SN+ individuals developing a manifest PD with regard to the prevalence of 10% of SN+ subjects and 1–2% of PD cases within the normal population. This underlines the significance of the echofeature in indicating a high risk for the development of PD in some of these SN+ individuals.

V. Longitudinal Clinical Follow-Up Studies

The complete understanding of the significance of TCS findings in the premotor diagnosis of PD depends on longitudinal studies with follow-up investigations of SN+ and SN- individuals. There is an ongoing large multicenter prospective trial examining the conversion rate from SN+ healthy control subjects to manifest PD in more than 1800 individuals and the relation with other putative premotor markers assessed at baseline and during follow-up visits. In this cohort, 11 initially healthy subjects converted to a clinically manifest PD after a mean follow-up time of 3 years (Berg *et al.*, submitted). While one of these PD converters had no sufficient temporal bone window for insonation, 8 of the remaining 10 exhibited SN+ at baseline (80%) resulting in a relative risk of 17.37 for SN+ extrapyramidally healthy subjects to develop PD within this relatively short time period. When combined with abnormalities in baseline smell tests and the presence of subtle signs of motor retardation, the relative risk for PD increased to 48.39. With a specificity of 99.6% and a positive predictive value of 25%, this combination of premotor markers including the SN echostatus seems to be a most valuable tool in the identification of a high-risk cohort for PD.

VI. Conclusion

Up to now, numerous lines of evidence have shown the association of SN+ with putative premotor signs of PD. Neuroimaging studies and first longitudinal data on SN+ healthy controls have further corroborated the hypothesis that SN+ indicates a higher risk for the development of PD. The combination of TCS with defined PD premotor markers may possibly allow the identification of a subgroup of PD patients *before* crossing the clinical—or better—motor threshold. This is crucial since future studies on putative disease-modifying drugs may only be successful when realized in very early disease stages before extended irreversible neuronal damage has taken place. Final data analysis and further longitudinal follow-up of these extensively examined study populations will give further insight into the impact of SN+ on the individual risk to develop PD later in life.

References

Abbott, R. D., Petrovitch, H., White, L. R., Masaki, K. H., Tanner, C. M., Curb, J. D., Grandinetti, A., Blanchette, P. L., Popper, J. S., and Ross, G. W. (2001). Frequency of bowel movements and the future risk of Parkinson's disease. *Neurology* **57**, 456–462.

Abbott, R. D., Ross, G. W., White, L. R., Sanderson, W. T., Burchfiel, C. M., Kashon, M., Sharp, D. S., Masaki, K. H., Curb, J. D., and Petrovitch, H. (2003). Environmental, lifestyle, and precursors of clinical Parkinson's disease: recent findings from the Honolulu-Asia Aging study. *J. Neurol.* **250**, 30–39.

Becker, G., Seufert, J., Bogdahn, U., Reichmann, H., and Reiners, K., (1995a). Degeneration of substantia nigra in chronic Parkinson's disease visualized by transcranial color-coded real-time sonography. *Neurology* **45**, 182–184.

Becker, G., Becker, T., Struck, M., Lindner, A., Burzer, K., Retz, W., Bogdahn, U., and Beckmann, H., (1995b). Reduced echogenicity of brainstem raphe specific to unipolar depression: a transcranial color-coded real-time sonography study. *Biol. Psychiatry* **38**, 180–184.

Becker, G., Müller, A., Braune, S., Büttner, T., Benecke, R., Greulich, W., Klein, W., Mark, G., Rieke, J., and Thümler, R., (2002). Early diagnosis of Parkinson's disease. *J. Neurol.* **249**, 40–48.

Behnke, S., Double, K. L., Duma, S., Broe, G. A., Guenther, V., Becker, G., and Halliday, G. M., (2007). Substantia nigra echomorphology in the healthy very old: Correlation with motor slowing. *Neuroimage* **34**, 1054–1059.

Behnke, S., Wollenweber, F. A., Probst, M., Hiry, T., Schneider, V., Dillmann, U., and Fassbender, K., (2009a). Substantia nigra hyperechogenicity in healthy controls: correlation with neuropsychological deficits. *Mov. Disord.* **24**(Suppl. 1), S182.

Behnke, S., Schroeder, U., Dillmann, U., Buchholz, H. G., Schreckenberger, M., Fuss, G., Reith, W., Berg, D., and Krick, C. M., (2009b). Hyperechogenicity of the substantia nigra in healthy controls is related to MRI changes and to neuronal loss as determined by f-dopa PET. *Neuroimage* **47**, 1237–1243.

Berendse, H. W., and Ponsen, M. M., (2006). Detection of preclinical Parkinson's disease along the olfactory trac(t). *J. Neural. Transm.* **70**, 321–325.

Berg, D., Becker, G., Zeiler, B., Tucha, O., Hofmann, E., Preier, M., Benz, P., Jost, W., Reiners, K., and Lange, K. W., (1999a). Vulnerability of the nigrostriatal system as detected by transcranial ultrasound. *Neurology* **53**, 1026–1031.

Berg, D., Supprian, T., Hofmann, E., Zeiler, B., Jager, A., Lange, K. W., Reiners, K., Becker, T., and Becker, G., (1999b). Depression in Parkinson's disease: Brainstem midline alteration on transcranial sonography and magnetic resonance imaging. *J. Neurol.* **246**, 1186–1193.

Berg, D., Jabs, B., Merschdorf, U., Beckmann, H., and Becker, G., (2001a). Echogenicity of substantia nigra determined by transcranial ultrasound correlates with severity of parkinsonian symptoms induced by neuroleptic therapy. *Biol. Psychiatry* **50**, 463–467.

Berg, D., Siefker, C., Ruprecht-Dorfler, P., and Becker, G., (2001b). Relationship of substantia nigra echogenicity and motor function in elderly subjects. *Neurology* **56**, 13–17.

Berg, D., Roggendorf, W., Schroder, U., Klein, R., Tatschner, T., Benz, P., Tucha, O., Preier, M., Lange, K. W., Reiners, K., Gerlach, M., and Becker, G., (2002). Echogenicity of the substantia nigra: Association with increased iron content and marker for susceptibility to nigrostriatal injury. *Arch. Neurol.* **59**, 999–1005.

Braak, H., Del Tredici, K., Rub, U., de Vos, R. A., Jansen Steur, E. N., and Braak, E., (2003). Staging of brain pathology related to sporadic Parkinson's disease. *Neurobiol. Aging* **24**, 197–211.

Brooks, D. J., and Doder, M., (2001). Depression in Parkinson's disease. *Curr. Opin. Neurol.* **14**, 465–470.

Chan-Palay, V., and Asan, E., (1989). Quantitation of catecholamine neurons in the locus coeruleus in human brains of normal young and older adults and in depression. *J. Comp. Neurol.* **287**, 357–372.

De Lau, L. M., Koudstaal, P. J., Hofman, A., and Breteler, M. M., (2006). Subjective complaints precede Parkinson disease: the Rotterdam study. *Arch. Neurol.* **63**, 362–365.

Doty, R. L., Shaman, P., Kimmelman, C. P., and Dann, M. S., (1984). University of Pennsylvania Smell Identification Test: a rapid quantitative olfactory function test for the clinic. *Laryngoscope* **94**, 176–178.

Dujardin, K., Duhamel, A., Becquet, E., Grunberg, C., Defebvre, L., Destee, A., (1999). Neuropsychological abnormalities in first degree relatives of patients with familial Parkinson's disease. *J. Neurol. Neurosurg. Psychiatr.* **67**, 323–328.

Emre, M., (2003). Dementia associated with Parkinson's disease. *Lancet Neurol.* **2**, 229–237.

Fearnley, J. M., and Lees, A. J., (1991). Ageing and Parkinson's disease: Substantia nigra regional selectivity. *Brain* **114**, 2283–2301.

Haehner, A., Hummel, T., Hummel, C., Sommer, U., Junghanns, S., and Reichmann, H., (2007). Olfactory loss may be a first sign of idiopathic Parkinson's disease. *Mov. Disord.* **22**, 839–842.

Holthoff, V. A., Vieregge, P., Kessler, J., Pietrzyk, U., Herholz, K., Bönner, J., Wagner, R., Wienhard, K., Pawlik, G., and Heiss, W. D., (1994). Discordant twins with Parkinson's disease: positron emission tomography and early signs of impaired cognitive circuits. *Ann. Neurol.* **36**, 176–182.

Hornykiewicz, O., and Kish, S. J., (1987). Biochemical pathophysiology of Parkinson's disease. *Adv. Neurol.* **45**, 19–34.

Iranzo, A., Molinuevo, J. L., Santamaría, J., Serradell, M., Martí, M. J., Valldeoriola, F., and Tolosa, E., (2006). Rapid-eye-movement sleep behaviour disorder as an early marker for a neurodegenerative disorder: A descriptive study. *Lancet Neurol.* **5**, 572–577.

Jabs, B. E., Bartsch, A. J., and Pfuhlmann, B. (2003). Susceptibility to neuroleptic-induced parkinsonism—age and increased substantia nigra echogenicity as putative risk factors. *Eur. Psychiatry* **18**, 177–181.

Leentjens, A. F., Van den Akker, M., Metsemakers, J. F., Lousberg, R., and Verhey, F. R. (2003). Higher incidence of depression preceding the onset of Parkinson's disease: A register study. *Mov. Disord.* **18**, 414–418.

Lees, A. J. (1992). When did Ray Kennedy's Parkinson's disease begin? *Mov. Disord.* **7**, 110–115.

Liepelt, I., Wendt, A., Schweitzer, K. J., Wolf, B., Godau, J., Gaenslen, A., Bruessel, T., and Berg, D. (2008). Substantia nigra hyperechogenicity assessed by transcranial sonography is related to neuropsychological impairment in the elderly population. *J. Neural. Transm.* **115**, 993–999.

Liepelt, I., Behnke, S., Schweitzer, K., Wolf, B., Godau, J., Wollenweber, F., Dillmann, U., Gaenslen, A., Di Santo, A., Maetzler, W., and Berg, D. (2009). Pre-motor signs of PD are related to SN hyperechogenicity assessed by TCS in an elderly population. *Neurobiol. Aging (in press)*.

Markopoulou, K., Larsen, K.W, Wszolek, E. K., Denson, M. A., Lang, A. E., Pfeiffer, R. F., and Wszolek, Z. K. (1997). Olfactory dysfunction in familial parkinsonism. *Neurology* **49**, 1262–1267.

Muslimovic, D., Post, B., Speelman, J. D., and Schmand, B. (2005). Cognitive profile of patients with newly diagnosed Parkinson disease. *Neurology* **65**, 1239–1245.

Ponsen, M. M., Stoffers, D., Twisk, J. W., Wolters, E., and Berendse, H. W. (2009). Hyposmia and executive dysfunction as predictors of future Parkinson's disease: A prospective study. *Mov. Disord.* **24**, 1060–1065.

Prettyman, R. (1998). Extrapyramidal signs in cognitively intact elderly people. *Age Ageing* **27**, 557–560.

Riederer, P., and Wuketich, S. (1976). Time course of nigrostriatal degeneration in Parkinson's disease: A detailed study of influential factors in human brain amine analogues. *J. Neural Transm.* **38**, 227–301.

Ross, G. W., Petrovitch, H., Abbott, R. D., Tanner, C. M., Popper, J., Masaki, K., Launer, L., White, L. R. (2008). Association of olfactory dysfunction with risk for future Parkinson's disease. *Ann. Neurol.* **63**, 167–173.

Ruprecht-Dorfler, P., Berg, D., Tucha, O., Benz, P., Meier-Meitinger, M., Alders, G. L., Lange, K. W., and Becker, G. (2003). Echogenicity of the substantia nigra in relatives of patients with sporadic Parkinson's disease. *Neuroimage* **18**, 416–422.

Ruprecht-Dorfler, P., Klotz, P., Becker, G., and Berg, D. (2007). Substantia nigra hyperechogenicity correlates with subtle motor dysfunction in tap dancers. *Parkinsonism Relat. Disord.* **13**, 362–364.

Schenck, C. H., Bundlie, S. R., and Mahowald, M. W. (1996). Delayed emergence of a parkinsonian disorder in 38% of 29 older men initially diagnosed with idiopathic rapid eye movement sleep behaviour disorder. *Neurology* **46**, 388–393.

Schroeder, U., Behnke, S., Buchholz, H.G., Landvogt, C., Fassbender, K., Schreckenberger, M., and Berg, D. (2010). Substantia nigra hyperechogenicity in healthy controls: A [^{18}Fluoro] Dopa-PET Follow-up study (submitted).

Schweitzer, K. J., Behnke, S., Liepelt, I., Wolf, B., Grosser, C., Godau, J., Gaenslen, A., Bruessel, T., Wendt, A., Abel, F., Muller, A., Gasser, T., and Berg, D. (2007). Cross-sectional study discloses a positive family history for Parkinson's disease and male gender as epidemiological risk factors for substantia nigra hyperechogenicity. *J. Neural Transm.* **114**, 1167–1171.

Sommer, U., Hummel, T., Cormann, K., Mueller, A., Frasnelli, J., Kropp, J., and Reichmann, H. (2004). Detection of presymptomatic Parkinson's disease: Combining smell tests, transcranial sonography, and spect. *Mov. Disord.* **19**, 1196–1202.

Stocchi, F., Vacca, L., Valente, M., and Ruggieri, S. (2001). Sleep disorders in Parkinson's disease. *Adv. Neurol.* **86**, 289–293.

Stockner, H., Iranzo, A., Seppi, K., Serradell, M., Gschliesser, V., Sojer, M., Valldeoriola, F., Molinuevo, J. L., Frauscher, B., Schmidauer, C., Santamaria, J., Hogl, B., Tolosa, E., and Poewe, W. (2009). Midbrain hyperechogenicity in idiopathic REM sleep behavior disorder. *Mov. Disord.* **24**, 1906–1909.

Tolosa, E., Compta, Y., and Gaig, C. (2007). The premotor phase of Parkinson's disease. *Parkinsonism Relat. Disord.* **13**, S2–S7.

Walter, U., Hoeppner, J., Prudente-Morrissey, L., Horowski, S., Herpertz, S. C., and Benecke, R. (2007). Parkinson's disease-like midbrain sonography abnormalities are frequent in depressive disorders. *Brain* **130**, 1799–807.

PATHOPHYSIOLOGY OF TRANSCRANIAL SONOGRAPHY SIGNAL CHANGES IN THE HUMAN SUBSTANTIA NIGRA

Kay L. Double,* Gabrielle Todd,† and Stephen R. Duma*

*Neuroscience Research Australia and the University of New South Wales, Sydney, NSW, Australia

†Discipline of Physiology, School of Medical Sciences, The University of Adelaide, Adelaide, SA, Australia

The characteristic change in ultrasound signal seen in the substantia nigra in Parkinson's disease and related disorders is known as hyperechogenicity and is suggested to be a risk or state marker for these disorders. The underlying cellular basis of this change, however, is unknown. Imaging studies on Parkinson's disease patients, experimental studies on animal models, and human postmortem studies support the hypotheses that an alteration in regional iron and/or change in regional cell composition within the substantia nigra underlie this echo feature. Future quantitative postmortem studies incorporating cellular and metal analyses of the substantia nigra, in Parkinson's disease patients and healthy controls of known echogenic status during life, will be important to clarify the cellular basis of this change.

As described in the previously, and also in the following chapters, transcranial sonography (TCS) has been used to identify changes in signal intensity in the substantia nigra (SN) in a variety of neurological disorders, particularly those marked by clinical parkinsonism. The alteration in signal, seen as an area of increased signal intensity and extent, is termed hyperechogenicity and the presence of SN hyperechogenicity in Parkinson's disease (PD) is suggested to be

INTERNATIONAL REVIEW OF NEUROBIOLOGY, VOL. 90
DOI: 10.1016/S0074-7742(10)90008-1

clinically useful in diagnosing this disorder, especially in the early stages (see chapter on Early Diagnosis of PD), when clinical diagnosis is notoriously difficult. Further evidence from a number of studies, as described in see chapter on Premotor Diagnosis of PD, suggests that TCS may also be useful in detecting preclinical PD. Moreover, studies described in other chapters within this volume suggest that TCS may assist with the differential diagnosis of a variety of other related motor disorders. Whilst data to date suggest that this simple and inexpensive imaging modality may be useful in improving diagnosis of a variety of disorders, and enabling the identification and study of other facets of movement dysfunction in individuals without clinically identifiable movement disorders, the cellular basis of SN hyperechogenicity is currently unknown. This chapter will review what is known about the pathophysiology of this echofeature and discuss the possible cellular basis of this change with a focus on the role of TCS in PD, the disorder in which the majority of research has been carried out.

SN hyperechogenicity is observed in 78–90% of patients with PD (Berg, 2007b; Doepp *et al.*, 2008; Spiegel *et al.*, 2006; Tsai *et al.*, 2007). Thus, we (Todd *et al.*, 2010) and others (Berg et al. 1999a; Walter, 2009) have suggested that this feature represents a risk or state marker for PD. It must be kept in mind, however, that a hyperechogenic nigra does not occur in all individuals with PD, although the reason for this is unknown and will be an interesting area for future study. Nevertheless, hyperechogenicity can also be observed in a proportion of individuals without parkinsonian signs. To date, the cellular basis of the hyperechogenic signal has not been definitively identified. The question is therefore asked: What are the possible cellular changes that could underlie this change?

I. Does Hyperechogenicity Reflect Death of Substantia Nigra Dopaminergic Neurons?

The most significant pathological change in the SN in PD seen at postmortem is the dramatic death of dopaminergic pigmented neurons of the substantia nigra pars compacta (Fig. 1). Does a hyperechogenic nigra reflect the death of these dopaminergic pigmented neurons in PD? To date, no studies in PD patients have reported a relation between the presence of hyperechogenicity and disease severity during life time to nigral neuronal number in the postmortem brain. A number of imaging studies, however, have addressed these questions in living patients. Fluorine 18-dopa ([^{18}F]-dopa) PET scans have been used to study nigrostriatal integrity in PD patients, and in healthy older adults with parkinsonian motor signs. These studies have evaluated the uptake of [^{18}F]-dopa in the striatum, and consistently report an association between reduced [^{18}F]-dopa uptake in the caudate nucleus and putamen of PD patients with hyperechogenicity of the SN

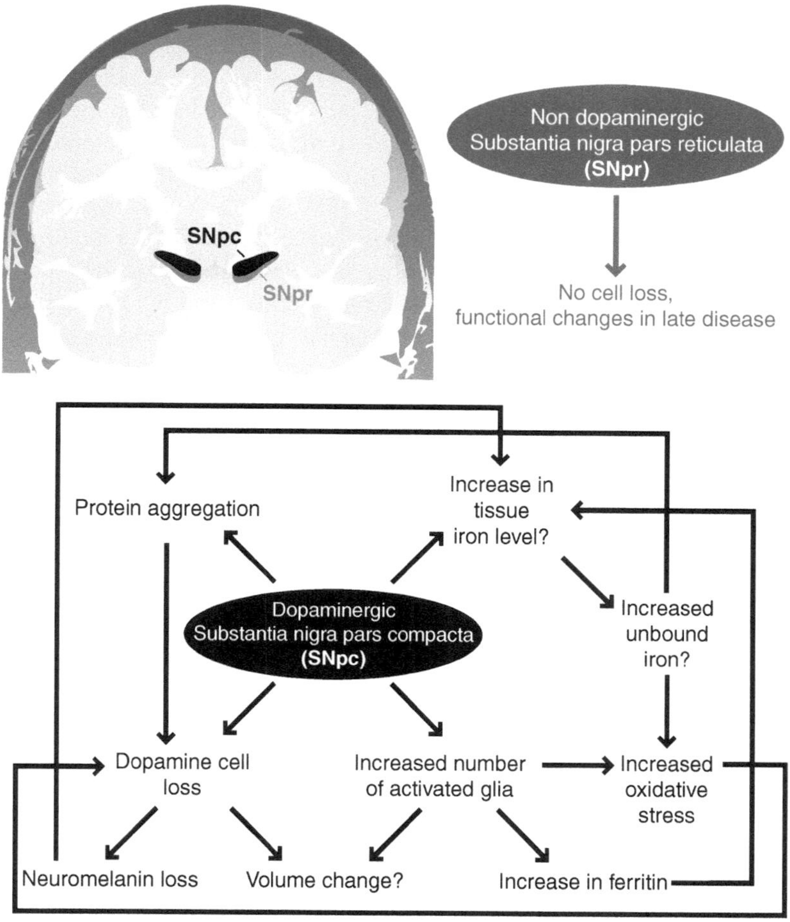

FIG. 1. Diagram showing the two subregions of the human substantia nigra, the dopaminergic substantia nigra pars compacta and the nondopaminergic substantia nigra pars reticulata and changes observed in these regions in Parkinson's disease, which may contribute to a hyperechogenic echosignal. It is unclear whether hyperechogenicity identified by TCS reflects features within the SN pars compata alone or within both the pars compata and the pars reticulata.

(Behnke *et al.*, 2009). Reduced [^{18}F]-dopa uptake is also observed in healthy aged individuals with SN hyperechogenicity (Behnke *et al.*, 2009; Berg *et al.*, 1999a, 2002; Ruprecht-Dörfler *et al.*, 2003). These data suggest that a structural change in the nigra, reflected as hyperechogenicity, is temporally associated with a functional change within the motor circuit, seen as reduced dopaminergic uptake in the striatum. A further imaging method used to assess functional changes within the nigrostriatal system is single photon emission computed tomography (SPECT).

SPECT is used to evaluate binding of dopamine transporters (DATs) in the striatum. Studies using ^{123}I-FP-CIT as a marker of DAT binding report a significant reduction in striatal DAT binding in PD patients, compared with controls (Doepp *et al.*, 2008). ^{123}I-FP-CIT uptake is particularly reduced in the putamen and is significantly lower in the striatum contralateral to the clinically more affected body side (Doepp *et al.*, 2008). Conflicting data have been reported, however, regarding the relation between ^{123}I-FP-CIT uptake and hyperechogencity in PD. One study reported no correlation between striatal ^{123}I-FP-CIT uptake and SN hyperechogenicity (Spiegel *et al.*, 2006), whilst another study reported a significant relation between striatal ^{123}I-FP-CIT uptake and SN hyperechogenicity (Weise *et al.*, 2009). These conflicting data may reflect compensatory mechanisms occurring within the nigrostriatal pathways in PD, which may increase ^{123}I-FP-CIT uptake in the remaining neurons following SN cell loss (Lee *et al.*, 2000). Nevertheless, a longitudinal study showing that the area of SN hyperechogencity does not change with progressive clinical progression of the disease, which is assumed to reflect progressive neuronal loss within the SN, also argues against the idea that this feature only reflects neuronal loss in the SN (Berg *et al.*, 2005).

A characteristic feature of PD is the unusually long preclinical period, calculated from postmortem and imaging data to be 5–10 years (Fearnley and Lees, 1991; Morrish *et al.*, 1998). At least 60% of the dopaminergic neurons in the SN must be lost prior to the development of motor dysfunction (Fearnley and Lees, 1991; Morrish *et al.*, 1998). Thus, the disease process is significantly advanced prior to clinical diagnosis. Using standard clinical tools to assess gross movement function in persons with clinical disease, such as the Unified Parkinson's Disease Rating Scale (UPDRS), movement appears to be normal in healthy SN hyperechogenic individuals carrying a mutation associated with the development of PD, such as PINK1 (Hagenah *et al.*, 2008) and Parkin (Walter *et al.*, 2004). However, there are also reports of subtle signs of motor dysfunction, not reaching the classification of "parkinsonism" in these individuals, suggesting that movement in this population may be compromised. Studies of healthy individuals in the normal population, i.e., individuals who do not carry a PD-associated gene mutation but nevertheless exhibit a hyperechogenic SN, may also exhibit subtle motor impairment (Behnke *et al.*, 2007; Berg *et al.*, 2002; Ruprecht-Dörfler *et al.*, 2007). It must be noted, however, that studies to date have utilized tests designed specifically for patients with PD, such as the UPDRS, rather than those designed to assess movement in healthy individuals, and the majority of studies have looked only at gross motor functions. An investigation of gross and fine motor functions in healthy SN hyperechogenic individuals using motor tests designed for the healthy population would clearly be of benefit to detect subtle motor dysfunction.

Non-PD individuals who exhibit a hyperechogenic nigra represent approximately 9% of healthy individuals aged 20–70 years (for review see Berg, 2007a), and the appearance of this feature increases with age, with 20% of healthy adults aged over 85 years exhibiting this feature (Behnke *et al.*, 2007). Clearly, the proportion of healthy SN hyperechogenic individuals is markedly higher than the reported incidence of PD in most populations (1–2% of individuals aged 60 or older; Olanow *et al.*, 2009), thus it would be impossible for all SN hyperechogenic individuals to be candidates for preclinical PD. This feature also occurs in individuals in young adulthood and in children (Iova *et al.*, 2004; Walter, 2009), far earlier than the reported 5- to 10-year preclinical period prior to the characteristic age of diagnosis of PD during the sixth decade of life. In addition, SN echogenicity is also changed—however, in the opposite direction, i.e., it is decreased—in another movement disorder, restless legs syndrome, which is not associated with altered nigral neuronal number (Godau *et al.*, 2007). Together, these data suggest the simple explanation of SN hyperechogenicity representing purely disease-associated SN cell death is unlikely to be the case.

II. Does Hyperechogenicity Reflect a Low Substantia Nigra Cell Number?

Is it then that the hyperechogenic signal reflects a low nigral neuron count, which might occur in some individuals as a result of natural variation, and in other individuals as a result of a (Parkinson's) disease-related process? This hypothesis suggests that the number of cells within the SN is highly variable and that a proportion of the population, e.g., 1 in 10 or even 2 in 10 older adults (Behnke *et al.*, 2007; Berg, 2007b), has a naturally low cell number count, which may be reflected as a change in the ultrasound echosignal. Most publications refer to a hyperechogenic signal associated with the anatomical region of the "substantia nigra", yet the human substantia nigra consists of two distinct nuclei: (1) the substantia nigra pars compacta and (2) the substantia nigra pars reticulata, which are anatomically and neurochemically distinct (Fig. 1). The neuromelanin-pigmented dopaminergic neurons of the SN project primarily to the caudate and putamen, while the γ-aminobutyric acid (GABA)-containing neurons of the pars reticulata project to the tectum and thalamus (McRitchie *et al.*, 1996). The pigmented neurons of the SN pars compacta are dramatically reduced in PD (although not all of the dopaminergic neurons within this region are equally vulnerable to the disease process), while the nondopaminergic neurons of the pars reticulata are preserved in PD (Hardman *et al.*, 1996). Whether the hyperechogenic signal in PD reflects the entire SN or just the vulnerable SN pars compacta is unclear. Detailed studies of the cytoarchitecture of the dopaminergic neurons of the human SN do not report, however, the presence of individuals with unusually low neuron numbers within the SN pars compacta. This suggests that a low neuronal count in this brain region cannot account for the altered echosignal in some healthy individuals, as a result of a wide normal variation in neuron number in this brain region, or in persons who have a low neuron count in this region as a result of disease-related neurodegeneration (McRitchie *et al.*, 1995, 1996). Another factor that might influence SN neuron number is age, and this is of particular interest as the incidence of hyperechogenicity in healthy individuals appears to be greater in the very old population (Behnke *et al.*, 2007), although the incidence of this feature does not change in adults from 20 to 70 years of age (Berg *et al.*, 1999). It is difficult to determine whether SN neuronal loss occurs naturally with age, as neuronal number can only be assessed at a single time point post mortem. Published studies of changes in SN neuronal numbers have primarily assessed neuronal density, a measure dependent upon tissue volume, and in these studies up to a 10% loss of neurons per decade have been reported. Other studies have quantified absolute neuronal number in the nigra, a measure corrected for tissue volume, and report no loss of neuron number with age. Further, the subpopulation of neurons quantified often differs, in that some studies quantify loss of pigmented neurons while other studies quantify loss of dopamine neurons. A comparative review of these studies has been published by Stark and Pakkenberg (Stark and Pakkenberg, 2004). In nonhuman primates, there is no loss of SN dopamine neurons with age, although functional changes, including neuronal depigmentation, occur (Collier *et al.*, 2007). The literature available on the human brain (Stark and Pakkenberg, 2004) suggests that functional and morphological changes occur within dopamine neurons with age but that only subtle neuronal degeneration results from increasing age. This then suggests that the increasing incidence of hyperechogenicity observed in the older population does not reflect nigral dopamine neuron loss, but rather that other age-related changes may play a role.

III. Does Hyperechogenicity Reflect a Change in Nigra Volume or Tissue Composition?

The loss of 60% of dopaminergic neurons in the SN pars compacta prior to development of motor symptoms in PD suggests that this region may be smaller in volume in this disorder. Neuronal loss, and possible subsequent changes in regional volume in the PD brain, has been investigated using T1-weighted MR imaging. This technique identifies SN neuronal loss in PD (Hutchinson and Raff, 2000, 2008; Hutchinson *et al.*, 2003; Michaeli *et al.*, 2007; Raff *et al.*, 2006) and is significantly associated with clinical severity, measured using the UPDRS (Hutchinson and Raff, 2000; Raff *et al.*, 2006), even in the earliest stages of the disease process (Hutchinson and Raff, 2000; Raff *et al.*, 2006). T1-weighted images are also used to study volumetric changes in the substantia nigra. Although the number of studies is small, these imaging studies suggest that the volume of the SN in PD patients is reduced, compared to that in healthy individuals (Menke *et al.*, 2009; Minati *et al.*, 2007). Interestingly, when using TCS, the area of the SN in PD appears larger, not smaller, and thus it is unclear if volumetric changes in this brain region contribute to this altered signal (Fig. 1). A further change in the parkinsonian SN is an increase in the number of glial cells present in this region as a result of the infiltration of these cells following neuronal degeneration (Fig. 1). The loss of neurons and the increase in glial cells effectively changes the neuron:glia ratio in this brain region and might thus affect the properties of the tissue, resulting in an alteration in reflected ultrasound waves. The increase in glia in the SN seen in idiopathic PD is also seen in other disorders in which marked degeneration of the SN can occur, such as multisystem atrophy, progressive supranuclear palsy, diffuse Lewy body disease, corticobasal degeneration, and dementia with Lewy bodies. An abnormal SN ultrasound signal is most common in idiopathic PD, but it also occurs in other disorders in which degeneration of this region occurs (Vlaar *et al.*, 2009). Interestingly, the studies reviewed by Vlaar *et al.* (2009) report that an abnormal ultrasound signal is less common in atypical parkinsonian syndromes (30% of atypical Parkinson's syndromes compared with 87% of patients with idiopathic PD in the studies reviewed), yet the variability of nigral pathology is also more highly variable in these disorders. It is possible that a threshold of neuronal loss or change in tissue composition must be present for the development of the hyperechogenic signal, which may or may not be present in persons with atypical parkinsonian syndromes. Indeed, this review found that the number of inconclusive ultrasound examinations reported for the parkinsonian syndromes was higher than that for idiopathic disease. Another study reported that all patients with dementia with Lewy bodies, another disorder with variable SN neuronal loss, had SN hyperechogenicity compared with 94–97% of patients with PD with or without dementia. It should, however, be noted that only 12 patients with dementia with Lewy bodies were assessed (Walter *et al.*, 2006). Again, the ability to directly correlate abnormalities on ultrasound signals during life with postmortem pathology will be invaluable to determine the cellular basis of this altered ultrasound signal.

IV. Does Hyperechogenicity Reflect a Change in Iron in the Substantia Nigra? Evidence from Animal Models and Human Postmortem Studies

One hypothesis widely discussed in the literature as the underlying cellular basis of SN hyperechogenicity is that this phenomenon reflects changes in iron levels and/or iron distribution in this area of the brain (Fig. 1). The basis of this hypothesis comes from a number of studies reporting alterations in iron levels in the SN in PD. It must, however, be noted that the type and cellular location of this posited iron change is unclear (Snyder and Connor, 2009), and not all studies found alterations in iron in this disorder. Experimental testing of the hypothesis that changes in regional iron may underlie the hyperechogenic signal has been limited to date, but the small amount of data available supports this idea. A study by Berg and colleagues experimentally manipulated SN iron levels in rats and found that injection of iron into this brain region resulted in hyperechogenicity in a dose-dependent manner, an effect also seen for zinc, i.e., another transition metal (Berg *et al.*, 1999b). Further, injection of 6-hydroxydopamine, a toxin known to release iron from ferritin stores, had a similar effect, whereas injection of ferritin alone or the iron chelator desferrioxamine did not (Berg *et al.*, 1999b). These data suggest that a hyperechogenic signal is associated with nonchelated iron whereas iron bound to chelating proteins does not produce a hyperechogenic signal. Ferritin in the SN is primarily located in glia, rather than in neurons, and in neurons it is only present in the H-form (Snyder and Connor, 2009). Another iron-binding molecule is present in the human (but not the rat; Fedorow *et al.*, 2005) SN: neuromelanin. Neuromelanin is a catecholamine-based polymer pigment, which is concentrated in the neurons of the human SN pars compacta, although it is also found in other catecholaminergic nuclei, such as the ventral tegmental area and the locus ceruleus. Neuromelanin binds iron at a high- and low-affinity binding site and may regulate cellular levels of free iron (Double *et al.*, 2003). In PD, the neuromelanin-containing neurons of the SN pars compacta degenerate, suggesting that a loss of this pigment, and the subsequent release of free iron (Fig. 1), may underlie the hyperechogenic signal. This hypothesis has been tested in one study in the postmortem human brain. In this study, ultrasound examinations were carried out on 40 control and 3 PD brains at postmortem, and the amounts of iron, neuromelanin, and ferritin subsequently quantified in the entire SN in each brain and the relationships between these variables and SN echogenicity were determined. SN iron and SN H- and L-ferritin levels were positively associated with the extent of the echogenic signal in this brain region, although there was also a significant relation between the extent of hyperechogenicity and iron associated with a reduction in neuromelanin tissue content when considered together, but no significant association between echogenicity and neuromelanin *per se* (Zecca *et al.*, 2005). Given the incidence of hyperechogenicity in the healthy population and the number of subjects in this study, it is likely that a proportion of the brains examined would have exhibited an abnormal hyperechogenic signal. Indeed, it was reported that 18 of the bilateral SNs investigated in the 40 control brains (80 in total, 23%) exhibited an echogenic signal of greater than 0.25 cm^2, a signal defined in many *in vivo* studies as the criterion for an abnormal hyperechogenic signal. As the postmortem imaging in this study was performed on brains removed from the skull, it can be assumed that the resultant echosignals are likely to differ from those found in the same brain during life (i.e., no penetration of ultrasound waves through the skull, probe closer to the SN), however, the variance of echogenic signals in this study suggests that at least a proportion of subjects with large areas of SN echogenicity postmortem may have displayed SN hyperechogenicity during life. Unfortunately, no quantitative analysis of SN neuron number was completed in this study, so it is not possible to conclude whether the extent of the echogenic signal was associated with variation in SN neuronal number. Thus, the question of whether the echogenic signal reflects, in part, a change in SN neuron number remains unanswered. The data strongly support the hypothesis, however, that the echogenic signal reflects an iron-associated mechanism. A simple relationship between increased iron and an enhanced echogenic signal is unlikely, as other regions of the healthy brain (such as the globus pallidus) exhibit higher iron loads, yet do not exhibit an enhanced echogenic signal on ultrasound examination (Snyder and Connor, 2009). This argues that the iron-mediated signal is state dependent. The association between H- and L-ferritin levels and echogenic signal in the human brain, although statistically significant, were not as robust as that of iron and echogenicity, suggesting that ferritin-bound iron may not play a significant role in the signal, in agreement with the animal data (Berg *et al.*, 1999b). Further, variations in the ceruloplasmin gene, an iron transport protein, have been linked with PD and with hyperechogenicity (Hochstrasser *et al.*, 2004), further supporting the idea that changes in iron-handling mechanisms are linked to this echofeature. The posited increase in free radicals thought to contribute to neurodegeneration in the SN in PD could be stimulated by changes in iron-handling pathways, and it has been suggested that an environment of increased oxidative load may underlie the development of hyperechogenicity (Berg, 2007a). This is supported by a recent finding in a postmortem study that the number of activated microglia in the SN are significantly associated with the extent of hyperechogenicity (D. Berg, personal communication, Fig. 1).

V. Does Hyperechogenicity Reflect a Change in Iron in the Substantia Nigra? Evidence from Imaging Studies

Reports of increased iron content in the brains of PD patients (i.e., early reports by Dexter *et al.*, 1987 and by Sofic *et al.*, 1988) stimulated investigations using neuroimaging modalities to determine whether iron accumulation could be seen in vivo. The most widely used MRI technique to measure brain iron content is T2 relaxometry—a method that measures T2 relaxation times, or the inversely related T2 relaxation rates. In brief, this method is based on the fact that iron accumulation causes local magnetic field inhomogeneity, which results in a loss of

T2 signal, and thus a shorter T2 relaxation time (or a longer T2 relaxation rate). The theory behind the relation of T2 relaxation time and iron accumulation is explained in greater detail by Ordidge and colleagues (Ordidge *et al.*, 1994). Early T2 relaxometry studies yielded conflicting results, but more recent publications report similar findings. The majority of studies report reduced T2 relaxation times (Antonini *et al.*, 1993; Behnke *et al.*, 2009; Kosta *et al.*, 2006; Michaeli *et al.*, 2007) or increased T2 relaxation rates (Bartzokis *et al.*, 1999; Gorell *et al.*, 1995; Graham *et al.*, 2000; Martin *et al.*, 2008; Wallis *et al.*, 2008) in the SN in PD patients compared with healthy individuals. Reduced relaxation times are equivalent to longer relaxation rates, and therefore these studies support the hypothesis that patients with PD have an enhanced iron content in the SN. Two further studies which support this hypothesis report that PD patients exhibit SN hypointensity, on T2-weighted images, compared with healthy individuals (Atasoy *et al.*, 2004; Sasaki *et al.*, 2006). Hypointensity on T2-weighted images equates to reduced T2 relaxation times, thus increased iron content, but is not as objective as T2 relaxometry. Two earlier studies reported that nigral T2 relaxation times did not significantly differ between PD patients and healthy individuals (Mondino *et al.*, 2002; Vymazal *et al.*, 1999); however, the latter did report a trend toward shorter T2 relaxation times in PD, again suggestive of increased iron content (Vymazal *et al.*, 1999). Although the relationship between hyperechogenicity and iron accumulation has not been proven, SN hyperechogenicity was recently shown to be related to reduced T2 relaxation times (Behnke *et al.*, 2009), even in individuals without PD, supporting the link between these findings. Further, a greater reduction in T2 relaxation time is reported to be associated with the more affected side in PD patients than in healthy individuals (Kosta *et al.*, 2006), suggesting that increased iron content is more marked in the SN most affected by the disease process. These data, combined with the finding that SN hyperechogenicity is greater in the brain hemisphere contralateral to the body side with more severe symptoms (Berg *et al.*, 2001a), also support the link between nigral iron content and hyperechogenicity. An association between increased T2 relaxation rates and motor symptom severity (Gorell *et al.*, 1995; Martin *et al.*, 2008; Wallis *et al.*, 2008), but not disease duration (Antonini *et al.*, 1993; Bartzokis *et al.*, 1999; Wallis *et al.*, 2008), has also been reported. Although not using T2 relaxometry, a related study found that hypointensity in the substantia nigra, observed on T2-weighted images, is associated with a higher score on the UPDRS (Atasoy *et al.*, 2004, see also Antonini *et al.*, 1993). In addition, Bartzokis *et al.* found that, compared to age-matched controls, older-onset PD patients had greater T2 relaxation rates, whereas young-onset patients had reduced T2 relaxation rates even following correction for disease duration, suggesting that iron content is related to age at disease onset (Bartzokis *et al.*, 1999). Hyperechogenicity is also reported to be associated with symptom severity in healthy older adults with subtle parkinsonism (Berg *et al.*, 2001b; Behnke *et al.*, 2007) but does not change in PD with disease duration (Berg *et al.*, 2005). Overall, these results suggest that increased iron is associated with greater severity of parkinsonian motor symptoms, but not disease duration, in PD patients.

In summary, imaging studies in PD and healthy individuals with SN hyperechogenicity, experimental studies in animal models, and postmortem studies in the human brain support the hypothesis that SN hyperechogenicity is associated with cellular changes and iron accumulation within this brain region. The relative contributions of these changes to the abnormal echosignal and the specific changes underlying this feature remain unclear and represent an important area for future studies.

References

Antonini, A., Leenders, K. L., Meier, D., Oertel, W. H., Boesiger, P., and Anliker, M. (1993). T2 relaxation time in patients with Parkinson's disease. *Neurology* **43**, 697–700.

Atasoy, H. T., Nuyan, O., Tunc, T., Yorubulut, M., Unal, A., and Inan, L. (2004). T2-weighted MRI in Parkinson's disease; Substantia nigra pars compacta hypointensity correlates with the clinical scores. *Neurol. India* **52**, 332–337.

Bartzokis, G., Cummings, J. L., Markham, C. H., Marmarelis, P. Z., Treciokas, L. J., Tishler, T. A., Marder, S. R., and Mintz, J. (1999). MRI evaluation of brain iron in earlier- and later-onset Parkinson's disease and normal subjects. *Magn. Reson. Imaging* **17**, 213–222.

Behnke, S., Double, K. L., Duma, S., Broe, G. A., Guenther, V., Becker, G., and Halliday, G. M. (2007). Substantia nigra echomorphology in the healthy very old: Correlation with motor slowing. *NeuroImage.* **34**, 1054–1059.

Behnke, S., Schroeder, U., Dillmann, U., Buchholz, H. G., Schreckenberger, M., Fuss, G., Reith, W., Berg, D., and Krick, C. M. (2009). Hyperechogenicity of the substantia nigra in healthy controls is related to MRI changes and to neuronal loss as determined by F-Dopa PET. *NeuroImage.* **47**, 1237–1243.

Berg, D. (2007a). Disturbance of iron metabolism as a contributing factor to SN hyperechogenicity in Parkinson's disease: Implications for idiopathic and monogenetic forms. *Neurochem. Res.* **32**, 1646–1654.

Berg, D. (2007b). Ultrasound in the (premotor) diagnosis of Parkinson's disease. *Parkinsonism Rel. Disord.* **13**, S429–S433.

Berg, D., Becker, G., Zeiler, B., Tucha, O., Hofmann, E., Preier, M., Benz, P., Jost, W., Reiners, K., and Lange, K. W. (1999a). Vulnerability of the nigrostriatal system as detected by transcranial ultrasound. *Neurology* **53**, 1026–1031.

Berg, D., Grote, C., Rausch, W. D., Mäurer, M., Wesemann, W., Riederer, P., and Becker, G. (1999b). Iron accumulation in the substantia nigra in rats visualized by ultrasound. *Ultrasound Med. Biol.* **25**, 901–904.

Berg, D., Merz, B., Reiners, K., Naumann, M., and Becker, G. (2005). Five-year follow-up study of hyperechogenicity of the substantia nigra in Parkinson's disease. *Mov. Disord.* **20**, 383–385.

Berg, D., Roggendorf, W., Schroder, U., Klein, R., Tatschner, T., Benz, P., Tucha, O., Preier, M., Lange, K. W., Reiners, K., Gerlach, M., and Becker, G. (2002). Echogenicity of the substantia nigra: Association with increased iron content and marker for susceptibility to nigrostriatal injury. *Arch. Neurol.* **59**, 999–1005.

Berg, D., Siefker, C., and Becker, G. (2001a). Echogenicity of the substantia nigra in Parkinson's disease and its relation to clinical findings. *J. Neurol.* **248**, 684–689.

Berg, D., Siefker, C., Ruprecht-Dorfler, P., and Becker, G. (2001b). Relationship of substantia nigra echogenicity and motor function in elderly subjects. *Neurology* **56**, 13–17.

Collier, T. J., Lipton, J., Daley, B. F., Palfi, S., Chu, Y., Sortwell, C., Bakay, R. A., Sladek, J.R.J., and Kordower, J. H. (2007). Aging-related changes in the nigrostriatal dopamine system and the response to MPTP in nonhuman primates: Diminished compensatory mechanisms as a prelude to Parkinsonism. *Neurobiol Dis.* **26**, 56–65.

Dexter, D. T., Wells, F. R., Agid, F., Agid, Y., Lees, A. J., Jenner, P., and Marsden, C. D. (1987). Increased nigral iron content in postmortem Parkinsonian brain. *Lancet.* **2** (8569), 1219–1220.

Doepp, F., Plotkin, M., Siegel, L., Kivi, A., Gruber, D., Lobsien, E., Kupsch, A., and Schreiber, S. J. (2008). Brain parenchyma sonography and ^{123}I-FP-CIT SPECT in Parkinson's disease and essential tremor. *Mov. Disord.* **23**, 405–410.

Double, K. L., Gerlach, M., Schünemann, V., Trautwein, A. X., Zecca, L., Gallorini, M., Youdim, M. B., Riederer, P., and Ben-Shachar, D. (2003). Iron-binding characteristics of neuromelanin of the human substantia nigra. *Biochem. Pharmacol.* **66**, 489–494.

Fearnley, J. M., and Lees, A. J. (1991). Ageing and Parkinson's disease: Substantia nigra regional selectivity. *Brain* **114**, 2283–2301

Fedorow, H., Tribl, F., Halliday, G., Gerlach, M., Riederer, P., and Double, K. L. (2005). Neuromelanin in human dopamine neurons: Comparison with peripheral melanins and relevance to Parkinson's disease. *Prog. Neurobiol.* **75**, 109–124

Godau, J., Schweitzer, K. J., Liepelt, I., Gerloff, C., and Berg, D. (2007). Substantia nigra hypoechogenicity: Definition and findings in restless legs syndrome. *Mov. Disord.* **22**, 187–192.

Gorell, J. M., Ordidge, R. J., Brown, G. G., Deniau, J. C., Buderer, N. M., and Helpern, J. A. (1995). Increased iron-related MRI contrast in the substantia nigra in Parkinson's disease. *Neurology* **45**, 1138–1143.

Graham, J. M., Paley, M.N.J., Grunewald, R. A., Hoggard, N., and Griffiths, P. D. (2000). Brain iron deposition in Parkinson's disease imaged using the PRIME magnetic resonance sequence. *Brain* **123**, 2423–2431.

Hagenah, J. M., Becker, B., Brüggemann, N., Djarmati, A., Lohmann, K., Sprenger, A., Klein, C., and Seidel, G. (2008). Transcranial sonography findings in a large family with homozygous and heterozygous PINK1 mutations. *J Neurol. Neurosurg. Psychiatry.* **79**, 1071–1074.

Hardman, C. D., McRitchie, D. A., Halliday, G. M., Cartwright, H. R., and Morris, J. G. (1996). Substantia nigra pars reticulata neurons in Parkinson's disease. *Neurodegeneration* **5**, 49–55.

Hochstrasser, H., Bauer, P., Walter, U., Behnke, S., Spiegel, J., Csoti, I., Zeiler, B., Bornemann, A., Pahnke, J., Becker, G., Riess, O., and Berg, D. (2004). Ceruloplasmin gene variations and substantia nigra hyperechogenicity in Parkinson disease. *Neurology* **63**, 1912–1917.

Hutchinson, M., and Raff, U. (2000). Structural changes of the substantia nigra in Parkinson's disease as revealed by MR imaging. *Am. J. Neuroradiol.* **21**, 697–701.

Hutchinson, M., and Raff, U. (2008). Detection of Parkinson's disease by MRI: Spin–lattice distribution imaging. *Mov. Disord.* **23**, 1991–1997.

Hutchinson, M., Raff, U., and Lebedev, S. (2003). MRI correlates of pathology in Parkinsonism: Segmented inversion recovery ratio imaging (SIRRIM). *NeuroImage* **20**, 1899–1902.

Iova, A., Garmashov, A., Androuchtchenko, N., Kehrer, M., Berg, D., Becker, G., and Garmashov, Y. (2004). Postnatal decrease in substantia nigra echogenicity. *J. Neurol.* **251**, 1451–1454.

Kosta, P., Argyropoulou, M., Markoula, S., and Konitsiotis, S. (2006). MRI evaluation of the basal ganglia size and iron content in patients with Parkinson's disease. *J. Neurol.* **253**, 26–32.

Lee, C. S., Samii, A., Sossi, V., Ruth, T. J., Schulzer, M., Holden, J. E., Wudel, J., Pal, P. K., de la Fuente-Fernandez, R., Calne, D. B., and Stoessl, A. J. (2000). In vivo positron emission tomographic evidence for compensatory changes in presynaptic dopaminergic nerve terminals in Parkinson's disease. *Ann. Neurol.* **47**, 493–503.

Martin, W.R.W., Wieler, M., and Gee, M. (2008). Midbrain iron content in early Parkinson disease: A potential biomarker of disease status. *Neurology* **70**, 1411–1417.

McRitchie, D. A., Halliday, G. M., and Cartwright, H. (1995). Quantitative analysis of the variability of substantia nigra pigmented cell clusters in the human. *Neuroscience* **68**, 539–551.

McRitchie, D. A., Hardman, C. D., and Halliday, G. M. (1996). Cytoarchitectural distribution of calcium binding proteins in midbrain dopaminergic regions of rats and humans. *J. Comp. Neurol.* **364**, 121–150.

Menke, R. A., Scholz, J., Miller, K. L., Deoni, S., Jbabdi, S., Matthews, P. M., and Zarei, M. (2009). MRI characteristics of the substantia nigra in Parkinson's disease: A combined quantitative T1 and DTI study. *NeuroImage* **47**, 435–441.

Michaeli, S., Oz, G., Sorce, D. J., Garwood, M., Ugurbil, K., Majestic, S., and Tuite, P. (2007). Assessment of brain iron and neuronal integrity in patients with Parkinson's disease using novel MRI contrasts. *Mov. Disord.* **22**, 334–340.

Minati, L., Grisoli, M., Carella, F., De Simone, T., Bruzzone, M. G., and Savoiardo, M. (2007). Imaging degeneration of the substantia nigra in Parkinson disease with inversion-recovery MR imaging. *Am. J. Neuroradiol.* **28**, 309–313.

Mondino, F., Filippi, P., Magliola, U., and Duca, S. (2002). Magnetic resonance relaxometry in Parkinson's disease. *Neurol. Sci.* **23**, s87–s88.

Morrish, P. K., Rakshi, J. S., Bailey, D. L., Sawle, G. V., and Brooks, D. J. (1998). Measuring the rate of progression and estimating the preclinical period of Parkinson's disease with [^{18}F]dopa PET. *J. Neurol. Neurosurg. Psychiatry* **64**, 314–319.

Olanow, C. W., Stern, M. B., and Sethi, K. (2009). The scientific and clinical basis for the treatment of Parkinson disease. *Neurology* **72**, S1–S136.

Ordidge, R. J., Gorell, J. M., Deniau, J. C., Knight, R. A., and Helpern, J. A. (1994). Assessment of relative brain iron concentrations using T2-weighted and T2*-weighted MRI at 3 Tesla. *Magn. Reson. Med.* **32**, 335–341.

Raff, U., Hutchinson, M., Rojas, G. M., and Huete, I. (2006). Inversion recovery MRI in idiopathic Parkinson disease is a very sensitive tool to assess neurodegeneration in the substantia nigra: Preliminary investigation. *Acad. Radiol.* **13**, 721–727.

Ruprecht-Dörfler, P., Berg, D., Tucha, O., Benz, P., Meier-Meitinger, M., Alders, G. L., Lange, K. W., and Becker, G. (2003). Echogenicity of the substantia nigra in relatives of patients with sporadic Parkinson's disease. *NeuroImage* **18**, 416–422.

Ruprecht-Dörfler, P., Klotz, P., Becker, G., and Berg, D. (2007). Substantia nigra hyperechogenicity correlates with subtle motor dysfunction in tap dancers. *Parkinsonism Relat. Disord.* **13**, 362–364.

Sasaki, M., Shibata, E., Tohyama, K., Takahashi, J., Otsuka, K., Tsuchiya, K., Takahashi, S., Ehara, S., Terayama, Y., and Sakai, A. (2006). Neuromelanin magnetic resonance imaging of locus ceruleus and substantia nigra in Parkinson's disease. *Neuroreport* **17**, 1215–1218.

Snyder, A. M., and Connor, J. R. (2009). Iron, the substantia nigra and related neurological disorders. *Biochim. Biophys. Acta* **1790**, 606–614.

Sofic, E., Riederer, P., Heinsen, H., Beckmann, H., Reynolds, G. P., Hebenstreit, G., and Youdim, M. B. (1988). Increased iron (III) and total iron content in post mortem substantia nigra of Parkinsonian brain. *J. Neural Transm.* **74**, 199–205.

Spiegel, J., Hellwig, D., Mollers, M.-O., Behnke, S., Jost, W., Fassbender, K., Samnick, S., Dillmann, U., Becker, G., and Kirsch, C.-M. (2006). Transcranial sonography and [^{123}I]FP-CIT SPECT disclose complementary aspects of Parkinson's disease. *Brain* **129**, 1188–1193.

Stark, A. K., and Pakkenberg, B. (2004). Histological changes of the dopaminergic nigrostriatal system in aging. *Cell Tissue Res.* **318**, 81–92.

Todd, G., Taylor, J. L., Baumann, D., Butler, J. E., Duma, S. R., Hayes, M., Carew-Jones, F., Piguet, O., Behnke, S., Ridding, M. C., Berg, D., and Double, K. L. (2010). Substantia nigra echomorphology and motor cortex excitability. *NeuroImage* **50**, 1351–1356.

Tsai, C.-F., Wu, R.-M., Huang, Y.-W., Chen, L.-L., Yip, P.-K., and Jeng, J.-S. (2007). Transcranial color-coded sonography helps differentiation between idiopathic Parkinson's disease and vascular Parkinsonism. *J. Neurol.* **254**, 501–507.

Vlaar, A. M., Bouwmans, A., Mess, W. H., Tromp, S. C., and Weber, W. E. (2009). Transcranial duplex in the differential diagnosis of Parkinsonian syndromes: A systematic review. *J. Neurol.* **256**, 530–538.

Vymazal, J., Righini, A., Brooks, R. A., Canesi, M., Mariani, C., Leonardi, M., and Pezzoli, G. (1999). T1 and T2 in the brain of healthy subjects, patients with Parkinson disease, and patients with multiple system atrophy: Relation to iron content. *Radiology* **211**, 489–495.

Wallis, L. I., Paley, M.N.J., Graham, J. M., Grünewald, R. A., Wignall, E. L., Joy, H. M., and Griffiths, P. D. (2008). MRI assessment of basal ganglia iron deposition in Parkinson's disease. *J. Magn. Reson. Imaging* **28**, 1061–1067.

Walter, U. (2009). Transcranial brain sonography findings in Parkinson's disease: Implications for pathogenesis, early diagnosis and therapy. *Expert Rev. Neurother.* **9**, 835–846.

Walter, U., Dressler, D., Wolters, A., Wittstock, M., Greim, B., and Benecke, R. (2006). Sonographic discrimination of dementia with Lewy bodies and Parkinson's disease with dementia. *J. Neurol.* **253**, 448–454.

Walter, U., Klein, C., Hilker, R., Benecke, R., Pramstaller, P.P., and Dressler, D. (2004). Brain parenchyma sonography detects preclinical Parkinsonism. *Mov. Disord.* **19**, 1445–1449.

Weise, D., Lorenz, R., Schliesser, M., Schirbel, A., Reiners, K., and Classen, J. (2009). Substantia nigra echogenicity: A structural correlate of functional impairment of the dopaminergic striatal projection in Parkinson's disease. *Mov. Disord.* **24**, 1669–1675.

Zecca, L., Berg, D., Arzberger, T., Ruprecht, P., Rausch, W. D., Musicco, M., Tampellini, D., Riederer, P., Gerlach, M., and Becker, G. (2005). In vivo detection of iron and neuromelanin by transcranial sonography: A new approach for early detection of substantia nigra damage. *Mov. Disord.* **20**, 1278–1285.

TRANSCRANIAL SONOGRAPHY FOR THE DISCRIMINATION OF IDIOPATHIC PARKINSON'S DISEASE FROM THE ATYPICAL PARKINSONIAN SYNDROMES

Angela E. P. Bouwmans,* Annemarie M. M. Vlaar,† Karin Srulijes,‡ Werner H. Mess,§ and W. E. J. Weber*

*Department of Neurology, University Hospital Maastricht, Maastricht, The Netherlands
†Department of Neurology, Sint Lucas Andreas Hospital, Amsterdam, The Netherlands
‡Department of Neurodegeneration, Hertie Institute of Clinical Brain Research and German Center of Neurodegenerative Diseases (DZNE), Tübingen, Germany
§Department of Clinical Neurophysiology, University Hospital Maastricht, Maastricht, The Netherlands

We reviewed eight studies on transcranial sonography (TCS) as a tool for differentiating idiopathic Parkinson's disease (IPD) from atypical parkinsonian syndromes (APS) and included some first data on TCS findings in the subforms of PSP. Changes of specific structures on TCS like the substantia nigra (SN), lenticular nucleus (LN), and the third ventricle are discussed as well as how they can contribute to differentiate between IPD, multiple system atrophy (MSA), progressive supranuclear palsy (PSP), Lewy body disease (LBD), and corticobasal degeneration (CBD). We finish with an algorithm that may be used to employ TCS as a diagnostic instrument delineating IPD from the APS and discerning among the APS themselves. As TCS is at present the most

DOI: 10.1016/S0074-7742(10)90009-3

promising tool for this particular diagnostic problem, this algorithm might be a suitable hypothesis to study in future research.

I. Introduction

Despite the availability of sophisticated neuroimaging techniques, the diagnostic differentiation between idiopathic Parkinson's disease (IPD) and the atypical parkinsonian syndromes (APS) still remains a challenge, especially at the beginning of symptoms. In one study, one-third of the initial clinical diagnoses of parkinsonian patients was changed after follow-up; IPD in particular was overdiagnosed and the APS were underdiagnosed, up to a factor of six (Caslake *et al.*, 2008). These observations were reproduced in other studies (Hughes *et al.*, 2002; Katzenschlager *et al.*, 2003; Schrag *et al.*, 2000). All these studies do confirm that neuropathological examination remains unrivalled as the gold standard diagnostic method for the parkinsonian syndromes. As these syndromes differ drastically in terms of therapy and prognosis, this stresses the need for a highly specific and sensitive tool to help differentiate between the various syndromes during life, and especially in the earlier stages of disease.

Since 1995, there is an increasing interest for TCS as a cheap and easily applicable tool to help differentiate between movement disorders (Becker *et al.*, 1995; Berg *et al.*, 2008; Walter *et al.*, 2007c). The investigation time is short and the technique is noninvasive. In this context, the substantia nigra (SN) is one of the major sonographic landmarks. Firstly, the brainstem is an 'echo-poor', butterfly-shaped structure, and the SN is not visible. When the SN is pathologically changed, it becomes hyperechogenic. Secondly, the lenticular nucleus (LN) can be studied through the TCS. The discriminating capacity of TCS also allows visual assessment of other structures in the brain, such as the thalami, caudate nuclei, raphe nuclei, third ventricle, and frontal horns.

This chapter discusses briefly the diagnostic criteria of the APS, including multiple system atrophy (MSA), progressive supranuclear palsy (PSP), diffuse Lewy body disease (LBD), and corticobasal degeneration (CBD). In all these disorders, there are serious difficulties with the accuracy of diagnosis, especially at the early stage of the disease. Subsequently, it reviews the literature on the use of transcranial sonography (TCS) as a diagnostic instrument in these patient populations. After summarizing all these data, an algorithm is presented for the use of and the interpretation of TCS findings on patients with parkinsonian syndromes.

II. Diagnostic Criteria of Atypical Parkinsonism

A. Multiple System Atrophy

MSA is an adult-onset, sporadic, progressive disorder with parkinsonism, cerebellar ataxia, autonomic failure, urogenital dysfunction, and corticospinal dysfunction (Geser *et al.*, 2006). One of the first symptoms is bladder dysfunction, accompanied by initially erectile dysfunction in male patients (Kirchhof *et al.*, 2003). The first motor symptoms are parkinsonism with bradykinesia, rigidity, gait instability, and tremor, thus making the differentiation between MSA and IPD difficult. A subgroup of MSA patients has cerebellar ataxia as the initial motor feature (Gilman *et al.*, 1999) (reference Gilman 1999). The predominant motor feature determines whether the patients are classified as MSA with parkinsonism (MSA-P) or as MSA with cerebellar ataxia (MSA-C). It is important to take into account that the predominant motor feature, and with this the diagnostic classification, can change over time. Parkinsonism in MSA can be asymmetric like in IPD. The tremor in MSA patients is usually irregular and posture- or action-dependent and polyminimyoclonus can be seen, but a classic "pill-rolling" resting tremor is rarely found. This can help to differentiate clinically between MSA and IPD. Parkinsonian features worsen faster over time in MSA than in IPD patients: motor scores measured with the UPDRS-III increase on average with less than 10% per year in IPD, but by more than 20% in MSA patients. An MSA patient usually becomes wheelchair-dependent within 5 years after the onset of symptoms, which is much faster than the average IPD patient (Seppi *et al.*, 2005)

In a 2008 expert meeting, a new set of diagnostic criteria was formulated (Gilman *et al.*, 2008). The meeting recommended that the diagnosis of MSA be divided into three groups (Tables I and II): probable, possible, and definite MSA. The category "definite MSA" is only possible by postmortem examination (Trojanowski and Revesz, 2007). The underlying neuropathology of MSA consists of degeneration of striatonigral and olivopontocerebellar structures together with great numbers of distinctive glial cytoplasmic inclusions formed by fibrillized alpha-synuclein proteins (Ozawa *et al.*, 2006; Spillantini *et al.*, 1998). Like IPD, this disease seems to be the result of an accumulation of intracellular alpha-synuclein and is categorized as an alpha-synucleinopathy. In the various categories, one or more symptoms of autonomic failure are obligatory, e.g., orthostatic hypotension (a decrease of blood pressure within 3 min of standing by at least 30 mmHg systolic or 15 mmHg diastolic), urinary incontinence, erectile dysfunction, and constipation. One of the criteria for the group of probable and possible MSA diagnostic classes is that there has to be a poor response of the parkinsonism in MSA to the chronic levodopa therapy. In daily clinical

TABLE I
DIAGNOSTIC CRITERIA FOR MULTIPLE SYSTEM ATROPHY

Diagnostic categories MSA	A sporadic, progressive, adult (>30 years)-onset disease characterized by
	Neuropathologic features
Definite	Widespread and abundant CNS alpha-synuclein-positive cytoplasmic inclusions *AND*
	Neurodegenerative changes in striatonigral or olivopontocerebellar structures
Probable	Autonomic failure *AND*
	Poorly levodopa-responsive parkinsonism *OR*
	Cerebellar syndrome
Possible	Parkinsonism *OR*
	Cerebellar syndrome *AND*
	At least one feature suggesting autonomic dysfunction *AND*
	At least one of the additional features listed elsewhere (Table II)

TABLE II
ADDITIONAL FEATURES OF DIAGNOSTIC CRITERIA FOR MULTIPLE SYSTEM ATROPHY

Diagnostic categories possible MSA	Additional features possible MSA
MSA-P or MSA-C	Babinski sign with hyperreflexia
	Stridor
MSA-C	Rapidly progressive parkinsonism
	Poor response to levodopa
	Postural instability within 3 years of motor onset
	Gait ataxia, cerebellar dysarthria, limb ataxia, or cerebellar oculomotor dysfunction
	Dysphagia within 5 years of motor onset
	Atrophy on mri of putamen, middle cerebellar peduncle, pons, or cerebellum
	Hypometabolism on FDG-PET in putamen, brainstem, or cerebellum
MSA-P	Parkinsonism
	Atrophy on MRI of putamen, middle cerebellar peduncle, or pons
	Hypometabolism on FDG-PET in putamen
	Presynaptic nigrostriatal dopaminergic denervation on SPECT or PET

practice, this criterion can be difficult to fulfil: 30% of the patients with MSA have a significant, albeit only initial, response to levodopa (Hughes *et al.*, 1992).

A validation study of these criteria showed that these new criteria had higher sensitivity at the first clinic visit compared with the previous criteria, but they never reach 100% (Osaki *et al.*, 2009), emphasizing the need for additional diagnostic tools, as prognosis in MSA is far worse than in IPD.

B. Progressive Supranuclear Palsy

PSP, also called the Steele–Richardson–Olzewksi syndrome, is a parkinsonian disorder characterized by early, frequent falls, dysphasia, vertical gaze disturbance, and mild dementia (Williams and Lees, 2009). The most recent set of diagnostic criteria for PSP was formulated in an international workshop, organized by the National Institute for Neurological Disorders and Stroke (NINDS) and the Society for PSP (SPSP) in 1996 (Litvan *et al.*, 1996). These NINDS–SPSP criteria (Table III), which were formulated from an extensive review of the literature, comparison with other previously published sets of criteria, and the consensus of experts, were validated on a clinical data set from autopsy-confirmed cases of PSP. The criteria specify three degrees of diagnostic certainty: possible PSP, probable PSP, and definite PSP. Possible PSP requires the presence of a gradually progressive disorder with onset at age 40 years or later, either vertical supranuclear gaze palsy or both slowing of vertical saccades and prominent postural instability with falls in the first year of onset, as well as no evidence of other diseases that could explain these features. Probable PSP requires vertical supranuclear gaze palsy, prominent postural instability, and falls in the first year of onset, as well as the other features of possible PSP. Definite PSP requires a history of probable or possible PSP and histopathologic evidence of typical PSP. Because of the clinical heterogeneity, PSP has recently been divided into four subgroups (Table IV).

Whereas MSA is neuropathologically classified as an alpha-synucleinopathy, cortical tau protein accumulation is the typical feature in PSP (Dickson, 2006); it is mainly located in the palladium, subthalamic nucleus, red nucleus, SN, pontine tegmentum, striatum, oculomotor nucleus, medulla, and dentate

TABLE III
Diagnostic Criteria for Progressive Supranuclear Palsy

Mandatory inclusionary criteria PSP	Mandatory exclusionary criteria PSP
Progressive disease course	History of encephalitis
Vertical supranuclear palsy with downward gaze abnormalities	Hallucinations
Severe postural instability with unexplained falls	Early or prominent cerebellar signs
	Noniatrogenic dysautonomia
	Unilateral dystonia
	Alien hand syndrome
	Early cortical dementia
	Focal lesions on neurologic examination or CT

TABLE IV
DIAGNOSTIC CRITERIA FOR VARIANTS OF PROGRESSIVE SUPRANUCLEAR PALSY

Variants PSP	Neuropathologically criteria	Clinical criteria
PSP-parkinsonism	Less severe tau pathology More restricted distribution More affected substantia nigra	Most common variant Tremor Moderate responsiveness
PSP-pure akinesia with gait freezing	Severe atrophy and neuronal loss specific in the globus pallidus, substantia nigra, subthalamic nucleus Less severe tau pathology in motor cortex, striatum, pontine nuclei, cerebellum	Early gait disturbance Micrographia Hypophonia Gait freezing Better prognosis compared to PSP
PSP-corticobasal syndrome	More severe tau pathology in midfrontal and inferior parietal cortex Less severe tau pathology in motor cortex	Progressive, asymmetric dystonia Apraxia Cortical sensory loss
PSP-progressive non-fluent aphasia	More severe tau pathology in temporal cortex and superior frontal gyrus Less severe tau pathology in brainstem and subcortical regions	Apraxia of speech

nucleus (Hauw *et al.*, 1994). These regional differences in pathological severity determine the clinical differences. The SN pars compacta is affected in PSP as in IPD, but the severity of pathology is substantially higher in PSP than in IPD (Murphy *et al.*, 2008). This also is the case for the ventrotegmentale areas, which influence the motor and cognitive functions. It is thus assumed that the destruction of these regions is the cause of the levodopa nonresponsiveness and frontostriatal cognitive dysfunction in patients with PSP (Sanchez-Gonzalez *et al.*, 2005).

Due to these clinicopathological variations, the accuracy of the clinical diagnosis of PSP remains low (Tsuboi *et al.*, 2005; Williams *et al.*, 2005, 2007). In a study on 60 patients clinically diagnosed as PSP, the eventual diagnosis was confirmed pathologically in only 47 cases (78%) (Osaki *et al.*, 2004). IPD is the main differential diagnosis, and it is generally thought to be 30 times more prevalent than PSP, although PSP has been underdiagnosed in several studies (Nath *et al.*, 2001).

Analysis of clinical data of pathologically proven PSP cases led to a number of efforts to clinically divide PSP into several subtypes, such as pure akinesia with gait freezing (PAGF), PSP–CBD, and progressive nonfluent aphasia (PNFA) (Brusa *et al.*, 1980; Cambier *et al.*, 1985; Williams *et al.*, 2009). Besides these rare variants, PSP-parkinsonism (PSP-P) and Richardson's syndrome (RS) represent the most common subtypes of PSP (Williams *et al.*, 2005). In postmortem analyses, severity and extent of pathology have been shown to differ between RS and PSP-P. The retrospectively analyzed clinical pictures revealed that RS presents in the beginning with falls, early cognitive dysfunction, abnormality of gaze, and postural instability, whereas PSP-P presents in the beginning with an asymmetric onset of parkinsonian symptoms, tremor, bradykinesia, nonaxial dystonia, and often response to levodopa medication. As response to L-Dopa therapy, progression and prognosis differ between these subtypes affecting treatment and daily care; therefore it is of importance to differentiate these forms. However, so far, there exists no diagnostic instrument to differentiate PSP-P and RS *in vivo* (Litvan *et al.*, 2003).

C. Dementia with Lewy Bodies

Dementia with Lewy bodies (DLB) is a progressive disease characterized by cognitive, neuropsychiatric, autonomic, sleep, and, to a lesser extent, motor symptoms. Lewy bodies and Lewy neurites are pathologic aggregations of alfa-synuclein. The pathologic presence of Lewy bodies occurs, as in IPD, in the brainstem, limbic system, and, this more markedly than in IPD, diffusely in the cortex. The main substrate of the cognitive dysfunction in both groups is assumed to be the presence of Lewy bodies in certain cortical areas. DLB and PD with dementia (PDD) seem to be extremes on a spectrum of LBD because of the similar clinical features such as fluctuating cognition, visual hallucinations, and parkinsonism. The consensus guidelines favor the diagnosis of PDD if parkinsonian symptoms precede 1 year before the onset of dementia (McKeith *et al.*, 1996). Therefore, the clinical course suggests earlier cortical involvement in DLB than in PDD.

In DLB the exact regional deposition of Lewy bodies is probably more a determinant of clinical symptoms than the total Lewy bodies count (Walter *et al.*, 2006), as nearly 60% of Alzheimer's disease patients also have Lewy bodies (Hamilton, 2000). Especially in the early stages, DLB can be impossible to discern clinically from IPD or Alzheimer's disease.

The most recent adjustments to the consensus diagnostic criteria of DLB have been made in 2005, the DLB consortium, which defines two diagnostic categories: probable and possible DLB. The adjustments included supportive features, as severe neuroleptica sensitivity abnormal and scintigraphy scans, and more precise definitions of demential features (McKeith *et al.*, 2005) (Tables V and VI). The introduction of these new clinical criteria for DLB leads to an increase in patients diagnosed with probable DLB. In a 2008 study on 196 patients with mild dementia, 20% received the diagnosis of DLB

TABLE V
Diagnostic Criteria for Dementia with Lewy Bodies

Clinical diagnosis DLB	Clinical findings	Probable DLB	Possible DLB
Central feature	Dementia Prominent or persistent memory impairment Deficits on test of attention, executive function, and visuospatial ability	Presence of one or more central features essential	Presence of one or more central features essential
Core features	Fluctuating cognition Recurrent visual hallucinations Spontaneous features of parkinsonism	Presence of two or more core features needed	Presence of one or more core features needed
Suggestive features	REM sleep behavior disorder Severe neuroleptic sensitivity Low-dopamine transporter uptake in basal ganglia by SPECT or PET imaging	One or more features needed in the presence of one or more core features	One or more features needed with no presence of core features

TABLE VI
Additional Features of Dementia with Lewy Bodies

Supportive clinical features DLB	Supportive imaging features DLB	Less likely clinical features DLB
Repeated falls and syncope	Relative preservation of medial temporal lobe structure on CT/MRI scan	Presence of cerebrovascular disease
Transient unexplained loss of consciousness	Generalized low uptake on SPECT/PET perfusion scan with reduced occipital activity	Presence of any other physical illness or brain disorder
Severe autonomic dysfunction	Abnormal MIBG myocardial scintigraphy	Parkinsonism only appears for the first time at a stage of severe dementia
Hallucinations in other modalities	Prominent slow wave activity on EEG with temporal lobe transient sharp waves	
Systematized delusions		
Depression		

compared with 65% who received the diagnosis of Alzheimer's disease (Aarsland *et al.*, 2008). Therefore, DLB seems to be the second most common type of dementia.

D. Corticobasal Degeneration

Previous literature on CBD suggested it was a distinct clinico-pathological disease; however, recently strong evidence has shown that it is a more heterogeneous disorder (Mahapatra *et al.*, 2004; Wadia and Lang, 2007). The same pathology of CBD can lead to diverse syndromes such as corticobasal syndrome, frontotemporal dementia, primary speech apraxia, progressive nonfluent aphasia, PSP-like syndrome, and posterior cortical atrophy syndrome (Boeve *et al.*, 2003).

The pathological features include frontoparietal cortical atrophy, commonly asymmetrical. In most cases the SN is severely depigmented. Histological features are neuronal loss and gliosis, most prominent in the superficial cortical layers. Furthermore, balloon neurons can be seen, but these are not specific for CBD (Wadia and Lang, 2007), as they also appear in PSP, Alzheimer's disease, and Creutzfeldt–Jakob's disease. However, the quantity and distribution of the balloon neurons are of diagnostic relevance. CBD is a rare disorder that usually differs from PD in its clinical presentation and pathogenesis, because it is associated with tau rather than with alpha-synuclein pathology.

CBD and PSP are closely related parkinsonian syndromes that have many clinical symptoms in common, such as rapid disease progression, poor levodopa response, eye movement abnormalities, cognitive impairment, apraxia, pyramidal signs, and dystonia (Litvan *et al.*, 1999). PSP is therefore the most frequent pathologic diagnosis in cases misdiagnosed as CBD; in pathologically diagnosed CBD series, PSP was the most frequent clinical misdiagnosis (Wadia and Lang, 2007). In a 1999 study, clinical data were examined from a group of 51 patients pathologically diagnosed with PSP and CBD (Litvan *et al.*, 1999). As a result, two predictor models were created to help differentiate between the two diseases. CBD patients present usually with lateralized motor and cognitive signs whereas PSP patients often had severe postural instability, symmetric parkinsonism, vertical supranuclear gaze palsy, speech, and frontal lobe features. However, these clinical features are not specific, and they only increase the index of suspicion. Official consensus diagnostic criteria for CBD have not been formulated, but numerous researchers use those listed in a recent paper by Mahapatra *et al.* (2004) (Table VII).

III. Systematic Review of the Literature on TCS and Parkinsonism

Searches for this review were performed in PubMed, with the terms mentioned as 'keywords'. Only papers in English and Dutch were reviewed. References were also identified from relevant articles. We were able to retrieve eight relevant papers on the value of TCS to discriminate among the parkinsonian

TABLE VII
DIAGNOSTIC CRITERIA FOR CORTICOBASAL DEGENERATION

Inclusion criteria	Core features	Supportive features	Exclusion criteria
(Asymmetric) rigidity	Insidious onset with progressive course	Focal/lateralized cognitive dysfunction	Early dementia
Cortical signs: – Apraxia –Cortical sensory loss	Cortical dysfunction	Focal or asymmetric hypo perfusion SPECT/PET	Sustained responsiveness to levodopa
Dystonia	No identifiable cause	Focal or asymmetric atrophy CT/MRI	Severe autonomic disturbances
Focal reflex myoclonus	Extrapyramidal dysfunction		Rest tremor
			Early vertical gaze
			Lesions on imaging

syndromes: six papers with data on discriminating MSA/PSP versus IPD, one with data on DLB versus IPD, and one paper on CBD versus PSP. Moreover, we add our own recent and therefore yet unpublished findings on TCS data in different subtypes of PSP (Tables VIII and IX).

A. RESULTS: TCS DISCRIMINATING MSA/PSP FROM IPD

The first paper on this subject was published by Walter *et al.* (2003). These authors concluded that when SN hyperechogenicity was marked on TCS, APS could be excluded with a positive predictive value of 100% for IPD (Table X). They also found no correlation between the duration of symptoms and the degree of SN hyperechogenicity. The median SN echogenic size was significantly different between the IPD and the MSA groups, but was not different between the MSA and the PSP groups. Moderate or marked SN hyperechogenicity had a positive predictive value of 92% for the diagnosis of IPD and a negative predictive value of 95% for not having done the diagnosis of IPD, that is, being diagnosed as APS.

Hyperechogenicity of the LN was more frequent in the APS group compared with the IPD group (Table XI). Hyperechogenicity of the LN had a positive predictive value for the diagnosis APS of 77%. Width of the third ventricle was larger in the PSP group than in the IPD group (Table XII), and no differences in this width were found between the MSA and the IPD groups. Thus, in this study, a third ventricle width of 9 mm or smaller had a positive predictive value of 71% for the diagnosis of IPD. No significant TCS differences were found in the thalami, caudate nuclei, and widths of the frontal horns.

TABLE VIII
DATA STUDIES OF TRANSCRANIAL DUPLEX SCANNING IN ATYPICAL PARKINSONIAN SYNDROMES

Studies	Number blinded operators TCS	Number blinded investigators diagnosis	Disease stage	Criteria size SN quantification hyperechogenicity available	criteria size SN cutoff point hyperechogenicity in cm^2
Bartova (2007)	NA	NA	Advanced	+	0.25
Behnke (2005)	NA	NA	Advanced	+	*moderate:* between 0.11 and 0.18/0.20 *marked*: 0.18/0.20 and above
Gaenslen (2008)	1	2-3	Beginning	+	0.20
Okawa (2007)	2	NA	Advanced	–	NA
Walter (2003)	2	NA	Advanced	+	*moderate*: between 0.20 and 0.25 *marked:* 0.25 and above
Walter (2004)	2	NA	Advanced	+	*moderate*: between 0.20 and 0.25 *marked*: 0.25 and above
Walter (2006)	1	NA	Advanced	+	*moderate*: between 0.20 and 0.25 *marked*: 0.25 and above
Walter (2007)	2	NA	Advanced	+	*moderate*: between 0.20 and 0.25 *marked*: 0.25 and above

TABLE IX
DATA ATYPICAL PARKINSONIAN SYNDROMES AVAILABLE IN STUDIES

Studies	MSA vs. IPD	PSP vs. IPD	CBD vs. IPD	LBD vs. IPD	PSP vs. CBD
Bartova (2007)	+	–	–	–	–
Behnke (2005)	+	+	–	–	–
Gaenslen (2008)	+	+	+	–	–
Okawa (2007)	+	+	–	–	–
Walter (2003)	+	+	–	–	–
Walter (2004)	–	–	–	–	+
Walter (2006)	–	–	–	+	–
Walter (2007)	+	+	–	–	–

In a similar study with 88 patients, Behnke *et al.* in 2005 found that SN hyperechogenicity of the IPD group compared with the PSP or MSA-P group was statistically significant (Table X) (Behnke *et al.*, 2005). In their patients, a marked hyperechogenic SN on one or both sides predicted IPD with a positive predictive value of 83%. The other significant difference was once again found

TABLE X

DATA MSA/PSP VS. IPD CONCERNING SN

Studies	IPD SN+	IPD SN−	MSA SN+	MSA SN−	PSP SN+	PSP SN−
Bartova (2007)	86/101	15/101	4/8	4/8	NA	NA
Behnke (2005)	*Total* 87/88	1/88	*Total* 28/32	4/32	*Total* 17/21:	4/21
	Moderate: 9/88		*Moderate*: 20/32		*Moderate*: 10/21	
	unilateral 6		unilateral 32		unilateral 3	
	bilateral 3		bilateral 218		bilateral 7	
	Marked: 78/88		*Marked*: 8/32		*Marked*: 7/21	
	unilateral 37		unilateral 2		unilateral 5	
	bilateral 41		bilateral 6		bilateral 2	
Gaenslen (2008)	37/39	2/39	0/4	4/4	0/4	4/4
Okawa (2007)	52/63	11/63	1/11	10/11	1/13	12/13
Walter (2003)	*Total* 24/25:	1/25	NA	NA	NA	NA
	Moderate: 5/25					
	unilateral 3					
	bilateral 2					
	Marked: 19/25					
	unilateral 13					
	bilateral 6					
Walter (2007)	131/134	3/134	2/21	19/21	9/22	13/22
Total	417/450	33/450	35/76	41/76	27/60	33/60
	93%	**7%**	**46%**	**54%**	**45%**	**55%**

With respect to the total numbers it needs to be considered that the cut off for SN hyperechogenicity differed in different studies. Behnke et al., used a lower cut off than the other studies (see table X). This may account for the high percentage of SN+ also in MSA and PSP.

TABLE XI
DATA MSA/PSP VS. IPD CONCERNING LN

Studies	IPD LN+	IPD LN−	MSA LN+	MSA LN−	PSP LN+	PSP LN−
Bartova (2007)	NA	NA	NA	NA	NA	NA
Behnke (2005)	10/88	78/88	23/32	9/32	13/18	5/18
Gaenslen (2008)	11/35	24/35	NA	NA	NA	NA
Okawa (2007)	NA	NA	NA	NA	NA	NA
Walter (2003)	5/22	17/22	NA	NA	NA	NA
Walter (2007)	31/131	100/131	15/21	6/21	19/20	1/20
Total	57/276	219/276	38/53	15/53	32/38	6/38
	21%	**79%**	**72%**	**28%**	**84%**	**16%**

between the IPD group and MSA-P or PSP group concerning the hyperechogenicity of the LN (Table XI). Hyperechogenic LN had a positive predictive value for an APS of 78%. When Behnke *et al.* combined ultrasound features, the positive predictive value increased: the positive predictive value for IPD of a hyperechogenic SN combined with a regular echogenic LN is 91%. The combination of a hyperechogenic LN and at maximum score a moderate echogenic SN was indicative for an atypical parkinsonian syndrome; the positive predictive value was 96%. Like other studies, no significant correlations between the disease severity in IPD and the extent of the hyperechogenic signal had been found.

Two subsequent studies by Okawa *et al.* and Bartova *et al.* in 2007 yielded similar results: a hyperechogenic SN does make the diagnosis of IPD more probable than MSA or PSP, with predictive values in the same range as previous studies (Table X) (Bartova *et al.*, 2007; Okawa *et al.*, 2007).

Walter *et al.* studied the issue again in 2007 on 134 patients(Walter *et al.*, 2007c). These authors divided the patients with IPD in two subgroups: older than 60 years versus younger than 60 years. They obtained several sonographic features, beginning with the echogenicity of the SN (Table X). The marker for a normal SN was a hyperechogenicity of a size less than 0.20 cm^2. A marked hyperechogenic SN was defined, as the size of the area was 0.25 cm^2 or more. They found that the size of the area of hyperechogenicity of the SN in the group of patients with IPD was larger compared with the group of patients with MSA-P. No correlation of SN hyperechogenicity was found with age, disease severity, or disease duration in either group. Besides, no significant difference was found between the size of the SN in the IPD group with a PD duration of no longer than 5 years compared with the IPD group with longer than 5 years' duration of symptoms, confirming earlier observations of the stability of this feature over time.

The *a priori* hypothesis of this 2007 study by Walther *et al.* was that sonographic discrimination of the patients with IPD from the group of patients with APS might be clearer in patients who have the onset at younger than 60 years compared with the patients with later onset (Walter *et al.*, 2007c). The results showed that in patients with onset of parkinsonism younger than 60 years, normal SN alone indicates MSA-P or PSP rather than IPD, with a sensitivity of 75% and a positive predictive value of 100%. In the whole group of patients with IPD regardless of their age, normal SN echogenicity discriminated MSA-P from IPD with a sensitivity of 90% and a positive predictive value of 86%. The finding of a hyperechogenic LN in combination with normal echogenic SN indicated MSA-P or PSP rather than PD, with a positive predictive value of 94%.

The width of the third ventricle was a third sonographic finding examined by Walter *et al.* in 2007 (Table XII) (Walter *et al.*, 2007c). Measurement of the third ventricle resulting in a width more than 10 mm in combination with a hyperechogenic LN discriminated PSP from IPD with a sensitivity of 84% and a positive predictive value of 89%. The finding of a hyperechogenic LN in combination with third-ventricle dilatation of more than 10 mm indicated MSA-P or PSP rather than IPD, with a positive predictive value of 94%.

They concluded that the combined finding of normal SN with hyperechogenicity of the LN excludes the diagnosis of IPD, indicating MSA-P or PSP, with a positive predictive value of 100%. Using TCS they concluded that MSA-P is best discriminated from IPD by normal SN, whereas PSP is best discriminated from IPD by the combined finding of third-ventricle dilatation of more than 10 mm and LN hyperechogenicity. In patients under 60 years with parkinsonian symptoms, a normal SN echogenicity excluded the diagnosis of IPD.

The most recent publication on this topic was by Gaenslen *et al.* (2008). This is the first prospective study, the earlier studies all being retrospective. They did TCS on 60 patients who visited the outpatient clinic with first mild signs of parkinsonism, but excluding patients with mostly rest tremor. All the investigators who performed the TCS were blinded for clinical data. Follow-up time was 12 months. The final diagnosis was reached by an independent neurologist who was

TABLE XII
DATA MSA/PSP VS. IPD CONCERNING THIRD VENTRICLE

Studies	IPD third ventricle dilatated	IPD third ventricle normal	MSA third ventricle dilatated	MSA third ventricle normal	PSP third ventricle dilatated	PSP third ventricle normal
Bartova (2007)	NA	NA	NA	NA	NA	NA
Behnke (2005)	NA	NA	NA	NA	NA	NA
Gaenslen (2008)	NA	NA	NA	NA	NA	NA
Okawa (2007)	NA	NA	NA	NA	NA	NA
Walter (2003)	NA	NA	NA	NA	NA	NA
Walter (2007)	14/130	116/130	4/19	15/19	17/20	3/20
Total	14/130	116/130	4/19	15/19	17/20	3/20
	11%	**89%**	**21%**	**79%**	**85%**	**15%**

blinded for the TCS results. Patients with an uncertain diagnosis had additional investigations by SPECT or PET scans. If this also did not result in a definite clinical diagnosis, a consensus diagnosis of two independent blinded experts was taken and, if necessary, a third expert.

Twenty-two (37%) of the sixty patients had a clinical diagnosis at baseline. After 12 months of follow-up, with additional imaging at 19 patients, a diagnosis was made in 52 patients (88%). At the beginning of the study, agreement between the clinical examiner and the rater was only 72.6% on the diagnosis, which increased after 12 months' follow-up up to 98.1%. The results showed that SN hyperechogenicity is not present in all IPD patients, although it has an excellent positive predictive value. LN hyperechogenicity had a low predictive value for the diagnosis of atypical parkinsonian syndromes (APS); however, the high negative predictive value enables the exclusion of APS in a patient with parkinsonism when LN is normal (Table XI). Two forms of bias may be present in this study: the exclusion of the patients with resting tremor, because it is known that this feature can precede IPD, and the failure to establish a clear diagnosis in all patients at the end of the follow-up, which was the case in seven patients. Moreover, follow-up was not very long. In conclusion, Gaenslen found that the sensitivity of the TCS at baseline compared with the endpoint diagnosis was 90.7% and the specificity was 82.4%. The positive predictive value of TCS for IPD was 92.9% and the classification accuracy 88.3%.

Only recently a growing awareness has evolved concerning the subtypes of PSP. So far, there exists no instrument to discern *in vivo* whether a patient suffers from RS or from PSP-P. In a first cross-sectional study we investigated whether transcranial sonography (TCS) can detect differences and can be used to differentiate RS from PSP-P *in vivo*. A total of 17 PSP patients were clinically well separated into nine RS patients and eight PSP-P patients and examined with TCS according to a standardized protocol. Median age at study inclusion was 67 years in RS patients and 64.5 years in PSP-P patients. Median disease duration (until study inclusion) in RS patients was 2.9 years and in PSP-P patients 6.1 years. Gender did not differ significantly between these two subtypes. Two RS patients could not be studied due to insufficient acoustic temporal bone windows. There was a significant difference in SN echogenicity between both PSP-subgroups (larger SN of both sides: in RS = 0.18 cm^2; in PSP-P = 0.25 cm^2; $P = 0.02$) (Figs. 1 and 2). A marked dilatation of the third ventricle (width > 10 mm) was found in five (71%) of the seven RS patients (Fig. 3), but in none of the PSP-P patients ($p = 0.02$) (Fig. 4). Other TCS findings such as widths of the frontal horns or echogenicity of lentiform and caudate nuclei did not discriminate RS and PSP-P.

In this pilot study, we could therefore show for the first time that TCS indeed differentiates between PSP-P and RS. Having in mind the study's limitations due to small sample size, SN hyperechogenicity, a characteristic TCS marker for IPD, and normal of the third ventricle, are strongly associated with PSP-parkinsonism. This finding sheds new light on the diagnostic accuracy of TCS in the

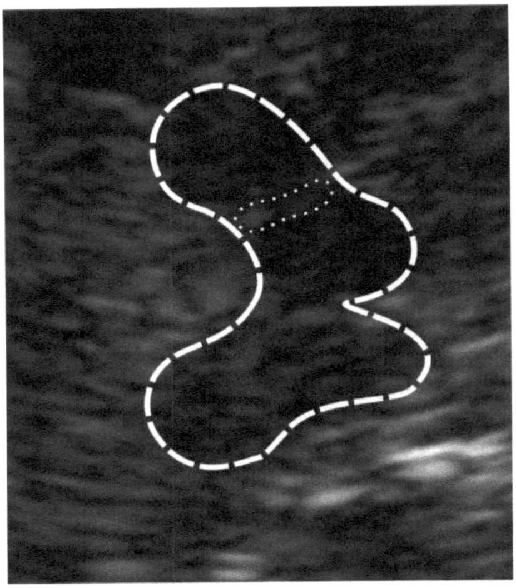

FIG. 1. Mesencephalic plane of a patient with Richardson syndrome: atrophic brainstem and normoechogenic substantia nigra.

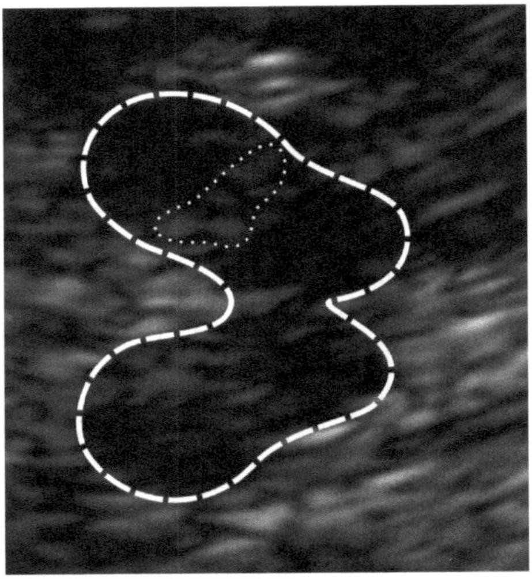

FIG. 2. Mesencephalic plane of a patient with PSP-parkinsonism: brainstem with hyperechogenic substantia nigra.

differentiation of IPD and PSP. It may well be that a diagnostic insecurity in former studies, that is, lower specificity of SN hyperechogenicity to discriminate for IPD, may have been due to the fact that subjects classified as having an APS

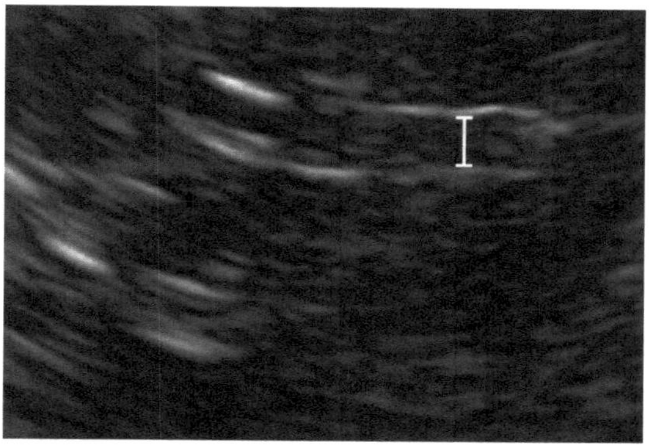

FIG. 3. Third ventricular plane of a patient with Richardson syndrom: normal anterior horn of the lateral ventricle but enlarged third ventricle.

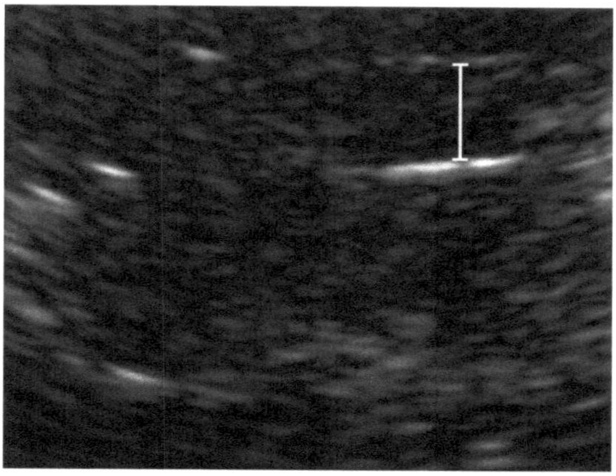

FIG. 4. Third ventricular plane of a patient with PSP-parkinsonism: normal anterior horn of the lateral ventricle and normal third ventricle.

may now be grouped into PSP-P and might therefore have shown SN hyperechogenicity. Certainly, larger studies are needed to confirm this finding. Moreover, future studies are needed to explore whether additional markers may help to differentiate PSP-P from PD.

B. Results: TCS to Discern DLB vs IPD

In 2006, Walter *et al.* studied a group of patients with dementia with Lewy bodies (DLB) and a group with Parkinson's disease divided into two subgroups, with dementia (PDD) and without dementia (PDnD), by TCS (Walter *et al.*, 2006).

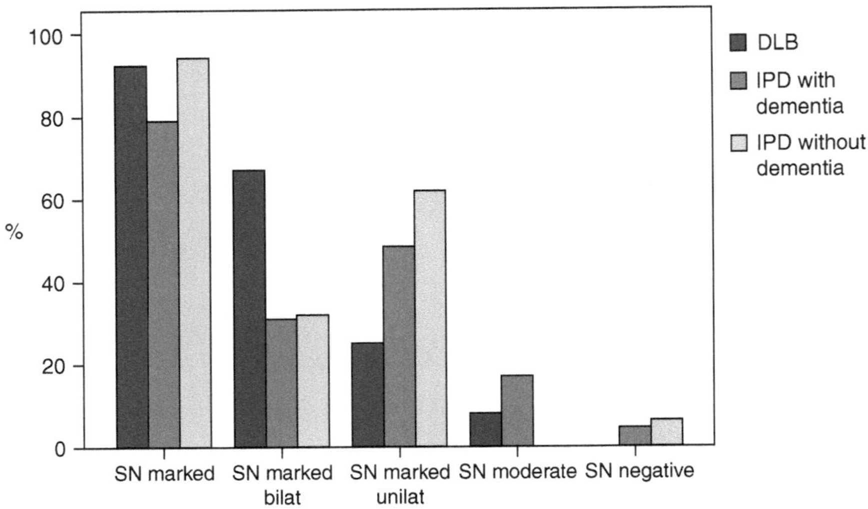

FIG. 5. Table showing the comparison between the group of patients with Dementia with Lewy Bodies and the group of patients with idiopathic Parkinson's disease concerning the echogenicity of the substantia nigra.

They found (Fig. 5) that the absence of bilateral marked SN hyperechogenicity discriminated PDD from DLB with a sensitivity of 65%, a specificity of 80%, and a positive predictive value of 89%. The mean right–left asymmetry index of SN echogenic size also helped to discriminate between DLB and PDD by a cutoff value of 1.15. An asymmetry index of 1.15 or above discriminated PDD from DLB with a sensitivity of 69%, a specificity of 80%, and a positive predictive value of 90%. PDD with bilateral marked SN hyperechogenicity had significantly earlier disease onset than PDD patients without these sonographic features. This was also found for the group of patients with PDnD but not for the group of patients with DLB. When the onset index is calculated by (age at disease onset × sum of bilateral echogenic sizes)/(asymmetry index), a cutoff value of 35.5 was found to discriminate best between DLB and PDD with bilateral hyperechogenicity.

When combining the criteria of (1) absence of bilateral marked SN hyperechogenicity and (2) onset index below 35.5 with bilateral SN hyperechogenicity, the discrimination of PDD from LBD increased with a sensitivity of 96%, a specificity of 80%, and a positive predictive value of 93%. Widths of third ventricle and of frontal horns correlated with UPDRS scores in PDD but not in PdnD or DLB patients. Neither widths of third ventricle nor widths of frontal horns correlated with scores on Addenbrooke's cognitive examination. The patients with DLB and PDD had significantly larger widths of third ventricle and of frontal horns of lateral ventricles, indicating subcortical atrophy, especially of the frontal brain (Fig. 6). Both disease entities share the characteristic feature of SN hyperechogenicity but a specific finding,

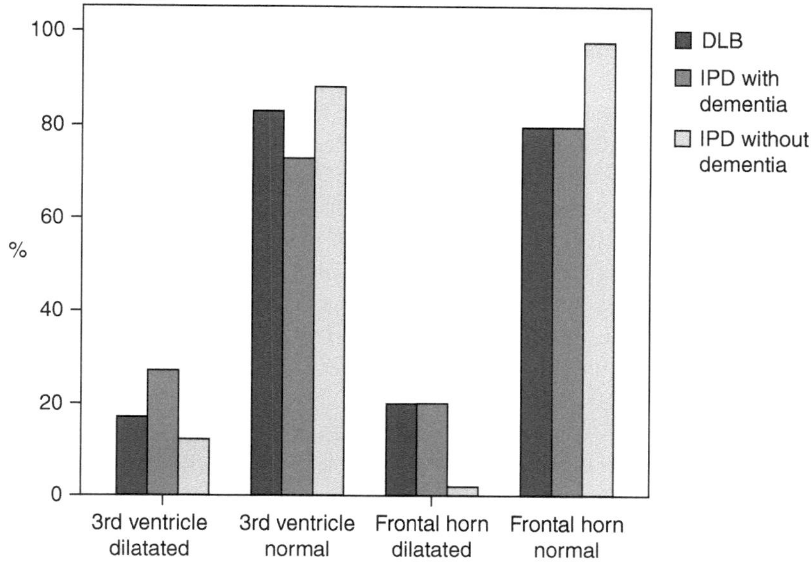

FIG. 6. Table showing the comparison between the group of patients with Dementia with Lewy Bodies and the group of patients with idiopathic Parkinson's disease concerning the size of the third ventricle and the frontal horn.

the more pronounced and symmetry of the TCS findings, distinguishes DLB from PDD and PdnD. This has also been shown in a comparison of clinical diagnosis, FP-CIT SPECT imaging, and autopsy of patients with DLB (Walker *et al.*, 2007). This is not related to earlier disease onset. It is remarkable that the SN echogenic sizes in DLB patients are comparable to those found in patients with CBD.

Because in this study all DLB patients had clinical symptoms of parkinsonism, which is an essential, but not obligate, feature for the probable diagnosis of DLB; further investigations are needed to know whether DLB patients without parkinsonism also exhibit SN hyperechogenicity. Although the difference between groups was not significant, the clinical observations that the parkinsonian features in DLB are more symmetrical than in IPD were actually reflected in the TCS findings. No significant differences were found between the groups concerning the TCS findings of the raphe, thalami, and caudate nuclei.

C. Results: TCS Delineating CBD from PSP

In 2004, Walter *et al.* studied a group of patients with CBD and PSP by TCS (Walter *et al.*, 2004). None of the CBD patients had dementia at symptom onset, early gait or speech disturbance, or vertical gaze palsy. Later, after neuropsychological investigation, four CBD patients (50%) and 11 PSP patients (84%) did have dementia. They all had symptoms of asymmetrical parkinsonism in addition to at least one cortical sign (apraxia, cortical sensory loss, or alien limb syndrome) or dystonia and focal reflex myoclonus. Marked SN hyperechogenicity had a positive value of 100% for the clinical diagnosis of CBD and a negative predictive value of 92% for not having the clinical diagnosis of CBD, indicating the diagnosis of PSP (see Fig. 7). The median SN echogenic size between the two groups was significantly different. Even the median echogenic size of 0.31 cm^2 in the CBD patients was larger than in IPD, where median SN sizes of 0.25 cm^2 were reported in several studies (Berg *et al.*, 2001; Walter *et al.*, 2003; Walter *et al.*, 2002). TCS findings of this study suggest that CBD and IPD share common mechanisms of nigrostriatal degeneration.

In PSP patients with dementia, the width of the third ventricle was negatively correlated with the score on Adenbrooke's cognitive examination. This is an interesting finding because it supports the hypothesis that in PSP patients a dysfunction of the midbrain ascending activating system is a relevant factor for the decline in cognitive abilities (Soliveri *et al.*, 2000). Furthermore, there was no correlation of third-ventricle width with age, age at disease onset, disease severity, or disease duration. The same observations hold for the widths of the frontal

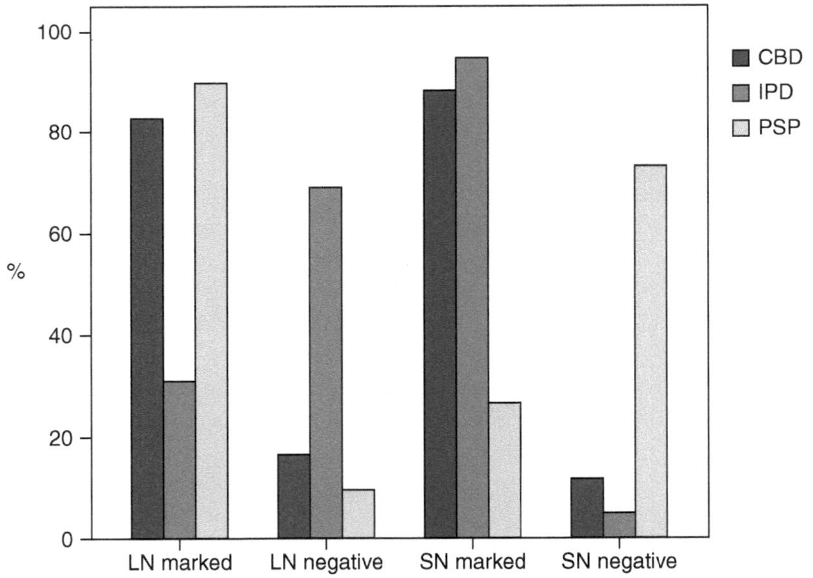

FIG. 7. Table showing the comparison between the group of patients with corticobasal degeneration and the group of patients with idiopathic Parkinson's disease and progressive supranuclear palsy concerning the echogenicity of the substantia nigra and the lenticular nucleus.

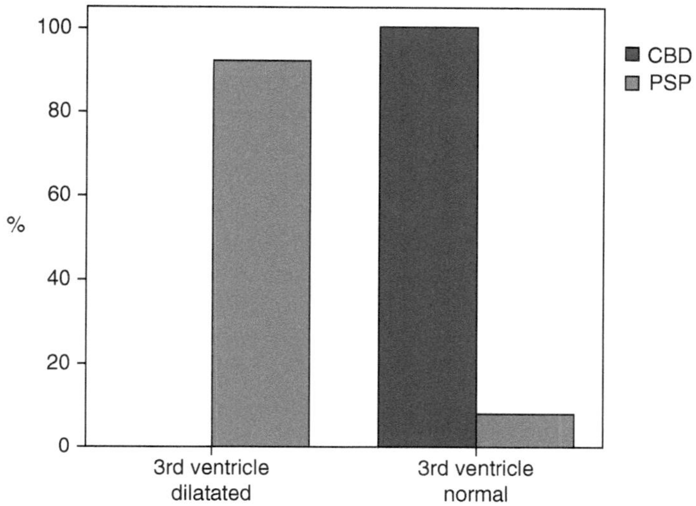

FIG. 8. Table showing the comparison between the group of patients with corticobasal degeneration and the group of patients with progressive supranuclear palsy concerning the size of the third ventricle.

horns. In all CBD patients, width of the third ventricle was below 10 mm (see Fig. 8). Widths of the third ventricle were significantly larger in the PSP group than in the CBD group. A third-ventricle width of more than 10 mm indicated PSP with a sensitivity of 83%, a specificity of 100%, and a positive predictive value of 100%.

The results showed that the presence of at least one of the TCS features, (1) marked SN hyperechogenicity and (2) third-ventricle width < 10 mm, indicated CBD with a sensitivity of 100%, a specificity of 83%, and a positive predictive value of 80%. Thus, SN hyperechogenicity, reported earlier as a characteristic TCS feature in IPD, is also typical for CBD and the presence of a hyperechogenic LN may help to differentiate.

In the same study, Walter *et al.* also examined other structures of the brainstem but in both groups no abnormalities were found in the thalami and brainstem raphe (Walter *et al.*, 2004). Although the caudate nucleus did show a percentage of abnormalities in both groups, this was not significant. The widths of the frontal horns did not differentiate between the two groups.

IV. Discussion

Before discussing the details of the studies, two points are to be noted on the research methodology. The field of TCS in movement disorders has been pioneered by the research groups of Berg and Walter (Berg *et al.*, 2008; Walter *et al.*, 2007b). Although their observations are currently being reproduced by other researchers, these two groups still largely dominate the field. Since TCS is a technique involving personal expertise and experience, we would caution against robust generalizations from the papers we reviewed before these results have been reproduced by others and before some consensus is reached on the exact use and interpretations of TCS in movement disorders. In the studies discussed earlier, different cutoff points concerning the hyperechogenic size of the SN have been used. As we pointed out before, this may give a false-positive effect of the numbers of positive SN (Vlaar *et al.*, 2009). We feel that an international workshop involving diagnosed and classified patients could serve a great purpose here.

Second, as we have stated earlier, most diagnostic studies involve later-stage, clearly diagnosed patients (Vlaar *et al.*, 2007b, 2009). As we have shown, the capabilities of a diagnostic test differ dramatically in different patient populations, especially so when the spectrum of these populations differ (Bachmann *et al.*, 2009). These later-stage patients are obviously not the patients for whom one needs additional diagnostic tools such as a TCS, as these patients already have an unequivocal clinical diagnosis. The challenge thus still remains to employ TCS (and ideally compare it with SPECT or PET) in patients with recent-onset symptoms. The study by Gaenslen represents such an approach and we have embarked on a similar diagnostic study several years ago (Gaenslen *et al.*, 2008; Vlaar *et al.*, 2007a).

Despite the above-mentioned objections, the problem remains that presently we do not have an accurate diagnostic tool to differentiate IPD from APS or differentiate among the various APS. IBZM-SPECT scans have been used to solve this clinical problem, but we found in a recent meta-analysis that it is rather inaccurate to answer this question (Vlaar *et al.*, 2007b). TCS thus remains one of the only promising techniques to differentiate IPD from APS, and to differentiate among the various APS (Walter *et al.*, 2007a).

Although the exact numbers differ from study to study, the data do point in one direction. From these we distilled a possible algorithm for the use of TCS to help differentiate between patients with IPD and APS (See algorithm). In summary, TCS may be used in the patient presenting with parkinsonian symptoms as follows. Should one find the combination of a hyperechogenic SN with a negative LN, then this would suggest the diagnosis of IPD with a high probability. Alternatively, the combination of a normal SN and a positive LN would favor the diagnosis of MSA or PSP. Additional investigations of the third ventricle would then differentiate between these two diseases, an increase in the third ventricle favoring the diagnosis of PSP. The other combinations of TCS findings probably are less reliable to reach a definitive diagnosis.

We feel that this provides a theoretical framework and hypothesis to test in future research. Ideally, this algorithm should be used to study and diagnostically classify by TCS patients with early-onset parkinsonian symptoms and tested against the gold standard of a clinical diagnosis after several years of follow-up.

V. Conclusion

TCS is a promising diagnostic technique to differentiate between different parkinsonian disorders. From the available data, the following possible scenario arises (a scenario that does need additional confirmation by prospective studies and consensus on TCS parameters): MSA and PSP can be distinguished from IPD by the absence of a hyperechogenic SN on TCS. A hyperechogenic LN indicates MSA or PSP in favor of IPD. Differentiation between MSA and PSP can be done by examining the third ventricle. If this is dilatated, PSP is the more likely diagnosis. In DLB, a hyperechogenic SN is found in general as well as a dilatated third ventricle, which can differentiate between DLB and IPD without dementia. Clinically it is often difficult to distinguish between CBD and PSP. In contrast to PSP patients with CBD generally show hyperechogenicity of the SN and a normal width of the third ventricle.

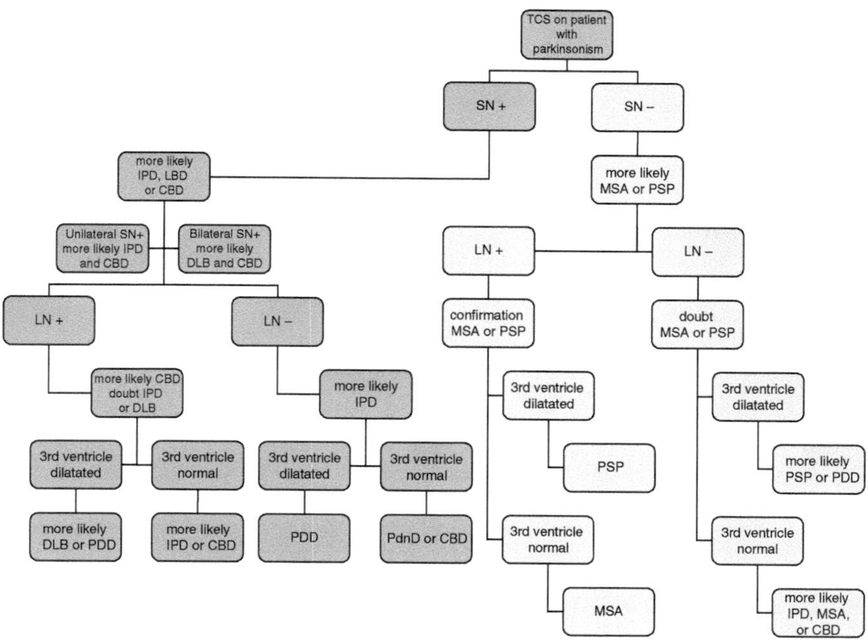

VI. Acknowledgments

We thank Drs. Anten, Vreeling, and Winogrodzka for their enthusiastic support and expertise contributing to our past and present studies.

References

Aarsland, D., Rongve, A., Nore, S. P., Skogseth, R., Skulstad, S., Ehrt, U., Hoprekstad, D., and Ballard, C. (2008). Frequency and case identification of dementia with Lewy bodies using the revised consensus criteria. *Dement. Geriatr. Cogn. Disord.* **26**, 445–452.

Bachmann, L. M., ter Riet, G., Weber, W. E., and Kessels, A. G. (2009). Multivariable adjustments counteract spectrum and test review bias in accuracy studies. *J. Clin. Epidemiol.* **62**, 357–361 .

Bartova, P., Skoloudik, D., Fadma, T., Ressner, P., Kanovsky, P., and Herzig, R. (2007). Substantia nigra evaluation in atypical parkinsonian syndromes and Parkinson's disease. *Mov. Disord.* **22**, S142.

Becker, G., Seufert, J., Bogdahn, U., Reichmann, H., and Reiners, K. (1995). Degeneration of substantia nigra in chronic Parkinson's disease visualized by transcranial color-coded real-time sonography. *Neurology* **45**, 182–184.

Behnke, S., Berg, D., Naumann, M., and Becker, G. (2005). Differentiation of Parkinson's disease and atypical parkinsonian syndromes by transcranial ultrasound. *J. Neurol. Neurosurg. Psychiatry* **76**, 423–425.

Berg, D., Godau, J., and Walter, U. (2008). Transcranial sonography in movement disorders. *Lancet Neurol.* **7**, 1044–1055.

Berg, D., Siefker, C., and Becker, G. (2001). Echogenicity of the substantia nigra in Parkinson's disease and its relation to clinical findings. *J. Neurol.* **248**, 684–689.

Boeve, B. F., Lang, A. E., and Litvan, I. (2003). Corticobasal degeneration and its relationship to progressive supranuclear palsy and frontotemporal dementia. *Ann. Neurol.* **54** (Suppl. 5), S15–19.

Caslake, R., Moore, J. N., Gordon, J. C., Harris, C. E., and Counsell, C. (2008). Changes in diagnosis with follow-up in an incident cohort of patients with parkinsonism. *J. Neurol. Neurosurg. Psychiatry* **79**, 1202–1207.

Dickson, D. W. (2006). Tauopathies. In Neurodegeneration: the Molecular Pathology of Dementia and Movement Disorders (D. Dickson, *et al.*, Eds.), pp. 82–154. ISN Neuropath Press, Basel

Gaenslen, A., Unmuth, B., Godau, J., Liepelt, I., Di Santo, A., Schweitzer, K. J., Gasser, T., Machulla, H. J., Reimold, M., Marek, K., and Berg, D. (2008). The specificity and sensitivity of transcranial ultrasound in the differential diagnosis of Parkinson's disease: a prospective blinded study. *Lancet Neurol.* **7**, 417–424.

Geser, F., *et al.*, (2006). Progression of multiple system atrophy (MSA): A prospective natural history study by the European MSA Study Group (EMSA SG). *Mov. Disord.* **21**, 179–186.

Gilman, S., *et al.* (1999). Consensus statement on the diagnosis of multiple system atrophy. *J. Neurol. Sci.* **163**, 94–98.

Gilman, S., *et al.* (2008). Second consensus statement on the diagnosis of multiple system atrophy. *Neurology* **71**, 670–676.

Hamilton, R. L. (2000). Lewy bodies in Alzheimer's disease: A neuropathological review of 145 cases using alpha-synuclein immunohistochemistry. *Brain Pathol.* **10**, 378–384.

Hauw, J. J., Daniel, S. E., Dickson, D., Horoupian, D. S., Jellinger, K., Lantos, P. L., McKee, A., Tabaton, M., and Litvan, I. (1994). Preliminary NINDS neuropathologic criteria for Steele–Richardson–Olszewski syndrome (progressive supranuclear palsy). *Neurology* **44**, 2015–2019.

Hughes, A. J., Daniel, S. E., Ben-Shlomo, Y., and Lees, A. J. (2002). The accuracy of diagnosis of parkinsonian syndromes in a specialist movement disorder service. *Brain* **125**, 861–870.
Hughes, A. J., Daniel, S. E., Kilford, L., and Lees, A. J. (1992). Accuracy of clinical diagnosis of idiopathic Parkinson's disease: A clinico-pathological study of 100 cases. *J. Neurol. Neurosurg. Psychiatry* **55**, 181–184.
Katzenschlager, R., Costa, D., Gerschlager, W., O'Sullivan, J., Zijlmans, J., Gacinovic, S., Pirker, W., Wills, A., Bhatia, K., Lees, A. J., and Brown, P. (2003). [^{123}I]-FP-CIT-SPECT demonstrates dopaminergic deficit in orthostatic tremor. *Ann. Neurol.* **53**, 489–496.
Kirchhof, K., Apostolidis, A. N., Mathias, C. J., and Fowler, C. J. (2003). Erectile and urinary dysfunction may be the presenting features in patients with multiple system atrophy: A retrospective study. *Int. J. Impot. Res.* **15**, 293–298.
Litvan, I., *et al.*, (1996). Accuracy of clinical criteria for the diagnosis of progressive supranuclear palsy (Steele–Richardson–Olszewski syndrome). *Neurology* **46**, 922–930.
Litvan, I., Grimes, D. A., Lang, A. E., Jankovic, J., McKee, A., Verny, M., Jellinger, K., Chaudhuri, K. R., and Pearce, R. K. (1999). Clinical features differentiating patients with postmortem confirmed progressive supranuclear palsy and corticobasal degeneration. *J. Neurol.* **246** (Suppl. 2), II1–5.
Mahapatra, R. K., Edwards, M. J., Schott, J. M., and Bhatia, K. P. (2004). Corticobasal degeneration. *Lancet Neurol.* **3**, 736–743.
McKeith, I. G., *et al.*, (2005). Diagnosis and management of dementia with Lewy bodies: Third report of the DLB Consortium. *Neurology* **65**, 1863–1872.
McKeith, I. G., *et al.* (1996). Consensus guidelines for the clinical and pathologic diagnosis of dementia with Lewy bodies (DLB): Report of the consortium on DLB international workshop. *Neurology* **47**, 1113–1124.
Murphy, K. E., Karaconji, T., Hardman, C. D., and Halliday, G. M. (2008). Excessive dopamine neuron loss in progressive supranuclear palsy. *Mov. Disord.* **23**, 607–610.
Nath, U., Ben-Shlomo, Y., Thomson, R. G., Morris, H. R., Wood, N. W., Lees, A. J., and Burn, D. J. (2001). The prevalence of progressive supranuclear palsy (Steele–Richardson–Olszewski syndrome) in the UK. *Brain* **124**, 1438–1449.
Okawa, M., Miwa, H., Kajimoto, Y., Hama, K., Morita, S., Nakanishi, I., and Kondo, T. (2007). Transcranial sonography of the substantia nigra in Japanese patients with Parkinson's disease or atypical parkinsonism: clinical potential and limitations. *Intern. Med.* **46**, 1527–1531.
Osaki, Y., Ben-Shlomo, Y., Lees, A. J., Daniel, S. E., Colosimo, C., Wenning, G., and Quinn, N. (2004). Accuracy of clinical diagnosis of progressive supranuclear palsy. *Mov. Disord.* **19**, 181–189.
Osaki, Y., Ben-Shlomo, Y., Lees, A. J., Wenning, G. K., and Quinn, N. P. (2009). A validation exercise on the new consensus criteria for multiple system atrophy. *Mov. Disord.* **24**, 2272–2276.
Ozawa, T., *et al.* (2006). The alpha-synuclein gene in multiple system atrophy. *J. Neurol. Neurosurg. Psychiatry* **77**, 464–467.
Sanchez-Gonzalez, M. A., Garcia-Cabezas, M. A., Rico, B., and Cavada, C. (2005). The primate thalamus is a key target for brain dopamine. *J. Neurosci.* **25**, 6076–6083.
Schrag, A., Good, C. D., Miszkiel, K., Morris, H. R., Mathias, C. J., Lees, A. J., and Quinn, N. P. (2000). Differentiation of atypical parkinsonian syndromes with routine MRI. *Neurology* **54**, 697–702.
Seppi, K., Yekhlef, F., Diem, A., Luginger Wolf, E., Mueller, J., Tison, F., Quinn, N. P., Poewe, W., and Wenning, G. K. (2005). Progression of parkinsonism in multiple system atrophy. *J. Neurol.* **252**, 91–96.
Soliveri, P., Monza, D., Paridi, D., Carella, F., Genitrini, S., Testa, D., and Girotti, F. (2000). Neuropsychological follow up in patients with Parkinson's disease, striatonigral degeneration-type multisystem atrophy, and progressive supranuclear palsy. *J. Neurol. Neurosurg. Psychiatry* **69**, 313–318.
Spillantini, M. G., Crowther, R. A., Jakes, R., Cairns, N. J., Lantos, P. L., and Goedert, M. (1998). Filamentous alpha-synuclein inclusions link multiple system atrophy with Parkinson's disease and dementia with Lewy bodies. *Neurosci. Lett.* **251**, 205–208.

Trojanowski, J. Q., and Revesz, T. (2007). Proposed neuropathological criteria for the post mortem diagnosis of multiple system atrophy. *Neuropathol. Appl. Neurobiol.* **33**, 615–620.

Tsuboi, Y., Josephs, K. A., Boeve, B. F., Litvan, I., Caselli, R. J., Caviness, J. N., Uitti, R. J., Bott, A. D., and Dickson, D. W. (2005). Increased tau burden in the cortices of progressive supranuclear palsy presenting with corticobasal syndrome. *Mov. Disord.* **20**, 982–988.

Vlaar, A. M., Bouwmans, A. E., Mess, W. H., Tromp, S. C., and Weber, W. E. (2009). Transcranial duplex in the differential diagnosis of parkinsonian syndromes: a systematic review. *J. Neurol.* **256**, 530–538.

Vlaar, A. M., Bouwmans, A. E., van Kroonenburgh, M. J., Mess, W. H., Tromp, S. C., Wuisman, P. G., Kessels, A. G., Winogrodzka, A., and Weber, W. E. (2007a). Protocol of a prospective study on the diagnostic value of transcranial duplex scanning of the substantia nigra in patients with parkinsonian symptoms. *BMC Neurol.* **7**, 28.

Vlaar, A. M., van Kroonenburgh, M. J., Kessels, A. G., and Weber, W. E. (2007b). Meta-analysis of the literature on diagnostic accuracy of SPECT in parkinsonian syndromes. *BMC Neurol.* **7**, 27.

Wadia, P. M., and Lang, A. E. (2007). The many faces of corticobasal degeneration. Parkinsonism *Relat. Disord.* **13** (Suppl. 3), S336–340.

Walker, Z., Jaros, E., Walker, R. W., Lee, L., Costa, D. C., Livingston, G., Ince, P. G., Perry, R., McKeith, I., and Katona, C. L. (2007). Dementia with Lewy bodies: a comparison of clinical diagnosis, FP-CIT single photon emission computed tomography imaging and autopsy. *J. Neurol. Neurosurg. Psychiatry* **78**, 1176–1181.

Walter, U., Behnke, S., Eyding, J., Niehaus, L., Postert, T., Seidel, G., and Berg, D. (2007a). Transcranial brain parenchyma sonography in movement disorders: State of the art. *Ultrasound Med. Biol.* **33**, 15–25.

Walter, U., Behnke, S., Eyding, J., Niehaus, L., Postert, T., Seidel, G., and Berg, D. (2007b). Transcranial brain parenchyma sonography in movement disorders: state of the art. *Ultrasound Med. Biol.* **33**, 15–25.

Walter, U., Dressler, D., Probst, T., Wolters, A., Abu-Mugheisib, M., Wittstock, M., and Benecke, R. (2007c). Transcranial brain sonography findings in discriminating between parkinsonism and idiopathic Parkinson disease. *Arch. Neurol.* **64**, 1635–1640.

Walter, U., Dressler, D., Wolters, A., Probst, T., Grossmann, A., and Benecke, R. (2004). Sonographic discrimination of corticobasal degeneration vs progressive supranuclear palsy. *Neurology* **63**, 504–509.

Walter, U., Dressler, D., Wolters, A., Wittstock, M., Greim, B., and Benecke, R. (2006). Sonographic discrimination of dementia with Lewy bodies and Parkinson's disease with dementia. *J. Neurol.* **253**, 448–454.

Walter, U., Niehaus, L., Probst, T., Benecke, R., Meyer, B. U., Dressler, D. (2003). Brain parenchyma sonography discriminates Parkinson's disease and atypical parkinsonian syndromes. *Neurology* **60**, 74–77.

Walter, U., Wittstock, M., Benecke, R., and Dressler, D. (2002). Substantia nigra echogenicity is normal in non-extrapyramidal cerebral disorders but increased in Parkinson's disease. *J. Neural. Transm.* **109**, 191–196.

Williams, D. R., de Silva, R., Paviour, D. C., Pittman, A., Watt, H. C., Kilford, L., Holton, J. L., Revesz, T., and Lees, A. J. (2005). Characteristics of two distinct clinical phenotypes in pathologically proven progressive supranuclear palsy: Richardson's syndrome and PSP-parkinsonism. *Brain* **128**, 1247–1258.

Williams, D. R., Holton, J. L., Strand, C., Pittman, A., de Silva, R., Lees, A. J., and Revesz, T. (2007). Pathological tau burden and distribution distinguishes progressive supranuclear palsy-parkinsonism from Richardson's syndrome. *Brain* **130**, 1566–1576.

Williams, D. R., and Lees, A. J. (2009). Progressive supranuclear palsy: Clinicopathological concepts and diagnostic challenges. *Lancet Neurol.* **8**, 270–279.

TRANSCRANIAL SONOGRAPHY IN THE DISCRIMINATION OF PARKINSON'S DISEASE VERSUS VASCULAR PARKINSONISM

Pablo Venegas-Francke

Hospital Clínico, Universidad de Chile,
Santiago de Chile

Cerebro-vascular disease is a well-known cause of parkinsonism. However, so far, there are no standardized clinical criteria that allow the diagnosis with accuracy and reliability. Rather, there is a great symptom overlap between idiopathic Parkinson's disease, other causes for parkinsonism, and vascular parkinsonism. Therefore, additional instruments are necessary to improve diagnostic accuracy. Transcranial sonography of brain parenchyma (TCS) has become a broadly applied tool in the diagnosis of Parkinson's disease, and secondary and atypical parkinsonian syndromes. In contrast to patients with idiopathic Parkinson's disease, patients with vascular parkinsonism in general show no hyperechogenicity of the substantia nigra. In contrast to a number of patients with atypical parkinsonian syndromes, also the basal ganglia are usually normoechogenic on TCS.

A more specific approach to vascular parkinsonism includes the Doppler or duplex technique in order to show stenosis of vessels. Therefore, the combination of TCS and Doppler/duplex imaging might help to improve diagnosis of vascular parkinsonism transcranial.

I. Introduction

Vascular parkinsonism (VP) is a well-known cause of secondary parkinsonism affecting primarily the elderly population in whom vascular occlusive diseases are common. In its classical form, VP affects the lower extremities to a more severe

INTERNATIONAL REVIEW OF
NEUROBIOLOGY, VOL. 90
DOI: 10.1016/S0074-7742(10)90010-X

degree than the upper ones, leading to early gait impairment and falls. VP is a predominantly symmetric, akinetic-rigid form of parkinsonism with, in general, a poor outcome on pharmacological treatment. For prognosis and evaluation of treatment options, it is therefore essential to get an early and correct diagnosis of VP. However, diagnostic accuracy is still a big challenge to clinicians (Benamer and Grosset, 2009).

This is mainly because classical clinical presentation of VP is rare. There is a huge overlap of symptoms and signs between the clinical presentations of idiopathic Parkinson's disease (PD) and VP (Okuda *et al.*, 2008). Structural neuroimaging with computed tomography (CT) or magnetic resonance imaging (MRI) are helpful to demonstrate vascular lesions. However, in the differential diagnosis of PD and VP, accuracy of diagnosis is low (Zijlmans *et al.*, 2004; Benamer and Grosset, 2009). This is mainly because there is no diagnostic marker for PD using these techniques. Therefore, it cannot be decided whether signs of vascular lesions on CT or MRI are the cause for the clinical picture of parkinsonism. Functional neuroimaging techniques depicting the nigrostriatal system like F-Dopa-PET or DATScan using SPECT imaging are, in general, helpful to establish the diagnosis of a primary affection of the nigrostriatal system, typical for PD. However, sometimes a reduced tracer uptake is also found in VP (Gerschlager *et al.*, 2002; Lorberboym *et al.*, 2004; Zijlmans *et al.*, 2007). Therefore, complete accuracy cannot be expected using functional neuroimaging. Moreover, these techniques are expensive, not broadly available, and use radioactivity.

Transcranial sonography of brain parenchyma (TCS) has been proven to be a useful technique in the diagnosis and differential diagnosis of movement disorders (Walter *et al.*, 2007; Godau and Berg, 2010). The characteristic finding of substantia nigra (SN) hyperechogenicity is already found in the early stages of PD, and there is growing evidence that it may even be detected before the onset of clinical symptoms (Tolosa *et al.*, 2007; Berg, 2008; Berg *et al.*, 2008). Because of the easy applicability and good diagnostic value, this technique is now being used in different populations all over the world (Miranda, 2006; Berg *et al.*, 2008).

Here, the role of TCS in the diagnosis of VP is reviewed, comparing ultrasound features of VP to those of PD. Moreover, TCS findings are compared to the data obtained by other neuroimaging techniques.

II. Vascular parkinsonism

A first description of parkinsonism caused by vascular lesions was done by Critchley (1929), establishing a disease entity with his observation (Critchley, 1929). In subsequent publications, this disease is referred to as arteriosclerotic parkinsonism, arteriosclerotic pseudo-parkinsonism, and lower body parkinsonism (Benamer and Grosset, 2009).

It has not been easy to define VP, since no widely accepted criteria are available. Vascular causality is very difficult to prove. On MRI, there exists no clear cut-off between the extension of white matter lesions in patients suffering from hypertension and those who really suffer from clinical signs of VP. Winikates and Jankovic (1999) suggest that VP should be suspected when a parkinsonian syndrome is detected in coexistence with clearly visible vascular lesions in the subcortical region (Winikates and Jankovic, 1999). Similarly, Zijlmans (2004) suggests criteria based on three steps: (i) A parkinsonian syndrome must be present as in idiopathic PD (bradykinesia plus at least one of the following: rigidity, postural instability, or resting tremor), (ii) signs of a cerebrovascular disease must be present too, and (iii) a relation between both is most probable (Zijlmans *et al.*, 2004; Benamer and Grosset, 2009). The latter refers to a temporal relation between either stroke affecting the basal ganglia and the beginning of parkinsonian symptoms or extensive white matter lesions affecting tracts from the basal ganglia or even the basal ganglia themselves in coexistence with parkinsonism and (very often) dementia.

Prevalence of VP is highly variable in different races and populations, ranging between 3% and 20% of all parkinsonian syndromes (Benamer and Grosset, 2009). The occurrence of this disease entity depends on the control of cerebrovascular risk factors and among those, most importantly, hypertension, genetic factors, and socioeconomic conditions. For example, the Japanese and Chilean populations have a high prevalence of arterial hypertension and subsequent VP; in Chile, nearly 20% of all parkinsonian syndromes are at least partly caused by a vascular disorder (Chilean Health Ministry, 2008).

Besides high blood pressure, diabetes mellitus, dyslipidemia, and other diseases known to contribute to a vascular risk are often associated with VP. Interestingly, some studies have shown that some of these diseases may constitute a protective factor for idiopathic PD (Jellinger, 2003; Scigliano *et al.*, 2006).

VP in its classical appearance can already be recognized in the clinical exam. The suspicion for VP in a patient is generally underlined by a long history of hypertension. When a neurological exam is performed, VP involves typically a stronger affection of the lower extremities compared to the upper ones, leading to the clinical picture of lower body parkinsonism with severe gait dysfunction, freezing, and often early falls due to gait apraxia.

Most patients do not have relevant resting tremor, motor functions of the arms are usually only mildly impaired, allowing often normal writing. By contrast, axial symptoms such as hypophonia, dysphagia, even pseudobulbar palsy and postural changes tend to be more pronounced (Benamer and Grosset, 2009).

Benamer and Grosset (2009) propose three clinical types of VP. According to their observations, the most frequent form is gait impairment, that is, the classical lower body parkinsonism; the second one is an insidious progressive and Levodopa-responsive form, resembling idiopathic PD, and the third form is a relatively rare acute onset of parkinsonism following acute basal ganglia lesions (for example, caused by ischemia) (Benamer and Grosset, 2009).

Therefore, although onset of VP may be sudden and severe, clinicians should be aware that most cases begin with minimal symptoms that progressively increase over time resembling a neurodegenerative pattern. The diagnostic difficulty becomes evident in postmortem analyses. There are several reports of subjects diagnosed and treated as PD during lifetime, in whom postmortem pathological exam revealed primarily vascular involvement of the basal ganglia (Colosimo *et al.*, 2000; Hughes *et al.*, 2001; Jain *et al.*, 2006).

In addition to the extrapyramidal signs, many patients have abnormal neurological findings with focal signs, such as extensor plantar response, hemiparesis, visual field deficits, and others. Cognitive assessment shows a frontal dysfunction pattern, most cases have low verbal fluency, impaired executive function, and low mental flexibility (Alonso-Prieto *et al.*, 2003; Owen, 2004).

Therefore, the typical VP could be summarized as lower extremities parkinsonism associated with frontal lobe dysfunction and focal signs in the neurological exam. However, this simplistic view usually does not reflect reality. In some patients, symptoms begin with asymmetric resting tremor of one hand, progressing slowly similar to PD. Other patients may have a good or excellent response to Levodopa treatment. The frequently associated neuropsychological impairment is by no means pathognomonic for VP, since patients with Parkinson's disease and other parkinsonian syndromes can also display these symptoms (Zijlmans *et al.*, 2007).

Differential diagnosis becomes even more difficult, when it is considered that VP and PD may co-occur. Here, it is of importance that vascular lesions tend to be mild in PD patients compared to age-matched non-PD subjects, meaning that strong microangiopathic white matter lesions or basal ganglia lacunar strokes are much less frequently found in idiopathic PD patients (Jellinger, 2003). Still, they do occur, making the assignment of symptoms to either VP or PD difficult.

As it is clinically difficult to differentiate PD and VD, auxiliary tests are needed. In VP, first-line morphologic neuroimaging examinations such as brain CT or MRT show a microangiopathic pattern and/or lacunar strokes of the basal ganglia. Unfortunately, there is no typical finding for PD on these structural neuroimaging techniques. Therefore, sometimes, SPECT or PET imaging is used to evaluate the functional integrity of the nigrostriatal system. However, also in VP, a decrease in presynaptic dopaminergic neurons can sometimes be observed, which is most likely due to lesions in the nigrostriatal pathway/basal ganglia. As both approaches have limitations, a technique which may show both—a typical feature for PD and the condition of brain vessels—would be of help.

III. TCS in the Differential Diagnosis of VP and PD

Transcranial ultrasound approaches for VP can be divided in two methods:

1. Transcranial sonography of the brain parenchyma using B-mode imaging (TCS)
2. Transcranial Doppler or color-coded duplex sonography of intracranial vessels (TCD, TCCS)

A. Two-Dimensional B-mode Transcranial Sonography (TCS)

The mesencephalic and basal ganglia scanning plane can be examined through the temporal bone window with a low-frequency transducer (1–4 Hz). Visualization of the substantia nigra, the lenticular nuclei, the midline raphe, the red nucleus, the thalamus, and the third and lateral ventricles is possible using this technique (Berg *et al.*, 2008) (see also the chapter 2 of this volume).

When hyperechogenicity of substantia nigra is evident, Parkinson's disease should be suspected; in cases of VP there is, in general, no hyperchogenicity of the substantia nigra (see Figs. 1 and 2).

Our experience using an ATL 500 Philips machine is very similar to those obtained with ultrasound machines from Siemens or others. To evaluate sensitivity and specificity of TCS in our center, we enrolled 116 cases (60 males and 56

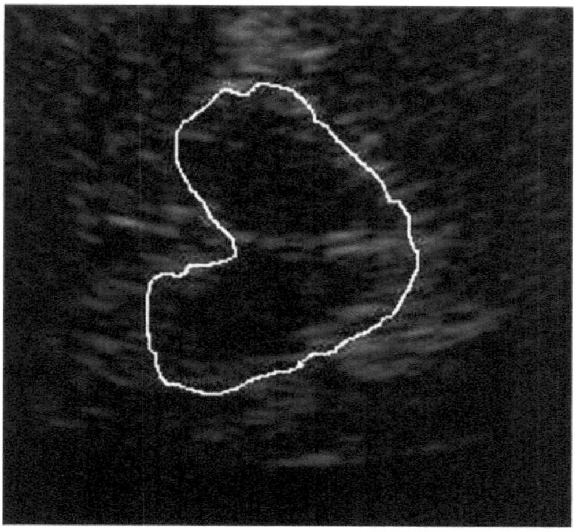

Fig. 1. Zoomed transcranial sonography picture of the butterfly-shaped mesencephalic brainstem of a patient with vascular parkinsonism. There is not hyperechogenicity of the substantia nigra; it is a "normal pattern" typical for secondary parkinsonism.

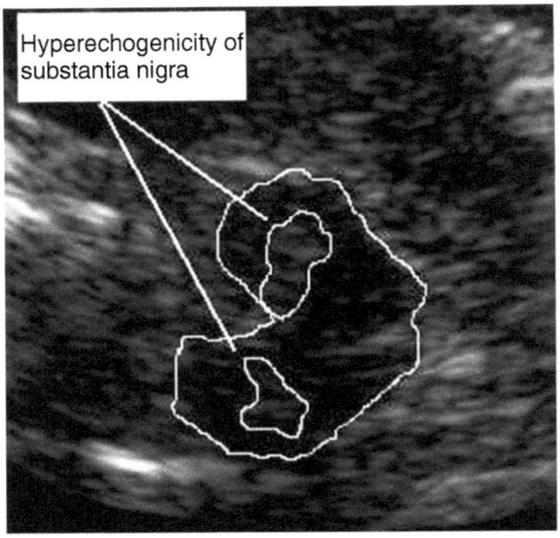

FIG. 2. Zoomed transcranial sonography picture of the mesencepahlic brainstem of a patient with Parkinson's disease. Large hyperechogenicities in the anatomical region of the substantia nigra are evident (encircled on both sides).

females, mean age 63 years (SD: 4.5 years) of parkinsonism from September 2008 to October 2009 who attended our movement disorder unit at the Clinical Hospital of the University of Chile. A detailed medical history, including concomitant diseases, concomitant medication, history of strokes and vascular diseases, was taken in addition to the neurological exam. All patients were also evaluated with the "sniffin' sticks screening battery". When a secondary cause of parkinsonism was suspected, a cerebral MRI was performed. Taking into consideration both medical history and neurological exam, the diagnosis of PD was made in 78 patients, VP was diagnosed in 12 patients, and 26 patients were diagnosed with parkinsonian syndromes due to other causes (atypical parkinsonian syndromes or parkinsonism induced by neurolopetic therapy or mimicked by depression). Transcranial sonography revealed substantia nigra hyperechogenicity in 90 patients; 65 of them were clinically diagnosed as idiophatic PD. Therefore, a sensitivity of 83% was calculated for diagnosis of PD in this patient cohort.

In cases of VP, normal echogenicity of the substantia nigra correlated in 95% with the clinical impression of VP, reflecting an excellent positive predictive value of 2D TCS for VP (Venegas, *et al.*, 2009a).

Taken together in VP, in general, normal midbrain and basal ganglia echogenicity are found. However, not seldom, the ventricular system may be enlarged.

Still, it is important to notice that it is possible that some cases of VP are associated with substantia nigra hyperechogenicity. This may reflect an overlapping of vascular and idiopathic PD. As both diseases are common in the elderly population, it is not surprising that in patients in whom VP is suspected clinically, additionally idiopathic PD may occur. In these cases, the second approach may help to differentiate VP from PD.

B. Evaluation of Vessels to Document a Basis for VP

As discussed above, VP occurs in patients in whom moderate-to-severe vascular disease has occurred, often for many years before manifestation of first motor signs. In these patients it is not rare to find intracranial stenosis. Tsai *et al.* (2007) published an article applying transcranial Doppler sonography of the main brain arteries to differentiate idiopathic PD from VP. They found significantly higher flow velocities in the anterior, middle, and posterior cerebral arteries in patients suffering from VP, compared to matched controls or in those suffering from idiopathic PD, indicating that intracranial stenosis is highly related to VP (see Fig. 3). Several studies indicate that PD patients less often suffer from arterial hypertension or diabetes mellitus. Besides, PD patients are known to have smoked

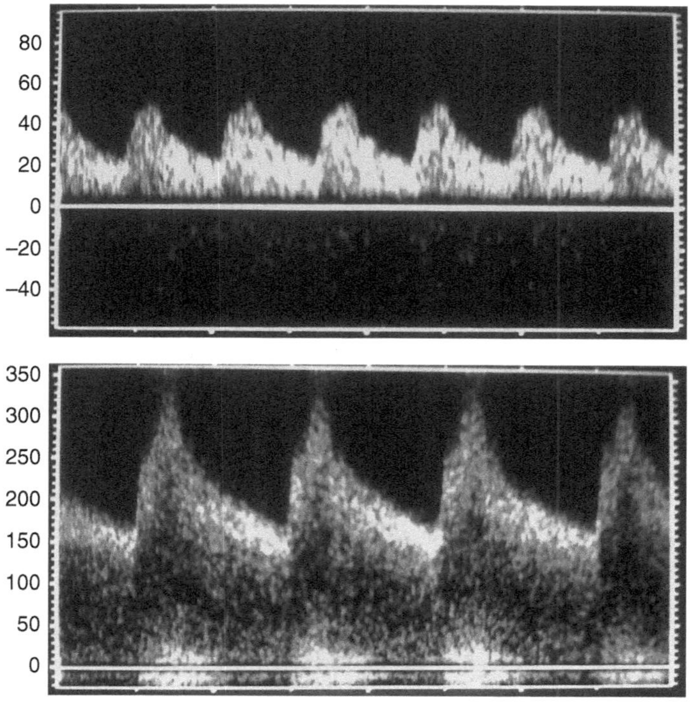

Fig. 3. (A) Middle cerebral artery normal hemodynamics of an idiopathic Parkinson disease patient measured by Doppler technique and (B) middle cerebral artery hemodynamics of a vascular parkinsonism patient. Here, an abnormal elevated blood flow velocity (measured as cm/sec) is evident.

less frequently during lifetime. Therefore, it is obvious that a method reflecting a disorder caused by vascular injury will show a marked difference between PD patients (with fewer cerebrovascular risk factors) and VP patients (with greater vascular risk factors) (Jellinger, 2003). It is also important to realize that not only extracranial but also intracranial vessels need to be examined, as there are some populations, like the Japanese or Chilean, in whom arteriosclerotic processes occur primarily intracranially rather than in the cervical carotid arteries (Diaz *et al.*, 2001).

Underlining the value of intracranial vessel assessment in the identification of VP, no statistical significant difference was found between PD and controls in the above-mentioned study (Tsai *et al.*, 2007).

Concerning the Doppler or duplex technique, assessment of the proximal segment of the middle cerebral artery is preferred, as little anatomical variation can be expected in this segment and it is generally easily accessible.

IV. Conclusion

VP is a cause of secondary parkinsonism. Its prevalence is variable and depends on vascular risk factors of specific populations. High prevalence rates are found in Asia and Latin America. In spite of its importance, VP is still clinically not clearly defined.

Medical history and physical exam may give rise to suspicion for VP. However, the typical findings on neuroimaging techniques like computed tomography cannot always be attributed to be the cause of the movement abnormalities, as there is no cut-off criterion that clearly defines age-related abnormalities and disease.

Transcranial ultrasound constitutes a unique technique to obtain supplementary information for the differential diagnosis of VP and PD. Using, on the one hand, a morphologic (TCS) view of midbrain and basal ganglia, substantia nigra hyperechogenicity as a typical feature for Parkinson's disease may be detected. On the other hand, evaluation of the intracerebral blood flow using Doppler and duplex techniques may help to define brain vascular pathology as a typical basis for VP. In some instances, an overlapping (vascular plus degenerative) parkinsonian syndrome may be present. In this case, hyperechogenicity of substantia nigra is usually found.

In summary, a typical VP patient does not have pathologically hyperechogenic areas at the midbrain, and the basal ganglia are normal (the latter is also found in PD). However, pathological findings are often revealed by middle cerebral artery Doppler or duplex sonography, showing higher blood flow velocities as a sign for stenosis. Further studies focusing on vessel hemodynamics are needed to improve evidence of Doppler techniques as useful complementary tools for the classical TCS approach in VP.

References

Alonso-Prieto, E., Esteban, E. M., Trujillo-Matienzo, C., Lara-Fernandez, G. E., Roussó-Viola, T. E., and Cordero-Eiriz, A. (2003). Specific alterations in attention in early stages of Parkinson's disease. *Rev. Neurol.* **36** (11), 1015–1018.

Becker, G., and Berg, D. (2001). Neuroimaging in basal ganglia disorders: Perspectives for transcranial ultrasound. *Mov. Disord.* **16**, 31–32.

Benamer, H., and Grosset, D. (2009). Vascular parkinsonism: A clinical review. *Eur. Neurol.* **61**, 11–15.

Berg, D. (2008). Biomarkers for the early detection of Alzheimer's and Parkinson's diseases. *Neurodegenerative Dis.* **5**, 133–136.

Berg, D., Godau, J., and Walter, U. (2008). Transcranial sonography in movement disorders. *Lancet Neurol.* **7**, 1044–1055.

Chilean Health Ministry. (2008). Chilean guidelines for treatment of Parkinson's disease. http://www.redsalud.gov.cl/archivos/guiasges/2008/GPC-Parkinson2008.pdf.

Colosito, C., Inghilleri, M., Chaudhuri, K. R. (2000). Parkinson's disease misdiagnosed as multiple system atrophy by sphincter electro-myography. *J. Neurol.* **247** (7), 559–561.

Critchley, M. (1929). Arterisclerotic parkinsonism. *Brain* **52**, 23–83.

Diaz, V., Plate, L., Erazo, S., Cumsille, M. A., and Venegas, P. (2001). Prevalence of carotid aterosclerosis in patients with cerebrovascular occlusive disease. *Rev. Med. Chile* **129**, 161–165.

Gaenslen, A., Unmuth, B., Godau, J., Liepelt, I., Di Santo, A., Schweitzer, K. J., et al. (2008). The specificity and sensitivity of transcranial ultrasound in the differential diagnosis of Parkinson's disease: A prospective blinded study. *Lancet Neurol.* **7**, 417–424.

Gerschlager, W., Bencsits, G., Pirker, W., Bloem, B. R., Asembaum, S., Prayer, D., et al. (2002). ^{123}I beta CIT SPECT distinguishes vascular parkinsonism from Parkinson's disease. *Mov. Disord.* **17**, 518–523.

Godau, J., and Berg, D. (2010). Role of transcranial ultrasound in the diagnosis of movement disorders. *Neuroimagin. Clin. N Am.* **20**, 87–101.

Hochstrasser, H., Bauer, P., Walter, U., Behnke, S., Spiegel, J., Csoti, B., et al. (2004). Ceruplasmin gene variations and substantia nigra hyperechogenicity in Parkinson's disease. *Neurology* **63**, 1912–1917.

Hughes, A. J., Daniel, S. E., and Lees, A. J. (2001). Improved accuracy of clinical diagnosis of Parkinson's disease. *Neurology* **57** (8), 1497–1499.

Jain, S., Lo, S. E., and Louis, E. (2006). Common misdiagnosis of common neurological disorders: How are we misdiagnosing essential tremor? *Arch. Neurol.* **63** (8), 1100–1104.

Jellinger, K. (2003). Prevalence of cerebrovascular lesions in Parkinson's disease: A postmortem study. *Acta Neuropathol.* **105**, 415–419.

Lorberboym, M., Djaldetti, R., Melamed, E., Sadeh, M., and Lampl, Y. (2004). ^{123}I FP-CIT SPECT imaging of dopamine transporters in patients with cerebrovascular disease and clinical diagnosis of vascular parkinsonism. *J. Nucl. Med.* **45**, 1688–1693.

Miranda, M. (2006). Transcranial ultrasonography: a new tool in the diagnosis of Parkinson's disease. *Rev. Med. Chile* **134**, 525–528.

Okuda, B., Kawabata, K., Tachibana, H., Kamogawa, K., and Okamoto, K. (2008). Primitive reflexes distinguish vascular parkinsonism from Parkinson's disease. *Clin. Neurol. Neurosurg.* **110**, 562–565.

Owen, A. (2004). Cognitive dysfunction in Parkinson's disease: The role of frontostraital circuitry. *Neuroscientist* **10** (6), 525–537.

Scigliano, G., Musicco, M., Soliveri, P., Piccolo, I., Ronchetti, G., and Girotti, F. (2006). Reduced risk factors for vascular disorders in Parkinson's disease patients: A case-control study. *Stroke* **37**, 1184–1188.

Tsai, C. F., Wu, R. M., Huang, Y. W., Chen, L. L., Yip, P. K., and Jen, J. S. (2007). Transcranial color-coded sonography helps differentiation between idiopathic Parkinson's disease and vascular parkinsonism. *J. Neurol.* **254**, 501–507

Tolosa, E., Compta, Y., and Gaig, C. (2007). The premotor phase of Parkinson's disease. *Parkinsonism Relat. Disord.* S2–S7.

Venegas, P., Miranda, G., Gomez, R., Hudson, L., and Sinning, M. (2009a). Sensitivity and specificity of transcranial sonography in Parkinson's disease. *Rev. Chil. Neuro-Psiquiat.* **47** (Suppl. 1): S22.

Venegas, P., Miranda, G., Sinning, M., and Galleguillos, L. (2009b). Transcranial sonography for diagnosis of Parkinson's disease and other movement disorders. Pilot study of 39 cases and clinicoradiological correlation. *Rev. Hosp. Clin. Univ. Chile* **20**, 5–13.

Walter, U., Hoeppner, J., Prudente-Morrisey, L., Horowski, S., Herpertz, S., and Benecke, R. (2007). Parkinson's disease-like midbrain sonography abnormalities are frequent in depressive disorders. *Brain* **130**, 1799–1807.

Winikates, J., and Jankovic, J. (1999). Clinical correlates of vascular parkinsonism. *Arch. Neurol.* **56**, 98–102.

Zecca, L., Berg, D., Arzberger, T., Ruprecht, P., Rausch, W., Mussico, M., et al. (2005). In vivo detection of iron and neuromelanin by transcranial sonography: A new approach for early detection of substantia nigra damage. *Mov. Disord.* **10**, 1278–1285.

Zijlmans, J., Daniel, S., Hughes, A., Revesz, T., and Lees, A. (2004). Clinicopathological investigation of vascular parkinsonism, including clinical criteria for diagnosis. *Mov. Disord.* **19**, 630–640.

Zijlmans, J., Evans, A., Fontes, F., Katzenschlager, R., Gacinovic, S., Lees, A., and Costa, D. (2007). (^{123}I)FP-CIT SPECT study in vascular parkinsonism and Parkinson's disease. *Mov. Disord.* **22**, 1278–1285.

TCS IN MONOGENIC FORMS OF PARKINSON'S DISEASE

Kathrin Brockmann* and Johann Hagenah†

*Department of Neurodegeneration, Hertie Institute of Clinical Brain Research and German Center of Neurodegenerative Diseases (DZNE), Tübingen, Germany
†Department of Neurology, University of Lübeck, Ratzeburger Allee 160, Lübeck, Germany

During the last decade, mutations in a growing number of genes have been found to cause monogenic forms of Parkinson's disease (PD). Moreover, many studies shed light on their contribution to sporadic variants of PD. Since hyperechogenicity of the substantia nigra (SN) represents a characteristic hallmark of sporadic PD questions arise concerning the echo pattern of monogenic forms of this disorder on TCS.

I. Introduction

There is an increasing plethora of genetic findings in Parkinson's disease (PD) with mutations in a growing number of genes found to cause monogenic forms of this disorder. To date, 15 loci (PARK1–15) and 12 genes are known to cause either autosomal dominantly or autosomal recessively inherited forms of PD (Table I). Point mutations in the gene for *α-synuclein (SNCA)*, as well as duplications and triplications of the wild-type gene, are responsible for a dominantly inherited form of this disorder. Mutations in the gene encoding *Leucine-rich repeat kinase 2 (LRRK2)* are the most frequent causes for autosomal dominant PD known so far. On the other hand, mutations in genes such as *Parkin* (*PRKN*), *PTEN-Induced Putative Kinase 1* (*PINK1*), and *DJ-1* are known to cause parkinsonism of early onset following an autosomal recessive pattern of inheritance. Since

INTERNATIONAL REVIEW OF
NEUROBIOLOGY, VOL. 90
DOI: 10.1016/S0074-7742(10)90011-1

0074-7742/10 $35.00

TABLE I

LOCI AND GENES ASSOCIATED WITH PARKINSON'S DISEASE

	Gene	Locus	Inheritance	Onset	Clinical features	TCS
PARK1	*SNCA*	4q21	AD	EO	Similar to idiopathic PD Cognitive/psychiatric impairment	Symptomatic MC: **SN+**
PARK2	*Parkin*	6q25	AR	EO	Frequent dystonia at beginning Early dopa-induced fluctuations Slow progression	Homozygous MC: **SN+** Heterozygous MC: **SN+**
PARK3	*?*	2p13	AD	LO	Similar to idiopathic PD	
PARK4	*SNCA*	4q21	AD	EO	Gene dosage effect More severe disease progression in triplications	Symptomatic MC: **SN+**
PARK5	*UCHL1*	4p14	AD	LO	Similar to idiopathic PD	TCS
PARK6	*PINK1*	1p35	AR	EO	Similar to PARK2	Homozygous MC: **SN+** Heterozygous MC: **SN+**
PARK7	*DJ-1*	1p36	AR	EO	Similar to PARK2	Homozygous MC: **SN+**
PARK8	*LRRK2*	12q12	AD	LO	Most frequent genetic cause for AD-LO PD Similar to idiopathic PD	Symptomatic MC: **SN+** Asymptomatic MC: **SN+**
PARK9	*ATP13A2*	1p36	AR	EO	Parkinsonism Pyramidal signs Supranuclear gaze palsy Cognitive dysfunction	Heterozygous MC: **SN−**
PARK10	*?*	1p32	AD	LO	Similar to idiopathic PD	
PARK11	*GIGYF2*	2q36	AD	LO	Similar to idiopathic PD	
PARK12	*?*	Xq	?	LO		
PARK13	*OMI/HTRA2*	2p13	AD	LO	Similar to idiopathic PD	
PARK14	*PLA2G6*	22q13.1	AR	EO	Parkinsonism Dystonia Pyramidal signs Cognitive impairment	
PARK15	*FBXO7*	22q12	AR	EO	Parkinsonian-pyramidal syndrome (PPS)	

EO = early onset. LO = late onset. PD = Parkinson's disease. LOPD = late-onset Parkinson's disease. PPS = Parkinsonian Pyramidal Syndrome.

both monogenic and sporadic forms of the disease tend to share many overlapping features that include, most importantly, parkinsonism due to nigrostriatal dopaminergic degeneration, the functions of these genes and their importance for the PD population as a whole are explored extensively to better understand the pathogenesis from which therapeutic options may be derived—for both monogenetic and eventually also sporadic PD.

Here, we highlight the advances of TCS in monogenic forms of PD, outline options for diagnostics, and discuss its importance for clinical and research applications.

II. Substantia Nigra Hyperechogenicity in First-Degree Relatives of Sporadic PD Patients

First-degree relatives of sporadic PD patients are about two times as likely to develop PD as subjects without a positive family history for PD (Marder *et al.*, 1996; Semchuk *et al.*, 1993). Furthermore, twin studies using positron emission tomography (PET) demonstrate subclinical alterations of the nigrostriatal system in healthy siblings of PD patients to be a frequent finding (Burn *et al.*, 1992; Piccini *et al.*, 1999). Therefore, the question arises whether SN hyperechogenicity may also be a common characteristic in the relatives of sporadic PD patients. Indeed, several studies provide evidence to confirm this hypothesis.

In one study, 58 first-degree relatives of 14 sporadic PD patients were examined, 26 (45%) of them exhibited hyperechogenicity of the SN on TCS (Ruprecht-Dorfler *et al.*, 2003). More interestingly, the relatives with SN hyperechogenicity more commonly had mild signs of hypokinesia as well as impaired executive functions compared with the relatives with a normal echo pattern of the SN. Furthermore, a cross-sectional study including 1120 subjects older than 50 years revealed an association of a positive family history for PD and SN hyperechogenicity (Schweitzer *et al.*, 2007).

Taking this into account, the prevalence of SN hyperechogenicity seems much higher in PD families than in the general population and might represent an indicator not only of PD vulnerability but also of subclinical nigrostriatal dysfunction, along with motor and executive impairment.

III. Substantia Nigra Hyperechogenicity in Monogenic Forms of PD

When evaluating monogenic forms of PD, several aspects are of particular interest: First, do symptomatic mutation carriers mimic the same SN echo pattern, which is typical for idiopathic PD? Second, do asymptomatic persons who carry a mutation within one of the PD-associated genes display hyperechogenicity of the SN? Third, what are the echo features of heterozygous mutation carriers of autosomal recessively inherited forms of PD on TCS in light of the fact that heterozygous mutations may increase the susceptibility for late-onset PD? In the following, we will review findings concerning these questions of the past 10 years and give a summarized description in Table I.

A. Genes Associated with Autosomal Recessively Inherited forms of PD

Mutations in the *Parkin* gene (*PRKN*) are known to be the most frequent cause of autosomal recessively inherited juvenile parkinsonism with age at onset before the age of 40 years. Hyperechogenicity of the SN was found in affected and asymptomatic mutation carriers of homozygous, compound heterozygous, and even single heterozygous mutations (Hagenah *et al.*, 2007; Walter *et al.*, 2004). The pathological TCS findings in the asymptomatic carriers of single heterozygous *Parkin* mutation are in line with pathological findings in nigrostriatal 18F-dopa PET (Walter *et al.*, 2004), functional MRI (fMRI) (Buhmann *et al.*, 2005), voxel-based morphometry (VBM) (Binkofski *et al.*, 2007) and TMS investigations (Baumer *et al.*, 2007), indicating a subclinical dysfunction of the nigrostriatal system.

Recent molecular studies support the assumption that *Parkin, PINK1* and *DJ-1*, the genes associated most commonly with autosomal recessively inherited juvenile PD, may act in a common pathway (Xiong *et al.*, 2009; Park *et al.*, 2009). Furthermore, the clinical phenotype recognized in patients carrying mutations in one of these genes is similar. Therefore, it is not surprising that similar SN echo patterns are found. In more detail, in a family with carriers of homozygous as well as of heterozygous mutations in the *PINK1* gene (Hagenah *et al.*, 2008) as well as in another study including also a symptomatic carrier of the *DJ-1* gene (Schweitzer *et al.*, 2007), TCS revealed hyperechogenicity of the SN similar to sporadic PD patients. Hence, a contribution of the mutations in these genes to the hyperechogenic signal can be supposed. However, there may also be additional factors involved in the genesis of the ultrasound feature, SN hyperechogenicity, as stated by the finding that also nonmutation carriers of *PINK1* mutation carriers can display SN hyperechogenicity similar to first-degree relatives of patients with idiopathic PD.

Importantly, not all recessive forms of monogenetic PD are associated with SN hyperechogenicity. Homozygous mutations in the *ATP13A2* gene turned out to be responsible for a complicated autosomal recessive form of early-onset parkinsonism known as the Kufor Rakeb syndrome (KRS), which was first described in 1994 in a consanguineous family living in Kufor Rakeb, Jordan (Najim al-Din *et al.*, 1994). Clinically, these patients present with early-onset parkinsonism combined with pyramidal degeneration, supranuclear gaze palsy,

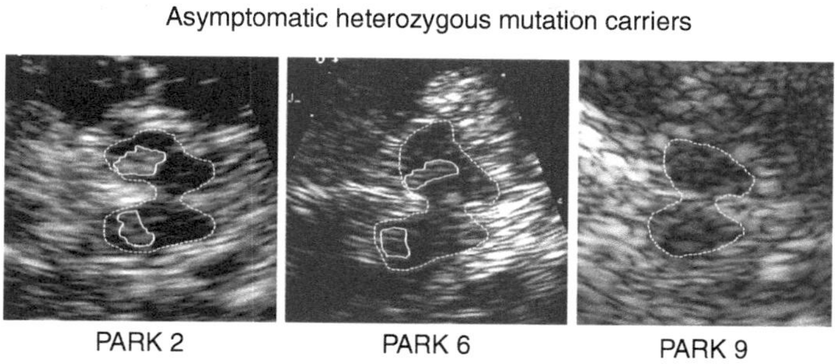

FIG. 1. Enlarged image of the hypoechogenic mesencephalic brainstem (encircled with dashed lines) of patients with different forms of recessively inherited monogenetic PD. In case of SN hyperechogenicity (PARK 2 and 6) the SN is encircled on both sides with full lines.

and cognitive dysfunction (Hampshire *et al.*, 2001; Najim al-Din *et al.*, 1994; Ramirez *et al.*, 2006; Williams *et al.*, 2005) suggesting a multisystem impairment. TCS data of one family showed no hyperechogenic SN in one compound heterozygous severely affected mutation carrier as well as in five heterozygous mutation carriers (Brüggemann *et al.*, submitted) (For comparison of SN echogenicity in recessive forms of PD, see also Fig. 1).

B. Genes Associated with Autosomal Dominantly Inherited forms of PD

Mutations in the *α-synuclein* (*SNCA*) gene were the first shown to cause familial PD (Polymeropoulos *et al.*, 1997). There are only two cases of TCS investigation in PARK1 published so far: One had a missense mutation (Schweitzer *et al.*, 2007) and the other PD patient carried a *de novo SNCA* duplication (Brueggemann *et al.*, 2008). Both of them showed abnormal SN hyperechogenicity. Regarding the recognized gene dosage effect of the phenotype due to different gene multiplications, it would be interesting to know whether mutation carriers with triplication differ from those who carry duplications of the *SNCA* gene in terms of brightness and SN area.

Mutations in the *LRRK2* gene are the most frequent genetic cause for late-onset autosomal dominant PD accounting not only for 10% of familial PD but also for 3% of sporadic PD, and even for up to 20–40% of all PS cases in select ethnic populations, for example, Ashkenazi Jews and North African Arabs. Clinical features associated with *LRRK2* mutations have been reported to be consistent with those of sporadic PD showing the typical asymmetric levodopa-responsive parkinsonism (Paisan-Ruiz *et al.*, 2005). The TCS findings in symptomatic *LRRK2*

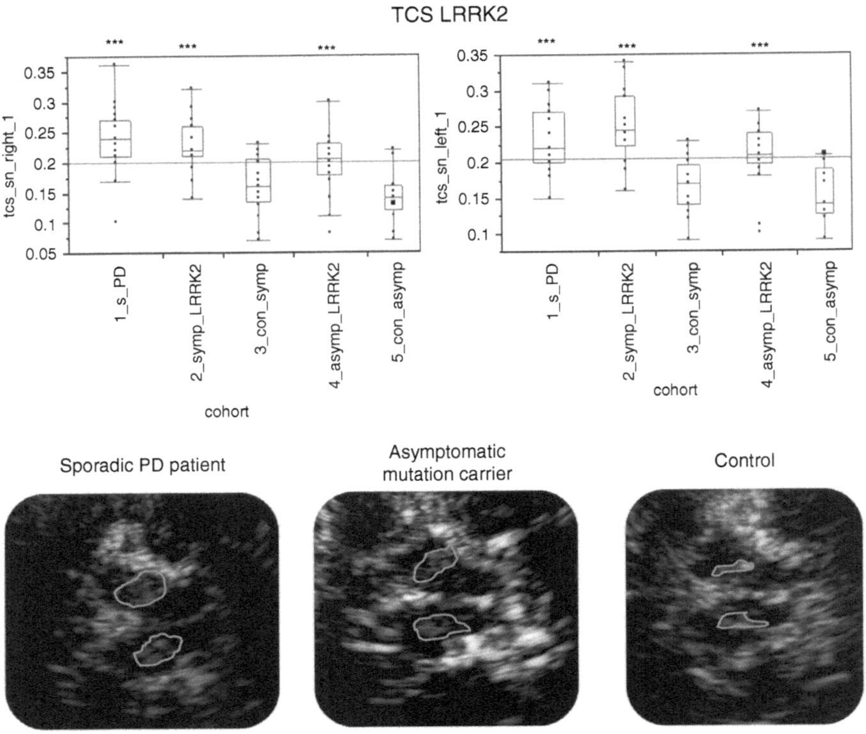

FIG. 2. Values of area of SN echogenicity in sporadic PD (sPD), symptomatic LRRK2 mutation carriers (symp_LRRK2) and asymptomatic LRRK2 mutation carriers (asymp_LRRK2) in comparison to corresponding age-machted healthy controls (con_symp, con_asymp) (upper picture). The SN in the hypoechogenic brainstem in a sporadic PD patient is similar to the area of hyperechogenicity in an asymptomatic mutation carrier and in both significantly larger than in a control person.

mutation carriers differ neither from the findings in symptomatic *SNCA, Parkin*, or *PINK1* mutation carriers revealing SN hyperechogenicity (Berg *et al.*, 2005; Hedrich *et al.*, 2006; Schweitzer *et al.*, 2007) nor from idiopathic PD. Recently, SN hyperechogenecity has also been found in asymptomatic mutation carriers of *LRRK2* mutations (Brockmann *et al.*, in preparation, Hagenah *et al.*, in preparation) (see also Fig. 2).

IV. Conclusions and Future Perspectives

Initial results from TCS studies on the most common forms of genetically based PD indicate that SN hyperechogenicity is also a frequent finding in monogenic forms of parkinsonism and, if prevalent, does not show a different pattern compared with sporadic PD. Asymptomatic mutation carriers not only of autosomal dominant but also of recessive forms of parkinsonism have been found to also present with the same echo pattern, suggesting SN hyperechogenicity as a possible indicator of PD vulnerability even in the presymptomatic stage of the disease. Multimodal follow-up studies in these individuals will help to learn more about the preclinical disease process. Moreover, a better understanding of the pathogenesis of these disorders may also contribute to elucidate the cause of SN hyperechogenicity.

References

Baumer, T., Pramstaller, P. P., Siebner, H. R., Schippling, S., Hagenah, J., Peller, M., Gerloff, C., Klein, C., and Munchau, A. (2007). Sensorimotor integration is abnormal in asymptomatic Parkin mutation carriers: A TMS study. *Neurology* **69** (21), 1976–1981.

Berg, D., Schweitzer, K., Leitner, P., Zimprich, A., Lichtner, P., Belcredi, P., Brussel, T., Schulte, C., Maass, S., and Nagele, T. (2005). Type and frequency of mutations in the LRRK2 gene in familial and sporadic Parkinson's disease. *Brain* **128** (Pt 12), 3000–11.

Binkofski, F., Reetz, K., Gaser, C., Hilker, R., Hagenah, J., Hedrich, K., van Eimeren, T., Thiel, A., Buchel, C., Pramstaller, P. P., Siebner, H. R., and Klein, C. (2007). Morphometric fingerprint of asymptomatic Parkin and PINK1 mutation carriers in the basal ganglia. *Neurology* **69** (9), 842–850.

Brueggemann, N., Odin, P., Gruenewald, A., Tadic, V., Hagenah, J., Seidel, G., Lohmann, K., Klein, C., and Djarmati, A. (2008). Re: Alpha-synuclein gene duplication is present in sporadic Parkinson disease. *Neurology* **71** (16), 1294.

Buhmann, C., Binkofski, F., Klein, C., Buchel, C., van Eimeren, T., Erdmann, C., Hedrich, K., Kasten, M., Hagenah, J., Deuschl, G., Pramstaller, P. P., and Siebner, H. R. (2005). Motor reorganization in asymptomatic carriers of a single mutant Parkin allele: A human model for presymptomatic parkinsonism. *Brain* **128** (Part 10), 2281–2290.

Burn, D. J., Mark, M. H., Playford, E. D., Maraganore, D. M., Zimmerman, Jr, T. R., Duvoisin, R. C., Harding, A. E., Marsden, C. D., and Brooks, D. J. (1992). Parkinson's disease in twins studied with 18F-dopa and positron emission tomography. *Neurology* **42** (10), 1894–1900.

Hagenah, J. M., Becker, B., Bruggemann, N., Djarmati, A., Lohmann, K., Sprenger, A., Klein, C., and Seidel, G. (2008). Transcranial sonography findings in a large family with homozygous and heterozygous PINK1 mutations. *J. Neurol. Neurosurg. Psychiatry* **79** (9), 1071–1074.

Hagenah, J. M., Konig, I. R., Becker, B., Hilker, R., Kasten, M., Hedrich, K., Pramstaller, P. P., Klein, C., and Seidel, G. (2007). Substantia nigra hyperechogenicity correlates with clinical status and number of Parkin mutated alleles. *J. Neurol.* **254** (10), 1407–1413.

Hampshire, D. J., Roberts, E., Crow, Y., Bond, J., Mubaidin, A., Wriekat, A. L., Al-Din, A., and Woods, C. G. (2001). Kufor–Rakeb syndrome, pallido-pyramidal degeneration with supranuclear upgaze paresis and dementia, maps to 1p36. *J. Med. Genet.* **38** (10), 680–682.

Hedrich, K., Winkler, S., Hagenah, J., Kabakci, K., Kasten, M., Schwinger, E., Volkmann, J., Pramstaller, P. P., Kostic, V., Vieregge, P., and Klein, C. (2006). Recurrent LRRK2 (Park8) mutations in early-onset Parkinson's disease. *Mov. Disord.* **21** (9), 1506–1510.

Marder, K., Tang, M. X., Mejia, H., Alfaro, B., Cote, L., Louis, E., Groves, J., and Mayeux, R. (1996). Risk of Parkinson's disease among first-degree relatives: A community-based study. *Neurology* **47** (1), 155–160.

Najim al-Din, A. S., Wriekat, A., Mubaidin, A., Dasouki, M., and Hiari, M. (1994). Pallido-pyramidal degeneration, supranuclear upgaze paresis and dementia: Kufor–Rakeb syndrome. *Acta Neurol. Scand.* **89** (5), 347–352.

Paisan-Ruiz, C., Saenz, A., Lopez de Munain, A., Marti, I., Martinez Gil, A., Marti-Masso, J. F., and Perez-Tur, J. (2005). Familial Parkinson's disease: Clinical and genetic analysis of four Basque families. *Ann. Neurol.* **57** (3), 365–372.

Park, J., Kim, Y., and Chung, J. (2009). Mitochondrial dysfunction and Parkinson's disease genes: Insights from Drosophila. *Dis. Model Mech.* **2** (7–8), 336–340.

Piccini, P., Burn, D. J., Ceravolo, R., Maraganore, D., and Brooks, D. J. (1999). The role of inheritance in sporadic Parkinson's disease: evidence from a longitudinal study of dopaminergic function in twins. *Ann. Neurol.* **45** (5), 577–582.

Polymeropoulos, M. H. Lavedan, C., Leroy, E., Ide, S. E., Dehejia, A., Dutra, A., Pike, B., Root, H., Rubenstein, J., Boyer, R., Stenroos, E. S., Chandrasekharappa, S. *et al.* (1997). Mutation in the alpha-synuclein gene identified in families with Parkinson's disease. *Science* **276** (5321), 2045–2047.

Ramirez, A., Heimbach, A., Grundemann, J., Stiller, B., Hampshire, D., Cid, L. P., Goebel, I., Mubaidin, A. F., Wriekat, A. L., Roeper, J., Al-Din, A., Hillmer, A. M. *et al.* (2006). Hereditary parkinsonism with dementia is caused by mutations in ATP13A2, encoding a lysosomal type 5 P-type ATPase. *Nat. Genet.* **38** (10), 1184–1191.

Ruprecht-Dorfler, P., Berg, D., Tucha, O., Benz, P., Meier-Meitinger, M., Alders, G. L., Lange, K. W., and Becker, G. (2003). Echogenicity of the substantia nigra in relatives of patients with sporadic Parkinson's disease. *Neuroimage* **18** (2), 416–422.

Schweitzer, K. J., Behnke, S., Liepelt, I., Wolf, B., Grosser, C., Godau, J., Gaenslen, A., Bruessel, T., Wendt, A., Abel, F., Muller, A., Gasser, T. *et al.* (2007). Cross-sectional study discloses a positive family history for Parkinson's disease and male gender as epidemiological risk factors for substantia nigra hyperechogenicity. *J. Neural. Transm.* **114** (9), 1167–1171.

Schweitzer, K. J., Brussel, T., Leitner, P., Kruger, R., Bauer, P., Woitalla, D., Tomiuk, J., Gasser, T., and Berg, D. (2007). Transcranial ultrasound in different monogenetic subtypes of Parkinson's disease. *J. Neurol.* **254** (5), 613–616.

Semchuk, K. M., Love, E. J., and Lee, R. G. (1993). Parkinson's disease: A test of the multifactorial etiologic hypothesis. *Neurology* **43** (6), 1173–1180.

Walter, U., Klein, C., Hilker, R., Benecke, R., Pramstaller, P. P., and Dressler, D. (2004). Brain parenchyma sonography detects preclinical parkinsonism. *Mov. Disord.* **19** (12), 1445–1449.

Williams, D. R., Hadeed, A., al-Din, A. S., Wreikat, A. L., and Lees, A. J. (2005). Kufor Rakeb disease: autosomal recessive, levodopa-responsive parkinsonism with pyramidal degeneration, supranuclear gaze palsy, and dementia. *Mov. Disord.* **20** (10), 1264–1271.

Xiong, H., Wang, D., Chen, L., Choo, Y. S., Ma, H., Tang, C., Xia, K., Jiang, W., Ronai, Z., Zhuang, X., and Zhang, Z. (2009). Parkin, PINK1, and DJ-1 form a ubiquitin E3 ligase complex promoting unfolded protein degradation. *J. Clin. Invest.* **119** (3), 650–660.

PART III

TRANSCRANIAL SONOGRAPHY IN OTHER MOVEMENT DISORDERS AND DEPRESSION

TRANSCRANIAL SONOGRAPHY IN BRAIN DISORDERS WITH TRACE METAL ACCUMULATION

Uwe Walter

Department of Neurology, University of Rostock, Rostock, Germany

Transcranial sonography (TCS) can detect trace metal accumulation in deep brain structures with higher sensitivity than conventional MRI. Especially, increased iron content in the substantia nigra in Parkinson's disease, increased copper content in the lenticular nucleus (LN) in Wilson's disease and idiopathic dystonia, and increased manganese content in the LN in manganese-induced Parkinsonism were detected with TCS, even in subjects with normal MRI. TCS, therefore, might be useful to detect an increased risk of developing neurological symptoms in relatives of patients with Parkinson's or Wilson's disease. The exact mechanism of how an elevated trace metal content leads to an increased echogenicity needs to be further elucidated.

I. Introduction

Transcranial sonography (TCS) is an alternative neuroimaging method that displays the brain through the intact skull. Compared to other well-established neuroimaging methods such as computed tomography (CT) and magnetic resonance imaging (MRI), the different imaging principle of TCS, that is, the reflection of ultrasound waves at tissue interfaces with diverse acoustic impedance, allows a different view into the brain (Becker and Berg, 2001). Moreover, contemporary TCS systems achieve today even a somewhat higher image resolution of echogenic deep brain structures compared with MRI under clinical conditions

INTERNATIONAL REVIEW OF
NEUROBIOLOGY, VOL. 90
DOI: 10.1016/S0074-7742(10)90012-3

TABLE I
MOVEMENT DISORDERS WITH ALTERATIONS OF TRACE METAL CONTENT IN DISTINCT DEEP BRAIN STRUCTURES

Disease	Altered brain metal	SN	RN	LN
Parkinson's disease	Iron	↑↑	–	(↑)
Hemochromatosis	Iron			↑
NBIA	Iron	–*	↑↑ *	(↑) *
Restless legs syndrome	Iron	↓↓	(↑)	–
Idiopathic dystonia	Copper, manganese	–	–	↑↑
Wilson's disease	Copper	(↑)	–	↑↑↑
Welding-related parkinsonism	Manganese	–	–	↑↑

SN, substantia nigra; RN, red nucleus; LN, lenticular nucleus; NBIA, neurodegeneration with brain iron accumulation (former: Hallervorden—Spatz disease); ↑↑, increased echogenicity of referring structure on TCS in more than 80% of patients; ↑, increased echogenicity of referring structure on TCS in 50—80% of patients; (↑), increased echogenicity of referring structure on TCS in less than 50% of patients; ↓↓, reduced echogenicity of referring structure on TCS in more than 80% of patients.

*No published studies; few cases studied.

(Walter *et al.*, 2008a). Subcortical brain structures, basal ganglia, as well as midbrain structures and central parts of the cerebellum, can well be assessed with TCS-provided adequate imaging conditions, especially a sufficient transtemporal acoustic bone window (Berg *et al.*, 2008).

A number of TCS studies have indicated that an increased echogenicity of distinct deep brain structures is likely to be caused by increased amounts of trace metals (iron, copper, manganese) within these structures (Table I). In this chapter, studies are summarized that led to the idea that trace metal accumulation may cause increased echogenicity on TCS, and TCS findings in metabolic disorders with primary trace metal accumulation in the brain are reviewed.

II. Substantia Nigra Echogenicity and Iron Metabolism

Becker *et al.* (1995) were the first to describe a characteristically increased echosignal ("hyperechogenicity") at the area of SN in patients with Parkinson's disease (PD) (Fig. 1). Meanwhile, this finding has been confirmed by a number of independent groups (Doepp *et al.*, 2008; Hagenah *et al.*, 2007; Huang *et al.*, 2007; Kim *et al.*, 2007; Kolevski *et al.*, 2007; Okawa *et al.*, 2007; Ressner *et al.*, 2007; Stockner *et al.*, 2007; Walter *et al.*, 2002). This TCS finding, present in about 90% of PD patients at cross section, is independent from PD duration and severity (Berg *et al.*, 2001a; Walter *et al.*, 2007a). Moreover, SN echogenicity was found to be stable

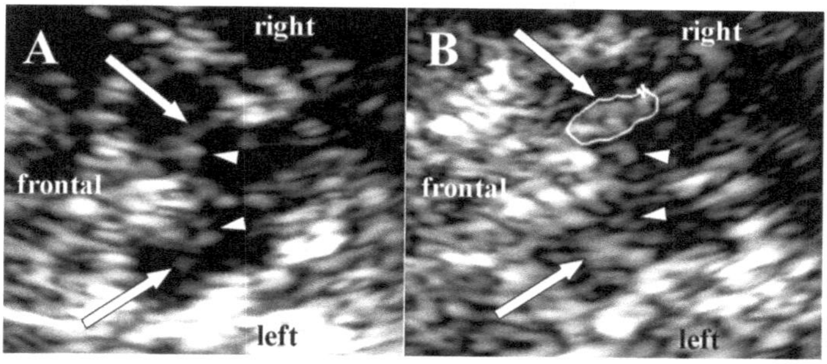

FIG. 1. Transcranial sonography (TCS) images of identical midbrain axial sections in two subjects. The butterfly-shaped midbrain is encircled for better visualization. (A) Normal, nearly invisible substantia nigra (arrows) (red nuclei: arrow heads). (B) Subject with increased echogenicity ("hyperechogenicity") of substantia nigra (arrows) (red nuclei: arrow heads). Echogenic area of the right substantia nigra was encircled for computerized measurement.

in a 5-year follow-up study of PD patients (Berg *et al.*, 2005). No correlation was found between the degree of SN hyperechogenicity and the striatal uptake of N-omega-fluoropropyl-2beta-carbomethoxy-3beta-4-[(123)I]iodophenyl-nortropane (FP-CIT) on SPECT, which is thought to represent a correlate for the degeneration of presynaptic dopaminergic neurons in PD (Spiegel *et al.*, 2006). These findings indicate that SN hyperechogenicity is not a correlate of the progressive degeneration of SN neurons. However, a close correlation between SN echogenicity and tissue iron content has been shown in animal studies and in postmortem studies of human brains (Berg *et al.*, 1999, 2002). These findings suggest that SN hyperechogenicity in PD is, at least in part, caused by an elevated iron content of the SN (Berg *et al.*, 2001b; Dexter *et al.*, 1991). On the other hand, increased iron content alone cannot be the only explanation for SN hyperechogenicity since iron accumulates over time in the SN of PD patients (Hirsch, 2006), and other iron-rich brain structures such as red nucleus or globus pallidus internus normally show no increased echogenicity on TCS (Walter *et al.*, 2007b). Therefore, additional factors such as abnormal iron–protein bindings were proposed to contribute to SN hyperechogenicity (Berg *et al.*, 2002). In an extensive postmortem study of 40 human brains, SN hyperechogenicity was related not only to increased iron content but also to a relatively decreased neuromelanin content of the SN (Zecca *et al.*, 2005). Neuromelanin is an iron-binding molecule of pigmented neurons and is thought to play a physiological role in intraneuronal iron homeostasis. In subjects with SN hyperechogenicity, where nigral iron levels are increased, saturation of high-affinity iron-binding sites on neuromelanin may overwhelm the protective capacity of this molecule, leading instead to an increase in redox-active iron, promoting subsequent cellular damage and neurodegeneration over a longer period of time. Interestingly, neuropathologic studies show that even a modest difference in iron concentration in the SN is sufficient to increase oxidative stress and cause neurodegeneration (Gałazka-Friedman and Friedman, 1997). Increased iron levels have been recently demonstrated in individual dopaminergic neurons and not only in glial cells, which strongly suggests that iron elevation in specific brain regions is due to primary changes in iron handling in PD (Oakley *et al.*, 2007). Iron that normally accumulates in the brain is primarily nonheme iron (Hallgren and Sourander, 1958). Although nonheme iron is generally unrelated to total body iron stores, the brains of subjects with PD have increased iron deposits of nonheme iron but lower levels of ferritin and transferrin, compared with controls (Connor *et al.*, 1995; Dexter *et al.*, 1991). This suggests that in PD there is an alteration of the system that regulates the synthesis of ferritin and other proteins involved in systemic iron metabolism.

The close correlation of the TCS finding of SN hyperechogenicity with increased iron content in the SN (Berg *et al.*, 2002; Zecca *et al.*, 2005), together with an apparently autosomal dominant inheritance of this echofeature in relatives of patients with idiopathic PD (Ruprecht-Dörfler *et al.*, 2003), supports the idea of a primary role of disturbed iron metabolism in PD. Subsequent mutation analyses of genes encoding for iron-transport and iron-regulatory proteins known to be associated with Parkinsonism led to the discovery of specific mutations in the *ferritin-H*, the *iron-regulatory protein 2*, and the *hemochromatosis* gene, respectively, in single PD patients with SN hyperechogenicity (Akbas *et al.*, 2006; Deplazes *et al.*, 2004; Akbas *et al.*, 2003).

The most striking association was found in the *ceruloplasmin* gene: of five exonic missense mutations, the I63T mutation was only found in one PD patient, the D544E and the R793H mutations in far more PD patients than in ethnically matched controls (Hochstrasser *et al.*, 2004). The *ceruloplasmin* gene mutations were clearly associated with the TCS finding of SN hyperechogenicity in PD patients and healthy control subjects. In subsequent cell culture studies, functional relevance of the *ceruloplasmin* mutations I63T and D544E for iron metabolism could be proven (Hochstrasser *et al.*, 2005). Ceruloplasmin oxidizes Fe^{2+} to Fe^{3+} without releasing radical oxygen species and is thought to play a role in regulating iron uptake into the central nervous system across the blood–brain barrier (Qian and Ke, 2001). The finding of parkinsonism present in some cases of aceruloplasminemia implies that sequence variations in the *ceruloplasmin* gene might cause iron-induced oxidative stress that could lead to or at least contribute to neurodegeneration in PD (Kohno *et al.*, 2000).

Further support for a relationship between SN echogenicity and a disturbance of iron metabolism came from TCS studies in patients with restless legs syndrome, a disease where a primary alteration of iron metabolism is discussed (Earley *et al.*, 2008), leading to decreased iron concentration within the SN and other brain structures (Astrakas *et al.*, 2008; Godau *et al.*, 2008). In contrast to PD, in RLS, an abnormal reduced echogenicity of the SN is typically found

(Schmidauer *et al.*, 2005; Godau *et al.*, 2007), which correlates with MRI measures indicating low iron content (Godau *et al.*, 2008). In patients with multiple sclerosis, the echogenicity of the SN was found to be clearly correlated with hypointensity on MRI T2-weighted images, which is thought to be a correlate of iron content (Walter *et al.*, 2009).

As yet, it remains open whether the changes of iron metabolism, reflected by SN hyperechogenicity, are a primary cause or a co-factor in the pathogenesis of PD. Nevertheless, influencing iron metabolism might become an interesting therapeutic target in PD (Arduino *et al.*, 2008; Xu *et al.*, 2008). In this context, TCS might be a useful tool for assessing the iron status of the SN.

III. Lenticular Nucleus Hyperechogenicity and Iron Accumulation

Berg *et al.* (2000a) reported on TCS findings in 14 patients with hemochromatosis, a hereditary metabolic disorder characterized by the deposition of iron-containing pigment in various organs, including the brain. They found in six patients a bilateral and in one a unilateral hyperechogenicity of the medial lenticular nucleus (LN) at the anatomic area of the globus pallidus interna (Berg *et al.*, 2000a). This corresponded to the CT finding of abnormal increased density of the globus pallidus interna in the referring patients.

Normally, the globus pallidus does not appear hyperechogenic (Naumann *et al.*, 1996), although histological and histochemical studies have shown that the globus pallidus is one of the brain regions containing the highest amounts of iron (Hallgren and Sourander, 1958; Morris *et al.*, 1992). Even in neurodegeneration with brain iron accumulation (NBIA, formerly Hallervorden–Spatz syndrome), a hereditary iron-storage disease with accumulation of iron-containing pigment especially in the globus pallidus and the SN, we found no clear abnormalities on TCS in the LN or the SN in patients with NBIA (Walter *et al.*, unpublished). This suggests that iron accumulation *per se* does not necessarily lead to an increased echogenicity. Other factors such as abnormal iron–protein bindings play a role in the generation of hyperechogenicity.

IV. Lenticular Nucleus Hyperechogenicity and Copper Accumulation

Naumann *et al.* (1996) were the first to report an abnormal increased echogenicity ("hyperechogenicity") of LN in patients with idiopathic dystonia (Fig. 2). This finding was surprising since with other neuroimaging techniques usually no

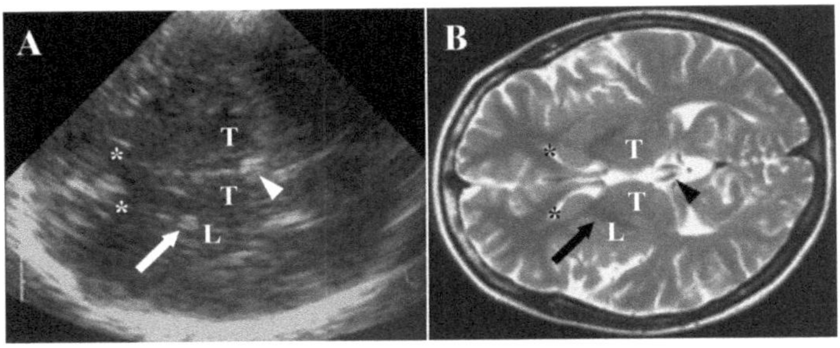

FIG. 2. Transcranial sonography (TCS) and magnetic resonance (MR) image of a patient with idiopathic dystonia. T indicates thalamus, L, lenticular nucleus, *frontal horn of lateral ventricle. (A) TCS image showing a dot-like hyperechogenicity of the lenticular nucleus. This finding is typical not only for idiopathic dystonia, but also for early stages of Wilson's disease. (B) MRI image corresponding to (A) with normal findings.

abnormalities are found in the brain of patients with idiopathic dystonia. Whereas it is well known from symptomatic forms of dystonia that lesions of LN or thalamus can result in contralateral focal or hemi-dystonia, histopathologic studies could not demonstrate consistent structural abnormalities in idiopathic dystonia. Nevertheless, electrophysiological and functional neuroimaging studies demonstrate a disinhibition of the thalamic and motor cortex neurons, pointing toward a key role of basal ganglia in the pathogenesis of idiopathic dystonia. TCS studies revealed that more than 75% of patients with cervical or upper-limb dystonia display LN hyperechogenicity, being most pronounced in the medial part representing the globus pallidus internus (Becker *et al.*, 1997; Naumann *et al.*, 1996). Patients with facial dystonia showed LN hyperechogenicity only in 31% of cases (Naumann *et al.*, 1996).

Driven by these TCS findings, postmortem examinations of brain tissue of patients with idiopathic dystonia were performed and yielded significantly increased copper (and manganese) levels in the globus pallidus and putamen whereas zinc, iron, and calcium concentrations turned out to be equal to those found in control brain samples (Becker *et al.*, 1999). Subsequent investigations found a reduction of the copper-transporting Menkes protein in the basal ganglia and in the leukocytes of patients with idiopathic dystonia (Berg *et al.*, 2000b; Kruse *et al.*, 2001). Menkes protein plays a key role in cellular copper homeostasis. As copper modulates synaptic function and acts as inhibitor of several receptor types, increase of copper levels in the LN could explain disinhibition of thalamus via disturbed function of neurons of the globus pallidus interna (Becker *et al.*, 2001). Hitherto, genetic studies failed to prove pathogenic mutations of genes related to copper-transporting enzymes in idiopathic dystonia (Bandmann *et al.*, 2002). Therefore, the disturbance of copper metabolism might also represent a secondary phenomenon. However, the suggestion that copper accumulation causes LN hyperechogenicity is supported by TCS findings in patients with Wilson's disease (Walter *et al.*, 2005). In clinical practice, the TCS finding of LN hyperechogenicity might be useful to support diagnosis of idiopathic dystonia in differentiation from tardive and psychogenic forms of dystonia, but this needs to be addressed in more detail by future studies.

Wilson's disease is an autosomal-recessive inherited disorder of hepatic copper metabolism caused by malfunction of a putative copper-transporting P-type ATPase, ATP7B, leading to toxic copper accumulation in hepatocytes and in extrahepatic tissues, including the cornea and the basal ganglia of the brain. Patients can present with hepatic symptoms, with neurological impairment, or with both. Neurological presentation begins in the second decade of life or later and includes dysarthria, akinesia, ataxia, tremor, and dystonia. MRI studies in Wilson's disease show frequent involvement of putamen, ventral thalamus, caudate head, globus pallidus, midbrain, pons, and cerebellum (King *et al.*, 1996; Oder *et al.*, 1993). Both genetic disorders of copper homeostasis, Wilson's and Menkes' diseases, are thought to cause neurodegeneration by uncontrolled redox reactivity of excessive copper (Rossi *et al.*, 2004).

In a study of 21 patients with Wilson's disease with ($n=18$) or without ($n=3$) neurologic symptoms TCS revealed LN hyperechogenicity in all assessable ($n=17$) symptomatic and in two of the three asymptomatic patients whereas 1.5-T MRI detected LN abnormalities only in 12 patients (Walter *et al.*, 2005). The correlation of LN hyperechogenicity with hypointense rather than hyperintense lesions on T2-weighted MR images suggested that LN hyperechogenicity in Wilson's disease is caused by copper accumulation rather than gliosis. Some patients with Wilson's disease also exhibited SN hyperechogenicity, hyperechogenic lesions of thalamus, and dilatation of third and lateral ventricles. LN echogenic sizes and ventricle widths correlated with disease severity (Walter *et al.*, 2005). The finding of LN hyperechogenicity even in neurologically asymptomatic patients with normal MRI suggests that TCS can detect copper accumulation in basal ganglia already in preclinical disease stages.

V. Lenticular Nucleus Hyperechogenicity and Manganese Accumulation

Manganese accumulation in the brain can be caused by chronic exposure to aerosolized manganese. Manganese-induced parkinsonism has been reported not only in Chilean miners (Mena *et al.*, 1967), but also in welders where they have worked in confined, unventilated spaces (Josephs *et al.*, 2005; Sadek *et al.*, 2003). Welding-related Parkinsonism has been suggested to be caused by neurotoxicity of aerosolized manganese, an elementary metal that is released in welding plumes

(Josephs *et al.*, 2005). Parkinsonism can occur after chronic exposure to high levels of manganese, usually above the permissible exposure limit at 5 mg/m^3 total dust set by the Occupational Safety and Health Administration (Jankovic, 2005). In a TCS study of two patients with welding-related Parkinsonism that was thought to be caused by chronic manganese intoxication, TCS revealed LN hyperechogenicity in both of them while SN echogenicity was normal (Walter *et al.*, 2008b). The MRI was normal in both subjects at the time of the study. Both subjects had stopped welding in unventilated spaces several years ago. These findings also support the idea that TCS is more sensitive than MRI in detecting certain types of trace metal accumulation in the basal ganglia.

VI. Conclusion

A number of studies have shown that TCS is suitable to detect trace metal accumulation in the basal ganglia and in the substantia nigra with higher sensitivity than conventional MRI. This may be useful for detecting preclinical disease stages with TCS in relatives of patients with PD and Wilson's disease (Berg, 2007; Walter, 2009b). A number of questions, however, remain to be clarified. First, the exact mechanisms leading to hyperechogenicity of a deep brain structure if trace metal content is increased are undisclosed yet. Furthermore, the exact time point, when the echogenicity increase occurs, is currently unclear.

References

Akbas, N., Hochstrasser, H., Deplazes, J., Tomiuk, J., Bauer, P., Walter, U., Behnke, S., Riess, O., and Berg, D. (2006). Screening for mutations of the HFE gene in Parkinson's disease patients with hyperechogenicity of the substantia nigra. *Neurosci. Lett.* **407**, 16–19.

Arduino, D., Silva, D., Cardoso, S. M., Chaves, S., Oliveira, C. R., and Santos, M. A. (2008). New hydroxypyridinone iron-chelators as potential anti-neurodegenerative drugs. *Front Biosci.* **13**, 6763–6774.

Astrakas, L. G., Konitsiotis, S., Margariti, P., Tsouli, S., Tzarouhi, L., and Argyropoulou, M. I. (2008). T2 relaxometry and fMRI of the brain in late-onset restless legs syndrome. *Neurology* **71**, 911–916.

Bandmann, O., Asmus, F., Sibbing, D., Grundmann, M., Schwab, S. G., Müller, J., Wildenauer, D. B., Poewe, W., Gasser, T., and Oertel, W. H. (2002). Copper genes are not implicated in the pathogenesis of focal dystonia. *Neurology* **59**, 782–783.

Becker, G., and Berg, D. (2001). Neuroimaging in basal ganglia disorders: Perspectives for transcranial ultrasound. *Mov. Disord.* **16**, 23–32.

Becker, G., Seufert, J., Bogdahn, U., Reichmann, H., and Reiners, K. (1995). Degeneration of substantia nigra in chronic Parkinson's disease visualized by transcranial color-coded real-time sonography. *Neurology* **45**, 182–184.

Becker, G., Naumann, M., Scheubeck, M., Hofmann, E., Deimling, M., Lindner, A., Gahn, G., Reiners, C., Toyka, K. V., and Reiners, K. (1997). Comparison of transcranial sonography, magnetic resonance imaging, and single photon emission computed tomography findings in idiopathic spasmodic torticollis. *Mov. Disord.* **12**, 79–88.

Becker, G., Berg, D., Rausch, W. D., Lange, H. K., Riederer, P., and Reiners, K. (1999). Increased tissue copper and manganese content in the lentiform nucleus in primary adult-onset dystonia. *Ann. Neurol.* **46**, 260–263.

Becker, G., Berg, D., Francis, M., and Naumann, M. (2001). Evidence for disturbances of copper metabolism in dystonia: From the image towards a new concept. *Neurology* 57, 2290–2294.

Berg, D., Grote, C., Rausch, W. D., Mäurer, M., Wesemann, W., Riederer, P., and Becker, G. (1999). Iron accumulation in the substantia nigra in rats visualized by ultrasound. *Ultrasound Med. Biol.* **25**, 901–904.

Berg, D., Hoggenmüller, U., Hofmann, E., Fischer, R., Kraus, M., Scheurlen, M., and Becker, G. (2000a). The basal ganglia in haemochromatosis. *Neuroradiology* **42**, 9–13.

Berg, D., Weishaupt, A., Francis, M. J., Miura, N., Yang, X. L., Goodyer, I. D., Naumann, M., Koltzenburg, M., Reiners, K., and Becker, G. (2000b). Changes of copper-transporting proteins and ceruloplasmin in the lentiform nuclei in primary adult-onset dystonia. *Ann. Neurol.* **47**, 827–830.

Berg, D., Siefker, C., and Becker, G. (2001a). Echogenicity of the substantia nigra in Parkinson's disease and its relation to clinical findings. *J. Neurol.* **248**, 684–689.

Berg, D., Gerlach, M., Youdim, M. B., Double, K. L., Zecca, L., Riederer, P., and Becker, G. (2001b). Brain iron pathways and their relevance to Parkinson's disease. *J. Neurochem.* **79**, 225–236.

Berg, D., Roggendorf, W., Schröder, U., Klein, R., Tatschner, T., Benz, P., Tucha, O., Preier, M., Lange, K. W., Reiners, K., Gerlach, M., and Becker, G. (2002). Echogenicity of the substantia nigra: Association with increased iron content and marker for susceptibility to nigrostriatal injury. *Arch. Neurol.* **59**, 999–1005.

Berg, D., Merz, B., Reiners, K., Naumann, M., and Becker, G, (2005). Five-year follow-up study of hyperechogenicity of the substantia nigra in Parkinson's disease. *Mov. Disord.* **20**, 383–385.

Berg, D. (2007). Ultrasound in the (premotor) diagnosis of Parkinson's disease. *Parkinsonism Relat. Disord.* **13** (Suppl. 3), S429–433.

Berg, D., Godau, J., and Walter, U. (2008). Transcranial sonography in movement disorders. *Lancet Neurol.* **7**, 1044–1055.

Connor, J. R., Snyder, B. S., Arosio, P., Loeffler, D. A., and LeWitt, P. (1995). A quantitative analysis of isoferritins in select regions of aged, parkinsonian, and Alzheimer's diseased brains. *J. Neurochem.* **65**, 717–724.

Deplazes, J., Schöbel, K., Hochstrasser, H., Bauer, P., Walter, U., Behnke, S., Spiegel, J., Becker, G., Riess, O., and Berg, D. (2003). Screening for mutations of the IRP2 gene in Parkinson's disease patients with hyperechogenicity of the substantia nigra. *J. Neural. Transm.* **111**, 515–521.

Dexter, D. T., Carayon, A., Javoy-Agid, F., Agid, Y., Wells, F. R., Daniel, S. E., Lees, A. J., Jenner, P., and Marsden, C. D. (1991). Alterations in the levels of iron, ferritin and other trace metals in Parkinson's disease and other neurodegenerative diseases affecting the basal ganglia. *Brain* **114**, 1953–1975.

Earley, C. J., Ponnuru, P., Wang, X., Patton, S. M., Conner, J. R., Beard, J. L., Taub, D. D., and Allen, R. P. (2008). Altered iron metabolism in lymphocytes from subjects with restless legs syndrome. *Sleep* **31**, 847–852.

Felletschin, B., Bauer, P., Walter, U., Behnke, S., Spiegel, J., Csoti, I., Sommer, U., Zeiler, B., Becker, G., Riess, O., and Berg, D. (2003). Screening for mutations of the ferritin light and heavy genes in Parkinson's disease patients with hyperechogenicity of the substantia nigra. *Neurosci. Lett.* **352**, 53–56.

Gałazka-Friedman, J., and Friedman, A. (1997). Controversies about iron in parkinsonian and control substantia nigra. *Acta Neurobiol. Exp. (Wars)* **57**, 217–225.

Godau, J., Schweitzer, K. J., Liepelt, I., Gerloff, C., and Berg, D. (2007). Substantia nigra hypoechogenicity: Definition and findings in restless legs syndrome. *Mov. Disord.* **22**, 187–192.

Godau, J., Klose, U., Di Santo, A., Schweitzer, K., and Berg, D. (2008). Multiregional brain iron deficiency in restless legs syndrome. *Mov. Disord.* **23**, 1184–1187.

Hagenah, J. M., König, I. R., Becker, B., Hilker, R., Kasten, M., Hedrich, K., Pramstaller, P. P., Klein, C., and Seidel, G. (2007). Substantia nigra hyperechogenicity correlates with clinical status and number of Parkin mutated alleles. *J. Neurol.* **254**, 1407–1413.

Hallgren, B., and Sourander, P. (1958). The effect of age on the non-haemin iron in the human brain. *J. Neurochem.* **3**, 41–51.

Hirsch, E. C. (2006). Altered regulation of iron transport and storage in Parkinson's disease. *J. Neural. Transm.* **71** (Suppl.), 201–204.

Hochstrasser, H., Bauer, P., Walter, U., Behnke, S., Spiegel, J., Csoti, I., Zeiler, B., Bornemann, A., Pahnke, J., Becker, G., Riess, O., and Berg, D. (2004). Ceruloplasmin gene variations and substantia nigra hyperechogenicity in Parkinson's disease. *Neurology* **63**, 1912–1917.

Hochstrasser, H., Tomiuk, J., Walter, U., Behnke, S., Spiegel, J., Krüger, R., Becker, G., Riess, O., and Berg, D. (2005). Functional relevance of ceruloplasmin mutations in Parkinson's disease. *FASEB J.* **19**, 1851–1853.

Huang, Y. W., Jeng, J. S., Tsai, C. F., Chen, L. L., and Wu, R. M. (2007). Transcranial imaging of substantia nigra hyperechogenicity in a Taiwanese cohort of Parkinson's disease. *Mov. Disord.* **22**, 550–555.

Jankovic, J. (2005). Searching for a relationship between manganese and welding and Parkinson's disease. *Neurology* **64**, 2021–2028.

Josephs, K. A., Ahlskog, J. E., Klos, K. J., Kumar, N., Fealey, R. D., Trenerry, M. R., Cowl, C. T. (2005). Neurologic manifestations in welders with pallidal MRI T1 hyperintensity. *Neurology* **64**, 2033–2039.

Kim, J. Y., Kim, S. T., Jeon, S. H., and Lee, W. Y. (2007). Midbrain transcranial sonography in Korean patients with Parkinson's disease. *Mov. Disord.* **22**, 1922–1926.

King, A. D., Walshe, J. M., Kendall, B. E., Chinn, R. J., Paley, M. N., Wilkinson, I. D., Halligan, S., and Hall-Craggs, M. A. (1996). Cranial MR imaging in Wilson's disease. *AJR Am. J. Roentgenol.* **167**, 1579–1584.

Kohno, S., Miyajima, H., Takahashi, Y., and Inoue, Y. (2000). Aceruloplasminemia with a novel mutation associated with parkinsonism. *Neurogenetics* **2**, 237–238.

Kolevski, G., Petrov, I., and Petrova, V. (2007). Transcranial sonography in the evaluation of Parkinson disease. *J. Ultrasound Med.* **26**, 509–512.

Kruse, N., Berg, D., Francis, M. J., Naumann, M., Rausch, W. D., Reiners, K., Rieckmann, P., Weishaupt, A., Becker, G. (2001). Reduction of Menkes mRNA and copper in leukocytes of patients with primary adult-onset dystonia. *Ann. Neurol.* **49**, 405–408.

Mena, I., Marin, O., Fuenzalida, S., and Cotzias, G. C. (1967). Chronic manganese poisoning. Clinical picture and manganese turnover. *Neurology.* **17**, 128–136.

Morris, C. M., Candy, J. M., Oakley, A. E., Bloxham, C. A., Edwardson, J. A. (1992). Histochemical distribution of non-haem iron in the human brain. *Acta Anat. (Basel).* **144**, 235–257.

Naumann, M., Becker, G., Toyka, K. V., Supprian, T., Reiners, K. (1996). Lenticular nucleus lesion in idiopathic dystonia detected by transcranial sonography. *Neurology* **47**, 1284–1290.

Oakley, A. E., Collingwood, J. F., Dobson, J., Love, G., Perrott, H. R., Edwardson, J. A., Elstner, M., and Morris, C. M. (2007). Individual dopaminergic neurons show raised iron levels in Parkinson disease. *Neurology* **68**, 1820–1825.

Oder, W., Prayer, L., Grimm, G., Spatt, J., Ferenci, P., Kollegger, H., Schneider, B., Gangl, A., and Deecke, L. (1993). Wilson's disease: evidence of subgroups derived from clinical findings and brain lesions. *Neurology* **43**, 120–124.

Okawa, M., Miwa, H., Kajimoto, Y., Hama, K., Morita, S., Nakanishi, I., and Kondo, T. (2007). Transcranial sonography of the substantia nigra in Japanese patients with Parkinson's disease or atypical parkinsonism: clinical potential and limitations. *Intern. Med.* **46**, 1527–1531.

Qian, Z. M., Ke, Y. (2001). Rethinking the role of ceruloplasmin in brain iron metabolism. *Brain Res. Brain Res. Rev.* **35**, 287–294.

Ressner, P., Skoloudik, D., Hlustik, P., and Kanovsky, P. (2007). Echogenicity of substantia nigra in Parkinson's disease—pilot study. *J. Neuroimaging.* **17**, 164–167.

Rossi, L., Lombardo, M. F., Ciriolo, M. R., and Rotilio, G. (2004). Mitochondrial dysfunction in neurodegenerative diseases associated with copper imbalance. *Neurochem. Res.* **29**, 493–504.

Ruprecht-Dörfler, P., Berg, D., Tucha, O., Benz, P., Meier-Meitinger, M., Alders, G. L., Lange, K. W., and Becker, G. (2003). Echogenicity of the substantia nigra in relatives of patients with sporadic Parkinson's disease. *Neuroimage* **18**, 416–422.

Sadek, A. H., Rauch, R., and Schulz, P. E. (2003). Parkinsonism due to manganism in a welder. *Int. J. Toxicol.* **22**, 393–401.

Schmidauer, C., Sojer, M., Seppi, K., Stockner, H., Högl, B., Biedermann, B., Brandauer, E., Peralta, C. M., Wenning, G. K., and Poewe, W. (2005). Transcranial ultrasound shows nigral hypoechogenicity in restless legs syndrome. *Ann. Neurol.* **58**, 630–634.

Stockner, H., Sojer, M., Seppi, K., Mueller, J., Wenning, G. K., Schmidauer, C., and Poewe, W. (2007). Midbrain sonography in patients with Essential Tremor. *Mov. Disord.* **22**, 414–417.

Spiegel, J., Hellwig, D., Möllers, M. O., Behnke, S., Jost, W., Fassbender, K., Samnick, S., Dillmann, U., Becker, G., and Kirsch, C. M. (2006). Transcranial sonography and [123I] FP-CIT SPECT disclose complementary aspects of Parkinson's disease. *Brain* **129**, 1188–1193.

Walter, U., Wittstock, M., Benecke, R., and Dressler, D. (2002). Substantia nigra echogenicity is normal in non-extrapyramidal cerebral disorders but increased in Parkinson's disease. *J. Neural. Transm.* **109**, 191–196.

Walter, U., Krolikowski, K., Tarnacka, B., Benecke, R., Czlonkowska, A., and Dressler, D. (2005). Sonographic detection of basal ganglia lesions in asymptomatic and symptomatic Wilson disease. *Neurology* **64**, 1726–1732.

Walter, U., Dressler, D., Wolters, A., Wittstock, M., and Benecke, R. (2007a). Transcranial brain sonography findings in clinical subgroups of idiopathic Parkinson's disease. *Mov Disord.* **22**, 48–54.

Walter, U., Behnke, S., Eyding, J., Niehaus, L., Postert, T., Seidel, G., and Berg, D. (2007b). Transcranial brain parenchyma sonography in movement disorders: state of the art. *Ultrasound Med Biol.* **33**, 15–25.

Walter, U., Kanowski, M., Kaufmann, J., Grossmann, A., Benecke, R., and Niehaus, L. (2008a). Contemporary ultrasound systems allow high-resolution transcranial imaging of small echogenic deep intracranial structures similarly as MRI: A phantom study. *Neuroimage.* **40**, 551–558.

Walter, U., Dressler, D., Lindemann, C., Slachevsky, A., and Miranda, M. (2008b). Transcranial sonography findings in welding-related Parkinsonism in comparison to Parkinson's disease. *Mov. Disord.* **23**, 141–145.

Walter, U., Wagner, S., Horowski, S., Benecke, R., and Zettl, U. K. (2009). Transcranial brain sonography findings predict disease progression in multiple sclerosis. *Neurology* **73**, 1010–1017.

Walter, U. (2009). Transcranial brain sonography findings in Parkinson's disease: implications for pathogenesis, early diagnosis and therapy. *Expert Rev. Neurother.* **9**, 835–846.

Xu, Q., Kanthasamy, A. G., and Reddy, M. B. (2008). Neuroprotective effect of the natural iron chelator, phytic acid in a cell culture model of Parkinson's disease. *Toxicology*. **245**, 101–108.

Zecca, L., Berg, D., Arzberger, T., Ruprecht, P., Rausch, W. D., Musicco, M., Tampellini, D., Riederer, P., Gerlach, M., and Becker, G. (2005). In vivo detection of iron and neuromelanin by transcranial sonography: a new approach for early detection of substantia nigra damage. *Mov. Disord.* **20**, 1278–1285.

TRANSCRANIAL SONOGRAPHY IN DYSTONIA

Alexandra Gaenslen

Department of Neurodegeneration, Hertie Institute of Clinical Brain Research and German Center of Neurodegenerative Diseases (DZNE), Tübingen, Germany

The cause of idiopathic dystonia is not entirely elucidated. In the pathophysiological model of dystonia, the basal ganglia play a major role, mainly putamen, globus pallidus internus, thalamus, and cortex. However, using conventional structural neuroimaging methods, no specific alterations could be detected in this area.

Using transcranial sonography (TCS) as a noninvasive, easy to perform, and side-effect-free method, it could be shown that in up to 75% of patients with cervical dystonia (CD), in a high percentage of other focal dystonias, but seldom in facial and genetically determined dystonia, hyperechogenicity of the medial part of the lentiform nuclei (LN) can be visualized in the third ventricular scanning plane. Based on these TCS findings an increased copper content of the LN could be verified in dystonia, opening new perspectives on possible pathophysiological aspects and future research. In clinical routine, this method may be used for early and differential diagnosis of primary dystonia.

I. Introduction

Dystonia is defined as a syndrome with twisting involuntary contractions of antagonistic muscles causing either repetitive movements or abnormal posture (Fahn *et al.*, 1998). Additionally, the term "dystonia" is used for dystonic symptoms

INTERNATIONAL REVIEW OF NEUROBIOLOGY, VOL. 90
DOI: 10.1016/S0074-7742(10)90013-5

occurring in other neurologic disorders. Dystonia is one of the common movement disorders in the population aged 50–89 years, with a prevalence of 1.8%, after different forms of tremor, restless legs syndrome, and Parkinsonian syndromes (Wenning *et al.*, 2005). Classification of dystonia is done either by etiology—that is, into primary (without observable cause) or secondary (known brain lesions) or by the affected body part: focal (one region), segmental (two or more neighboring regions), multifocal (two or more not neighboring regions), and generalized (muscles of the whole body involved). The prevalence of segmental and focal idiopathic dystonia fluctuates between 6 and 37 per 100,000 persons in different countries depending on the design of survey and data evaluation (Asgeirsson *et al.*, 2006).

The diagnosis of idiopathic dystonia is still clinical; other diseases, especially those associated with dystonia—that is, dystonia as a secondary form—need to be excluded by neuroimaging, blood and urine analyses (e.g., Wilson's disease (WD)), and lumbar puncture, if there are any suggestions of an inflammatory process. Clinically, it is important to define the involved muscles and regions for classification and evaluation of treatment options.

II. Pathophysiology

The cause of idiopathic dystonia is not entirely elucidated.

So far, 17 forms of monogenetically caused dystonia syndromes have been described, presenting either as dystonia only or dystonia plus syndromes (Mueller, 2009).

From secondary dystonia with known brain lesions, it can be derived that primarily the basal ganglia, especially the LN, seem to be causative for the involuntary muscle contractions (Kostic *et al.*, 1996; Lorenzana *et al.*, 1992; Molho and Factor, 1993; Kim *et al.*, 2007; Takahashi *et al.*, 2009), indicating that this anatomical region may also be pathophysiologically involved in primary dystonia. Moreover, lesions in the brainstem and cerebellum have been found to be associated with secondary focal dystonia (LeDoux and Brady, 2003; Loher and Krauss, 2009). Here, primarily networks involving the basal ganglia seem to be affected. In these regions, as well as in the premotor and motor cortex, hypometabolism could be demonstrated in patients with spasmodic torticollis investigated with [^{18}F]-fluoro-2-deoxy-D-glucose ([^{18}F]-FDG) positron emission tomography (PET) (Karbe *et al.*, 1992).

From the known circuits, a pathophysiological model of dystonia is derived, in which the basal ganglia play a major role. Overactivity of the putamen is proposed to lead to a reduction of inhibitory function of the globus pallidus internus (GPi) and increased thalamic input to the cortex (Berardelli *et al.*, 1998).

III. Neuroimaging in Dystonia

To exclude secondary forms of dystonia, conventional neuroimaging techniques (cCT or MRI) are necessary. In patients with idiopathic dystonia no specific morphologic changes can be visualized by conventional CT or MRI (Naumann *et al.*, 1996). Only with special imaging techniques it has been possible to identify differences between patients with dystonia and healthy controls. Using diffusion tensor imaging (DTI), changes in mean diffusivity and fractional anisotropy in patients with primary cervical dystonia and blepharospasm, mainly in the basal ganglia and motor cortex areas can be visualized (Fabbrini *et al.*, 2008). T_2 times have been shown to be prolonged in the LN in 22 patients with primary dystonic torticollis compared with sex- and age-matched healthy persons without structural lesions in the basal ganglia (Schneider *et al.*, 1994). Voxel-based morphometry (VBM) in patients with primary dystonia compared with healthy controls showed gray matter volume increase in the GPi and prefrontal cortex (Egger *et al.*, 2007).

Using FDG-PET, a reduction of glucose metabolism could be demonstrated in the basal ganglia, thalamus, and frontal cortex in 15 patients with primary dystonia (Karbe *et al.*, 1992). Although these techniques underline the role of the basal ganglia in dystonia, they cannot broadly be used to ascertain the diagnosis of idiopathic dystonia in the clinical routine.

IV. TCS in Primary Dystonia

In 1995, for the first time, the group of Georg Becker performed TCS in patients with dystonia. Knowing that no structural abnormalities are seen on standard CCT and MRI, already the first observations were surprising: In up to 75% of patients with cervical dystonia (CD) hyperechogenicity of the medial part (see Fig. 1) of the LNs could be depicted, scanning the third ventricular scanning plane with standard parameters (see also chapter on methods) (Naumann *et al.*, 1996; Becker *et al.*, 1997). The same feature was frequently seen in patients with upper limb dystonia (ULD) (83%), but rarely in facial dystonia (31%), and very rarely in healthy controls (12%) (Naumann *et al.*, 1996). The areas of hyperechogenicity were larger contralateral to the clinically affected side and the intensity of hyperechogenicity of the LN seemed to correlate with the severity of dystonic symptoms in CD and ULD patients (Naumann *et al.*, 1996).

So far, there are hardly any reports on TCS findings in monogenetic forms of dystonia. In one study, including five patients with Dyt 5 mutations, hyperechogenicity of the LN was only found in one patient (Hagenah *et al.*, 2006).

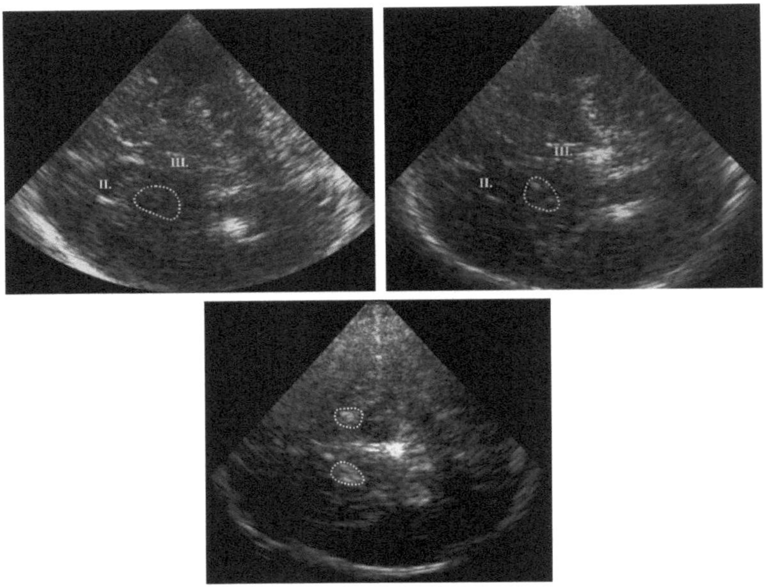

FIG. 1. Echogenicity of the basal ganglia. III.—3. ventricle, II. —frontal horn of second ventricle, encircled area—lentiform nuclei: (A) normal echogenicity, isoechogenic to surrounding brain tissue, (B) hyperechogenicity of the medial and dorsal part of the lentiform nucleus contralateral to the insonation side; (C) prominent hyperechogenicity bilaterally.

Moreover, from our own investigations we made the observation that in patients with dystonia caused by Dyt 1 mutations ($n = 3$ examined so far) normally no hyperechogenicity of the LN is found (own data, not yet published).

Other anatomical regions—such as the SN and the ventricular system-examined in the routine procedure with TCS are usually normal. In some dystonic patients, an interrupted or missing mesencephalic midline structure (raphe) can be seen. This echofeature typically correlates with major depression or depression in neurodegenerative diseases—not with dystonia itself (Naumann *et al.*, 1996; Becker *et al.*, 2001).

V. TCS in Secondary Dystonia

Disorders like Wilson's disease and others may manifest with dystonic movements and especially in the early disease course it may be difficult to differentiate these from primary dystonia, which, however, is essential to initiate the right treatment strategy. Dystonic symptoms may even be the first manifestation of early-onset Parkinsonism (EOP) (Lucking *et al.*, 2000). One pilot study compared five patients with dopa-responsive dystonia (DRD), five *parkin*-homozygous PD patients, and four heterozygous *parkin*-carriers with 10 healthy controls. In this study, only the echogenicity of the SN was evaluated. In all *parkin*-PD patients but only in one DRD patient hyperechogenicity of the SN was found, $P < 0.008$ (Hagenah *et al.,* 2007). Therefore, this study indicates that evaluation of the SN may be important to differentiate Parkinsonian syndromes associated with SN hyperechogenicity from dystonia, in which usually a normal SN is found. In another report, however, the same group stated that only one of these five patients with the monogenetic form of dystonia displayed LN hyperechogenicity (see above, Hagenah *et al.*, 2006) indicating that, in monogenetic forms of dystonia, evaluation of the echogenicity of the basal ganglia does not contribute to the differential diagnosis of monogenetic dystonia versus early-onset PD.

Another study examined 21 patients with WD (18 with neurological symptoms and 3 without) and found all of them to be neurologically symptomatic, and hyperechogenicity in the area of the LN on TCS in 66% of the asymptomatic WD patients. Importantly, the extension of the area of hyperechogenicity correlated with the WD rating scale score (Walter *et al.*, 2005), that is, the larger the area the more severe the symptoms. Of 19 patients additionally investigated with MRI, 63% showed hyper- or hypointense lesions of the LN on T_2-weighted imaging, of these 18 (95%) exhibited LN hyperechogenicity on TCS. Moreover, in 32% of these patients, hyperechogenicity of the thalamus was found on TCS corresponding to hyperintense lesions on MRI in the same regions. In 21% abnormalities were only seen on MRI and in 16% only on TCS. The sensitivity of TCS for morphological changes associated with WD was higher in the LN than in the thalamus, which may be due to the fact that the LN is more often affected by the disease than other brain regions (Mikol *et al.*, 2005).

Also other neurodegenerative disorders may—up to a certain percentage—be associated with LN hyperechogenicity. One of these is Huntington's disease (HD). Here, the most striking finding is an enlargement of the anterior horns (due to atrophy of the caudate nucleus) and hyperechogenicity of the caudate nucleus (see also chapter on Huntington's disease). However, also hyperechogenicity of the LN was found in 6% of 45 patients in the first study published on TCS in HD patients (Postert *et al.*, 1999).

Considering the possible etiologies of LN hyperechogenicity, it needs to be noted that accumulation of different forms of trace metals—that is, iron accumulation in different forms of neurodegeneration with brain iron accumulation (NBIA)—a substantial increase in calcium content like in Fahr's disease, and also slight calcification of the basal ganglia generally lead to an increase in echogenicity of the affected tissue (Berg *et al.*, 2008). Importantly, small calcifications can be more striking on TCS than on CCT images.

VI. TCS in Patients with Herniated Cervical Disk

In one of the early studies, patients with herniated cervical disks were used as control persons for dystonia (Becker *et al.*, 1997). In this study, it was observed that in a strikingly high percentage of patients with nuchal pain due to herniated cervical disks, especially those with recurrent events, hyperechogenicity of the LN was found. This observation led to a study comparing subjects with herniated cervical disks and healthy persons (Becker *et al.*, 2002). In this study, 13 patients with herniated cervical disks and additionally severe chronical nuchal pain were included. All of them were found to have hyperechogenicity of the LN (Becker *et al.*, 2002). As the reason for an atraumatic cervical disk herniation is still not clear and all patients included suffered from severe chronic nuchal pain, caused by hardening of muscles, it was suggested that at least in subjects affected severely a dystonic genesis (constant contraction of muscles) may be possible. In a more recent study involving 26 patients with herniated cervical disks, a hyperechogenic LN was found in 70% (data in preparation).

VII. TCS in Dystonia and Deep Brain Stimulation

In dystonia, typically, hyperechogenicity of the basal ganglia is detected in the medial part of LN, that is, the globus pallidus internus. This nucleus is also the target of deep brain stimulation in dystonia patients, untreatable by oral medication or injections with botulinum toxin (Gruber *et al.*, 2009; Jeong *et al.*, 2009; Oropilla *et al.*, 2009). So far, intraoperative monitoring of the placement of electrodes can hardly be done by neuroimaging. Here, a new and promising application field is opening for TCS. In a first study, Walter *et al.* (2009) described the monitoring of placement of electrodes for deep brain stimulation in two dystonic patients. Pre- and postoperatively patients were examined with MRI and intraoperatively with TCS. The placement of electrodes within the GPi was assessed correctly with both the methods and the clinical outcome was excellent in both patients (see also chapter on future applications). Moreover, it became obvious that the additional use of intraoperative color-code sonography of parenchymal vessels could be helpful to detect adjacent vessels and thereby prevent hemorrhages (Walter *et al.*, 2009).

VIII. Possible Causes of Hyperechogenicity of the LN in Dystonia

In a first postmortem study on brain tissue of three patients with adult-onset idiopathic dystonia and 10 healthy controls, trace metal content and calcium were investigated. In the GP and putamen, increased levels of copper and manganese were found in the dystonic patients in comparison to controls, whereas there was no difference in the levels of iron, zinc, or calcium (Becker *et al.*, 1999). Further analyses of copper-metabolizing proteins revealed that in the same brain tissue the level of the copper-metabolizing Menkes protein was reduced (Berg *et al.*, 2000). This led to the hypothesis that a deficit of Menkes protein could lead to increased levels of copper in patients with primary dystonia (Berg *et al.*, 2000). In a study of 14 patients with CD, decreased levels of Menkes mRNA were detected. Besides, copper concentrations in the lymphocytes in the peripheral blood were reduced compared to patients with blepharospasm and healthy controls (Kruse *et al.*, 2001). The fact that patients with blepharospasm were found to have similar levels of copper and Menkes mRNA as healthy controls reflects the generally normal TCS findings. Only in a minority (31%) of patients with blepharospasm hyperechogenicity of the LN can be detected, while in CD 75% of the affected subjects show this echofeature (Naumann *et al.*, 1996). Different underlying pathomechanisms may therefore be discussed with regard to these different forms of dystonic syndromes.

From the imaging and postmortem observations, interesting aspects can therefore be derived: Using TCS, it has become possible for the first time to draw a pathophysiological parallel between WD and primary dystonia—besides both presenting with dystonic movements, both show similar abnormalities on TCS. WD is a genetically determined disorder, caused by different mutations in the ATP7B gene. The effect is a defect in copper metabolism. Copper cannot be transported out of the liver into the gall and then be excreted. Rather it accumulates, mainly in the liver and the brain. This copper accumulation in brain tissue can be detected with MRI, showing increased signal intensity on T1-weighted images mainly in the putamen, GP, and thalamus (Kim *et al.*, 2006; van Wassenaer-van Hall *et al.*, 1996). Making use of copper-chelating therapy, signal alterations can be decreased, in parallel with the amelioration of clinical symptoms, especially in pediatric WD patients (Kim *et al.*, 2006).

In summary, there is evidence for an involvement of the basal ganglia in the pathogenesis of primary dystonia. Hyperechogenicity detected by TCS may at least, in part, be caused by an increased content of copper in this brain region. Still, further investigations are needed to prove this hypothesis.

IX. Conclusions

TCS is a cheap and easy-to-handle tool for clinical praxis. Importantly, it can also be used in moving patients without deteriorating the quality of images, which is of utmost importance in dystonic patients. It can be used to

(i) strengthen the clinical diagnosis of primary dystonia
(ii) get an impression of an affection of the basal ganglia by other trace metal or calcium-accumulating diseases
(iii) intraoperative control of the placement of electrodes for deep brain stimulation.

Moreover, since first new pathophysiological aspects could be derived from TCS observations and consecutive examination of brain tissue, TCS could lay the ground for future research activities in this area to get a better understanding of the pathophysiology of primary dystonia.

References

Asgeirsson, H., Jakobsson, F., Hjaltason, H., Jonsdottir, H., Sveinbjornsdottir, S. (2006). Prevalence study of primary dystonia in Iceland. *Mov. Disord.* **21**, 293–298.

Becker, G., Berg, D., Kruse, N., Schroder, U., Warmuth-Metz, M., Rieckmann, P., Naumann, M., Reiners, K. (2002). Evidence for shoulder girdle dystonia in selected patients with cervical disc prolapse. *Mov. Disord.* **17**, 710–716.

Becker, G., Berg, D., Lesch, K. P., Becker, T. (2001). Basal limbic system alteration in major depression: A hypothesis supported by transcranial sonography and MRI findings. *Int. J. Neuropsychopharmacol.* **4**, 21–31.

Becker, G., Berg, D., Rausch, W. D., Lange, H. K., Riederer, P., Reiners, K. (1999). Increased tissue copper and manganese content in the lentiform nucleus in primary adult-onset dystonia. *Ann. Neurol.* **46**, 260–263.

Berardelli, A., Rothwell, J. C., Hallett, M., Thompson, P. D., Manfredi, M., Marsden, C. D. (1998). The pathophysiology of primary dystonia. *Brain* **121** (Part 7), 1195–1212.

Berg, D., Godau, J., Walter, U. (2008). Transcranial sonography in movement disorders. *Lancet Neurol.* **7**, 1044–1055.

Berg, D., Weishaupt, A., Francis, M. J., Miura, N., Yang, X. L., Goodyer, I. D., Naumann, M., Koltzenburg, M., Reiners, K., Becker, G. (2000). Changes of copper-transporting proteins and ceruloplasmin in the lentiform nuclei in primary adult-onset dystonia. *Ann. Neurol.* **47**, 827–830.

Egger, K., Mueller, J., Schocke, M., Brenneis, C., Rinnerthaler, M., Seppi, K., Trieb, T., Wenning, G. K., Hallett, M., Poewe, W. (2007). Voxel based morphometry reveals specific gray matter changes in primary dystonia. *Mov. Disord.* **22**, 1538–1542.

Fabbrini, G., Pantano, P., Totaro, P., Calistri, V., Colosimo, C., Carmellini, M., Defazio, G., Berardelli, A. (2008). Diffusion tensor imaging in patients with primary cervical dystonia and in patients with blepharospasm. *Eur. J. Neurol.* **15**, 185–189.

Fahn, S., Bressman, S. B., Marsden, C. D. (1998). Classification of dystonia. *Adv. Neurol.* **78**, 1–10.

Gruber, D., Trottenberg, T., Kivi, A., Schoenecker, T., Kopp, U. A., Hoffmann, K. T., Schneider, G. H., Kuhn, A. A., Kupsch, A. (2009). Long-term effects of pallidal deep brain stimulation in tardive dystonia. *Neurology* **73**, 53–58.

Hagenah, J. M., Hedrich, K., Becker, B., Pramstaller, P. P., Seidel, G., Klein, C. (2006). Distinguishing early-onset PD from dopa-responsive dystonia with transcranial sonography. *Neurology* **66**, 1951–1952.

Jeong, S. G., Lee, M. K., Kang, J. Y., Jun, S. M., Lee, W. H., Ghang, C. G. (2009). Pallidal deep brain stimulation in primary cervical dystonia with phasic type: Clinical outcome and postoperative course. *J. Korean Neurosurg. Soc.* **46**, 346–350.

Karbe, H., Holthoff, V. A., Rudolf, J., Herholz, K., Heiss, W. D. (1992). Positron emission tomography demonstrates frontal cortex and basal ganglia hypometabolism in dystonia. *Neurology* **42**, 1540–1544.

Kim, T. J., Kim, I. O., Kim, W. S., Cheon, J. E., Moon, S. G., Kwon, J. W., Seo, J. K., Yeon, K. M. (2006). MR imaging of the brain in Wilson's disease of childhood: Findings before and after treatment with clinical correlation. *AJNR Am. J. Neuroradiol.* **27**, 1373–1378.

Kruse, N., Berg, D., Francis, M. J., Naumann, M., Rausch, W. D., Reiners, K., Rieckmann, P., Weishaupt, A., Becker, G. (2001). Reduction of Menkes mRNA and copper in leukocytes of patients with primary adult-onset dystonia. *Ann. Neurol.* **49**, 405–408.

LeDoux, M. S., Brady, K. A. (2003). Secondary cervical dystonia associated with structural lesions of the central nervous system. *Mov. Disord.* **18**, 60–69.

Loher, T. J., Krauss, J. K. (2009). Dystonia associated with pontomesencephalic lesions. *Mov. Disord.* **24**, 157–167.

Mikol, J., Vital, C., Wassef, M., Chappuis, P., Poupon, J., Lecharpentier, M., Woimant, F. (2005). Extensive cortico-subcortical lesions in Wilson's disease: Clinico-pathological study of two cases. *Acta Neuropathol.* **110**, 451–458.

Mueller, U. (2009). The monogenetic primary dystonias. *Brain* **132**, 2005–2025.

Naumann, M., Becker, G., Toyka, K. V., Supprian, T., Reiners, K. (1996). Lenticular nucleus lesion in idiopathic dystonia detected by transcranial sonography. *Neurology* **47**, 1284–1290.

Oropilla, J. Q., Diesta, C. C., Itthimathin, P., Suchowersky, O., Kiss, Z. H. (2009). Both thalamic and pallidal deep brain stimulation for myoclonic dystonia. *J. Neurosurg.*

Postert, T., Lack, B., Kuhn, W., Jergas, M., Andrich, J., Braun, B., Przuntek, H., Sprengelmeyer, R., Agelink, M., Buttner, T. (1999). Basal ganglia alterations and brain atrophy in Huntington's disease depicted by transcranial real time sonography. *J. Neurol. Neurosurg. Psychiatry* **67**, 457–462.

Schneider, S., Feifel, E., Ott, D., Schumacher, M., Lucking, C. H., Deuschl, G. (1994). Prolonged MRI T2 times of the lentiform nucleus in idiopathic spasmodic torticollis. *Neurology* **44**, 846–850.

Walter, U., Krolikowski, K., Tarnacka, B., Benecke, R., Czlonkowska, A., Dressler, D. (2005). Sonographic detection of basal ganglia lesions in asymptomatic and symptomatic Wilson disease. *Neurology* **64**, 1726–1732.

Walter, U., Wolters, A., Wittstock, M., Benecke, R., Schroeder, H. W., Muller, J. U. (2009). Deep brain stimulation in dystonia: Sonographic monitoring of electrode placement into the globus pallidus internus. *Mov. Disord.* **24**, 1538–1541.

TRANSCRANIAL SONOGRAPHY IN ESSENTIAL TREMOR

Heike Stockner* and Isabel Wurster†

*Medical University Innsbruck, Anichstrasse, Innsbruck
†Department of Neurodegeneration, Hertie Institute of Clinical Brain Research and German Center of Neurodegenerative Diseases (DZNE), Tübingen, Germany

Essential tremor (ET) is the most common adult movement disorder. Especially in early stages the clinical differentiation between essential and Parkinsonian tremor may be a diagnostic challenge. Recent studies have consistently found hyperechogenic alterations in the area of the substantia nigra (SN) in patients with Parkinson's disease (PD) using transcranial sonography (TCS). The present chapter summarizes five studies, which have been performed to investigate SN echogenicity in patients with ET compared with PD patients and healthy control subjects and also includes our own experience.

All of the studies published so far have shown that hyperechogenicity of the SN is a typical finding in about 90% of patients with PD, but not in patients with ET. In ET patients, the prevalence of hyperechogenicity is in the range of healthy control subjects or slightly above, which may indicate an increased risk for PD in the subgroup of ET patients with hyperechogenicity, consistent with the increased risk for PD that has been established in ET patient cohorts.

As TCS is a noninvasive and inexpensive method, it seems to be a valuable instrument in the differentiation between ET and PD. Future follow-up studies on ET patients with SN hyperechogenicity will reveal whether those with increased echosignal are indeed the ones who may develop PD later in life.

INTERNATIONAL REVIEW OF
NEUROBIOLOGY, VOL. 90
DOI: 10.1016/S0074-7742(10)90014-7

I. Introduction—Essential Tremor as a Diagnostic Challenge

Essential tremor (ET) is the most common adult movement disorder with a prevalence ranging from 2.8 to 4% in individuals aged older than 40 years (Dogu *et al.*, 2003), and a prevalence of 14% in people older than 65 years (Wenning *et al.*, 2005). According to the consensus statement of the Movement Disorder Society, definite ET is defined as abnormal action tremor of both upper limbs or isolated head tremor with no abnormal posturing, dystonia, or other neurologic signs (Deuschl *et al.*, 1998).

Up to now, a disease-specific diagnostic marker for ET is lacking and no specific pathological changes have been found in the brains of ET patients. Misdiagnosis of ET is very common and is estimated as an overdiagnosed condition (Schrag *et al.*, 2000). Despite different phenomenology, the differentiation of Parkinsonian and ET may be a diagnostic challenge. One study shows that about one in three patients with tremor was misdiagnosed as having ET, with the most frequent false diagnoses being Parkinson's disease (PD) and dystonia (Jain *et al.*, 2006). The extent to which other mild neurologic signs may occur in ET has also been a topic of debate (Jankovic, 2002). Cogwheeling in the absence of rigidity is common and unsteady gait is seen in some advanced patients. Recent studies suggest that some patients with long-standing ET later develop rest tremor and other Parkinsonian signs (Jankovic, 2002). There may be also some doubts in differentiating between "true" rest tremor and insufficient relaxation in patients with moderate postural tremor (Montgomery *et al.*, 2000; Schrag *et al.*, 2000). Additionally, postural tremor can be the initially presenting symptom in patients with PD (Lee *et al.*, 1999) and other movement disorders as dystonia and spinocerebellar ataxias. In some PD patients, tremor remains the predominant feature in the course of the disease and hypokinetic symptoms can be still mild and do not seem to progress. Recently, a study in 16 PD patients with tremor as the predominant feature showed only mild progression, calling this condition "benign tremulous Parkinsonism" or "isolated tremor with very mild Parkinsonian signs."(Josephs *et al.*, 2006)

Additionally, the incidence of both ET and PD increases with age and coexistence of these frequent movement disorders in a single patient may occur by chance, but several studies report an association of ET and PD (Geraghty, 1985; Shahed and Jankovic, 2007). Some studies have found ET patients to harbor a higher risk for PD than found in the general population, and studies in ET families have described an increased risk for PD also in first-degree relatives of patients with ET (Payami *et al.*, 1994; Jankovic *et al.* 1995; Louis *et al.*, 2003). A twin study showed that 70% of twins with postural tremor had PD or a twin with PD (Tanner *et al.*, 2001). In a study of 678 ET patients, 6.1% showed features of PD, indicating that the frequency of PD in ET is higher than expected and known in the general population (Koller *et al.*, 1994). A further link between a subset of ET patients and PD is supported by pathological findings in brains of patients with ET. Six of ten ET patients studied at the Essential Tremor Centralized Brain Repository exhibited a distinctive pattern of Lewy bodies in the locus ceruleus. This observation further supports that a proportion of ET patients could have a form of Lewy body disease (Louis *et al.*, 2005).

The use of DAT SPECT has been shown to facilitate the differential diagnosis in patients with isolated tremor symptoms (Scherfler, 2007). Loss of striatal dopamine nerve terminal function is a hallmark of neurodegenerative Parkinsonism and is strongly related to the decrease of dopamine transporter (DAT) density, which can be measured by single photon emission computed tomography (SPECT). Cases with possible ET without other neurological features consistently have normal DAT imaging (Asenbaum, 1998). Some SPECT studies are thought to be 90% sensitive in differentiating PD from normal subjects or ET. Brooks *et al.* (1992) showed that patients with isolated rest tremor had consistently abnormal striatal 18F-dopa uptake, whereas those with isolated postural tremor did not. Lee *et al.* performed a brain SPECT study in patients with long duration of isolated postural tremor, in whom rest tremor developed after the onset of postural tremor (Lee *et al.*, 1999). Many of these patients demonstrated reduced levels of striatal dopamine transporters comparable with the levels seen in typical PD. The authors concluded that later in their clinical courses, some patients with postural tremor may acquire rest tremor in association with mild substantia nigra neuronal loss. Chaudhuri *et al.* showed that longstanding isolated asymmetrical postural tremor may evolve to PD, supported in 5 of 13 reported cases (38%) by abnormal [123I]ß-CIT SPECT whereas uncertain Parkinsonian signs and normal [123I]FP-CIT led to a change of diagnosis to ET in 5 of 24 cases (Chaudhuri *et al.*, 2005).

II. Transcranial Sonography in the Differential Diagnosis of ET and PD

Transcranial sonography (TCS) has consistently shown hyperechogenic alterations in the substantia nigra in patients with PD (Becker *et al.* 1995; Berg, *et al.*, 2001). This hyperechogenicity has been related to increased tissue concentrations of iron (Zecca *et al.*, 2005). SN hyperchogenicity, which may be detected in about 10% of healthy controls, has been proposed to constitute a potential risk marker for PD in these subjects (Berg *et al.*, 1995), as indicated by several lines of evidence (see also Chapter on early and premotor diagnosis). For example, one study in healthy elderly subjects found reduced 18F-Dopa uptake measured by PET in a subgroup of healthy subjects with hyperechogenic SN in TCS compared with subjects with normal midbrain echogenicity (Berg *et al.*, 2001). Reports of midbrain hyperechogenicity in asymptomatic carriers of the *Parkin* gene mutation also suggest that midbrain echogenicity could be related to PD risk (Walter *et al.* 2004).

Up to now, five TCS studies have been performed to investigate SN echogenicity in patients with ET than in PD patients and healthy control subjects.

(1) In a study published by Niehaus *et al.* (2004) 10% of 20 patients with essential tremor showed SN hyperechogenicity in comparison to 93% of 30 tremor-dominant PD patients. There was no significant difference between SN echogenicity in patients with essential tremor and controls. The authors concluded that TCS enables a good prediction for the diagnosis of PD.

(2) The aim of the study of Budisic *et al.* (2009) was to determine the usefulness of TCS in the differential diagnosis of PD and ET. TCS was performed in 80 patients with PD, 30 patients with ET, and 80 control subjects by an investigator blinded to the clinical diagnosis. SN hyperechogenicity was found in 91% of PD patients, 10% of healthy subjects, and in 13% patients with ET. The authors concluded that SN hyperechogenicity is a highly specific finding of PD and TCS may differentiate between PD and ET patients.

(3) Doepp *et al.* (2008) investigated the accuracy of TCS for differentiation between idiopathic Parkinson's disease (PD) and essential tremor (ET) in comparison to (123)I-FP-CIT SPECT. They performed TCS and FP-CIT SPECT investigations in 71 patients (46 PD patients and 25 ET). Of the 42 patients with abnormal FP-CIT SPECT findings, 36 had an increased echogenicity of the SN whereas 27 of 29 patients with normal SPECT findings showed normal echogenicity of the SN. Referring to FP-CIT SPECT, the sensitivity of TCS for the detection of PD was 86% and the specificity 93%; the positive and negative predictive values were 95 and 82%, respectively. In accordance with a previous SPECT study in patients with PD, the extent of SN hyperechogenicity did not correlate with the degree of reduction in dopamine transporter binding. The authors concluded that TCS allows a reliable differentiation of PD and ET.

(4) The study of Okawa *et al.* (2007) was performed in Japan and included 98 patients with PD, 17 with progressive supranuclear palsy (PSP), 14 with multiple system atrophy (MSA), and 19 with ET, and 30 age-matched controls. In this study, SN hyperechogenicity was assessed semiquantitatively, using a four-point scale to define SN echogenicity (I = none or faint, II = equivocal, III = definite, IV = marked). Definite or marked SN hyperechogenicity was observed in 83% of the accessible PD patients (52 of 63), while in the ET group none of the included subjects and

in the MSA, PSP, and control groups only one of each group, respectively, displayed hyperechogenicity at the SN. The authors concluded that TCS provides a useful tool for the differential diagnosis between PD and ET as well as between PD and atypical parkinsonism.

In this study, the rate of recording failures due to an insufficient acoustic bone window was higher compared with other studies, in which the rate was reported to be about 10%. The SN could not be assessed by TCS in 59% of women older than 60 years. This may limit the potential role of TCS in Japanese patients.

(5) In our study (Stockner *et al.*, 2007) the prevalence of SN hyperechogenicity in normal controls was 3% and about 75% in patients with PD. Midbrain hyperechogenicity was seen in 16% of patients with ET, which was significantly more frequent than in healthy controls. Similar to other series (Cohen *et al.*, 2003), about 14% of the ET patients of this study had additional rest tremor. Although rest tremor in ET has been discussed to constitute a possible sign of comorbid tremor-dominant PD (Cohen *et al.*, 2003), hyperechogenicity in ET patients of this study was independent of the presence of concomitant rest tremor. Compared with previous studies (Berg *et al.*, 1999, 2001), we used a more conservative method to calculate the cut-off value defining hyperechogenicity (2 standard deviations from the mean of echogenic area of 100 normal individuals in the log-transformated data set corresponded to $0.24\,cm^2$), resulting in a lower overall frequency of hyperechogenicity in control subjects and in PD patients in this study. A post hoc analysis revealed that implementation of the 90th percentile of the normal population to define hyperechogenicity (i.e., $0.20\,cm^2$) in this data set led to findings consistent with previous studies with a prevalence of hyperechogenicity in the area of the SN in healthy controls of 9% and about 90% in patients with PD.

III. Conclusion

Especially in early stages the clinical differentiation between essential and Parkinsonian tremor may be a diagnostic challenge. Although the diagnostic accuracy of presynaptic PET and SPECT imaging in the differential diagnosis of essential tremor and tremor-dominant PD is more than 90% (Asenbaum *et al.*, 1998; Benamer *et al.*, 2000) the application of these investigation tools is still limited.

TCS is a noninvasive and inexpensive method that enables an easy and wide application. TCS studies have shown that hyperechogenicity of the SN is a typical finding in about 90% of patients with PD, but not in patients with ET. In ET patients the prevalence of hyperechogenicity is in the range of healthy control subjects or slightly above, which may indicate that there is an increased risk for PD in the subgroup of ET patients with SN hyperechogenicity. Future follow-up studies on ET patients with SN hyperechogenicity will reveal whether those with the increased echo signal are indeed the ones who may develop PD later in life.

A recent prospective study in a cohort of patients with different types of Parkinsonism has suggested a sensitivity of 90.7% and a specificity of 82.4% of midbrain hyperechogenicity on TCS for a diagnosis of idiopathic PD (Gaenslen *et al.*, 2008).

From the studies published so far, it can therefore be concluded that the finding of normal echogenicity in subjects with an unclear tremor syndrome including postural/action and a slight resting tremor in general allows the diagnosis of ET (Fig. 1). In patients with mixed tremor and SN hyperechogenicity, one may be prone to diagnose PD. Here additional clinical features are

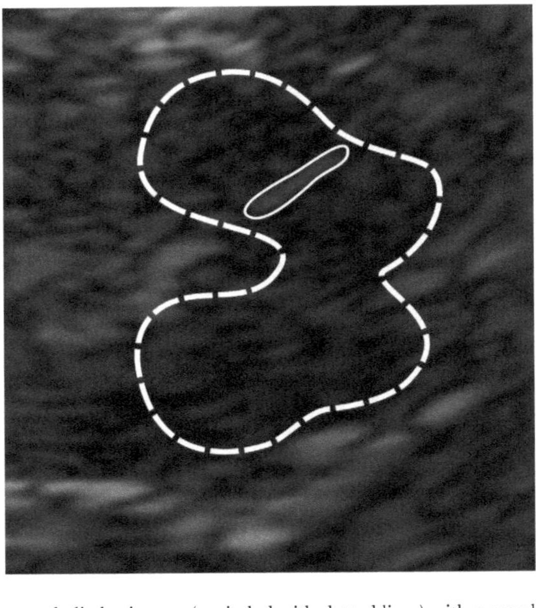

FIG. 1. (A) Mesencephalic brainstem (encircled with dotted lines) with normal echogenicity of the SN (encircled) of a patient with essential tremor (clinically predominant postural and action tremor and also slight resting tremor).

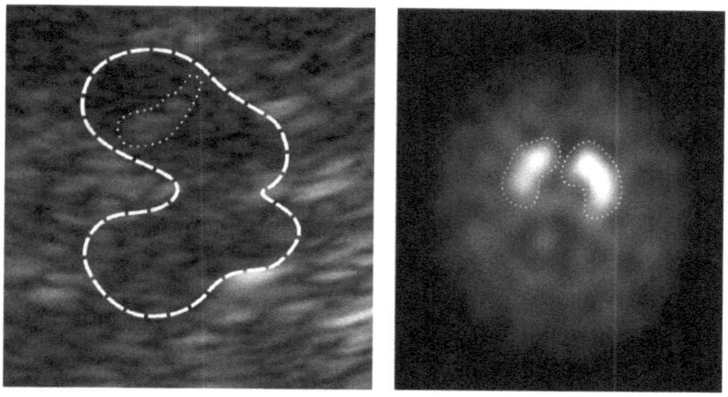

FIG. 2. (A) Mesencephalic brainstem (encircled with dotted lines) with hyperechogenicity of the SN (encircled) of a patient with a history of essential tremor and additional resting tremor since 2 months. (B) DATScan of the same patient. Slight reduction of radiotracer uptake contralateral to the side of the beginning resting tremor. Although the uptake was still in the normal range, the side difference could imply the beginning of PD in this patient, which is compatible with the ultrasound finding.

necessary to make the diagnosis, or functional neuroimaging of the presynaptic dopaminergic system (Figs. 2 and 3). Still, even if no further hints can be derived from these investigations, subjects with mixed tremor and SN hyperechogenicity may be more likely to develop PD in the future. This hypothesis needs to be proven in longitudinal investigations. So far, a regular follow-up of these patients is recommended.

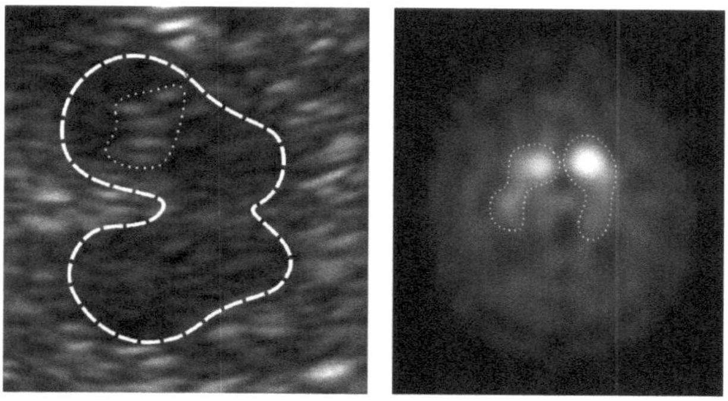

FIG. 3. (A) Mesencephalic brainstem (encircled with dotted lines) with hyperechogenicity of the SN (encircled) of a patient with a history of essential tremor, additional resting and slight bradykinesia. (B) DATScan of the same patient. Reduced tracer uptake on both sides, with less uptake, especially in the putamen, contralateral to the clinically more affected side.Both ultrasound and DATScan are compatible with the diagnosis of PD.

References

Asenbaum, S., Pirker, W., Angelberger, P., Bencsits, G., Pruckmayer, M., and Brucke, T. (1998). [123I]beta-CIT and SPECT in essential tremor and Parkinson's disease. *J. Neural. Transm.* **105**, 1213–1228.

Becker, G., Seufert, J., Bogdahn, U., et al. (1995). Degeneration of substantia nigra in chronic Parkinson's disease visualized by transcranial color-coded real time sonography. *Neurology* **45**, 182–184.

Benamer, T. S., Patterson, J., Grosset, D. G., et al. (2000). Accurate differentiation of parkinsonism and essential tremor using visual assessment of [123I]-FP-CIT SPECT imaging: The [123I]-FP-CIT study group. *Mov. Disord.* **15**, 503–510.

Berg, D., Becker, G., Zeiler, B., et al. (1999). Vulnerability of the nigrostriatal system as detected by transcranial ultrasound. *Neurology* **53**, 1026–1031.

Berg, D., Siefker, C., and Becker, G. (2001). Echogenicity of the substantia nigra in Parkinson's disease and its relation to clinical findings. *J. Neurol.* **248**, 684–689.

Berg, D., Siefker, C., Ruprecht-Dorfler, P., and Becker, G. (2001). Relationship of substantia nigra echogenicity and motor function in elderly subjects. *Neurology* **56**, 13–17.

Brooks, D. J., Playford, E. D., Ibanez, V., Sawle, G. V., Thompson, P. D., Findley, L. J., et al. (1992). Isolated tremor and disruption of the nigrostriatal dopaminergic system: An 18F-dopa PET study. *Neurology* **42**, 1554–1560.

Budisic, M., Trkanjec, Z., Bosnjak, J., Lovrencic-Huzjan, A., Vukovic, V., and Demarin, V. (2009). Distinguishing Parkinson's disease and essential tremor with transcranial sonography. *Acta Neurol. Scand.* **119** (1), 17–21.

Chaudhuri, K. R., Buxton-Thomas, M. Dhawan, V. Peng, R. Meilak, C. and Brooks, D. J. (2005). Long duration asymmetrical postural tremor is likely to predict development of Parkinson's disease and not essential tremor: Clinical follow up study of 13 cases. *J. Neurol. Neurosurg. Psychiatry* **76**, 115–117.

Cohen, O., Pullman, S., Jurewicz, E., Watner, D., Louis, E. D. (2003). Rest tremor in patients with essential tremor: Prevalence, clinical correlates, and electrophysiologic characteristics. *Arch. Neurol.* **60**, 405–410.

Deuschl, G., Bain, P., and Brin, M. (1998). Consensus statement of the Movement Disorder Society on Tremor. Ad Hoc Scientific Committee. *Mov. Disord.* **13** (Suppl. 3), 2–23.

Doepp, F., Plotkin, M., Siegel, L., Kivi, A., Gruber, D., Lobsien, E., Kupsch, A., and Schreiber, S. J. (2008). Brain parenchyma sonography and 123I-FP-CIT SPECT in Parkinson's disease and essential tremor. *Mov. Disord.* **23** (3), 405–410.

Dogu, O., Sevim, S., Camdeviren, H., et al. (2003). Prevalence of essential tremor: Door-to-door neurologic exams in Mersin Province, Turkey. *Neurology* **61**, 1804–1806.

Gaenslen, A., Unmuth, B., Godau, J., et al. (2008). The specificity and sensitivity of transcranial ultrasound in the diff erential diagnosis of Parkinson's disease: A prospective blinded study. *Lancet Neurol.* **7** (5), 417–424.

Geraghty, J. J., Jankovic, J., and Zetusky, W. J. (1985). Association between essential tremor and Parkinson's disease. *Ann. Neurol.* **17**, 329–333.

Jain, S., Lo, S. E., and Louis, E. D. (2006). Common misdiagnosis of a common neurological disorder: How are we misdiagnosing essential tremor? *Arch. Neurol.* **63** (8), 1100–1104.

Jankovic, J. (2002). Essential tremor: A heterogeneous disorder. *Mov. Disord.* **17** (4), 638–644.

Jankovic, J., Beach, J., Schwartz, K., and Contant, C. (1995). Tremor and longevity in relatives of patients with Parkinson's disease, essential tremor, and control subjects. *Neurology* **45**, 645–648.

Josephs, K. A., Matsumoto, J. Y., and Ahlskog, J. E. (2006). Benign tremulous parkinsonism. *Arch. Neurol.* **63**, 354–357.

Koller, W. C., Busenbark, K., and Miner, K. (1994). The relationship of essential tremor to other movement disorders: Report on 678 patients. Essential Tremor Study Group. *Ann. Neurol.* **35**, 717–3

Lee, M. S., Kim, Y. D., Im, J. H., Kim, H. J., Rinne, J. O., and Bhatia, K. P. (1999). 123I-IPT brain SPECT study in essential tremor and Parkinson's disease. *Neurology* **52**, 1422–1426.

Louis, E. D., Levy, G., Mejia-Santana, H., Cote, L., Andrews, H., Harris, J., et al. (2003). Risk of action tremor in relatives of tremor-dominant and postural instability gait disorder PD. *Neurology* **61**, 931–936.

Louis, E. D., Vonsattel, J. P., Hong, L. S., Ross, G. W., Lyons, K. E., and Pahwa, R. (2005). Essential tremor pathology: A case-control study from the essential tremor centralized brain repository. *Mov. Disord.* **20**, 1241.

Montgomery, Jr E. B., Baker, K. B., Lyons, K., and Koller, W. C. (2000). Motor initiation and execution in essential tremor and Parkinson's disease. *Mov. Disord.* **15**, 511–515.

Niehaus, L., Savyer, N., Weber, U., Hertel, R., Trottenberg, T., and Kupsch, A. (2004). Brain parenchyma sonography in patients with essential tremor and Parkinson's disease. *Cerebrovasc. Dis.* **17** (Suppl. 4).

Okawa, M., Miwa, H., Kajimoto, Y., Hama, K., Morita, S., Nakanishi, I., Kondo, T. (2007). Transcranial sonography of the substantia nigra in Japanese patients with Parkinson's disease or atypical parkinsonism: Clinical potential and limitations. *Intern. Med.* **46** (18), 1527–1531.

Payami, H., Larsen, K., Bernard, S., and Nutt, J. (1994). Increased risk of Parkinson's disease in parents and siblings of patients. *Ann. Neurol.* **36**, 659–661.

Scherfler, C., Schwarz, J., Antonini, A., Grosset, D., Valldeoriola, F., Marek, K., Oertel, W., Tolosa, E., Lees, A. J., and Poewe, W. Role of DAT-SPECT in the diagnostic work up of Parkinsonism. *Mov. Dis.* Published Online: 7 May 2007.

Schrag, A., Mnchau, A., Bhatia, K. P., Quinn, N. P., and Marsden, C. D. (2000). Essential tremor: An overdiagnosed condition? *J. Neurol.* **247** (12), 955–959.

Shahed, J., and Jankovic, J. (2007). Exploring the relationship between essential tremor and Parkinson's disease. *Parkinsonism Relat. Disord.* **13** (2), 67–76.

Stockner, H., Sojer, M., Seppi, K., Mueller, J., Wenning, G. K., Schmidauer, C., and Poewe, W. (2007). Midbrain sonography in patients with Essential Tremor. *Mov. Disord.* **22**, 414–417.

Tanner, C. M., Goldman, S. M., Lyons, K. E., Aston, D. A., Tetrud, J. W., Welsh, M. D., et al. (2001). Essential tremor in twins: An assessment of genetic vs. environmental determinants of etiology. *Neurology* **57**, 1389–1391.

Walter, U., Klein, C., Hilker, R., Benecke, R., Pramstaller, P. P., and Dressler, D. (2004). Brain parenchyma sonography detects preclinical parkinsonism. *Mov. Disord.* **19**, 1445–1449.

Wenning, G. K., Kiechl, S., Seppi, K., et al. (2005). Prevalence of movement disorders in men and women aged 50–89 years (Bruneck Study cohort): A population-based study. *Lancet Neurol.* **4**, 815–820.

Zecca, L., Berg, D., Arzberger, T., Ruprecht, P., Rausch, W. D., Musicco, M., Tampellini, D., Riederer, P., Gerlach, M., and Becker, G. (2005). In vivo detection of iron and neuromelanin by transcranial sonography: A new approach for early detection of substantia nigra damage. *Mov. Disord.* **20** (10), 1278–1285.

TRANSCRANIAL SONOGRAPHY IN RESTLESS LEGS SYNDROME

Jana Godau* and Martin Sojer†

*Department of Neurodegeneration, Hertie Institute of Clinical Brain Research and German Center of Neurodegenerative Diseases (DZNE), Tübingen, Germany
†University of Innsbruck, Innsbruck, Austria

Restless legs syndrome (RLS) is one of the most common neurological disorders. The diagnosis is mainly based on the patient's report of the characteristic symptoms and further on the response to dopaminergic treatment. In the past years, transcranial B-Mode sonography (TCS) of the mesencephalic brainstem has been introduced as a promising new method to assist the diagnostic approach to RLS. This review of literature describes the typical TCS findings in RLS including hypoechogenicity of the substantia nigra and raphe as well as hyperechogenicity of the red nucleus. The diagnostic parameters for patients with idiopathic and various forms of symptomatic RLS will be discussed as well as the current knowledge about the pathophysiology underlying these sonographical abnormalities.

I. Restless Legs Syndrome

Restless legs syndrome (RLS) with a prevalence of 5–10% is one of the most common neurological disorders. It is clinically characterized by an

INTERNATIONAL REVIEW OF NEUROBIOLOGY, VOL. 90
DOI: 10.1016/S0074-7742(10)90015-9

irresistible urge to move the legs and sometimes other body parts, which is accompanied by a sensory discomfort, primarily occurs at rest and can be relieved by moving around. Typically, these symptoms follow a circadian rhythm with most pronounced symptoms in the evening or at night, subsequently leading to a profound disturbance of sleep architecture resulting in daytime sleepiness. About 80% of the patients also suffer from involuntary periodic limb movements (PLMs) (Walters, 1995). In addition to the typical RLS symptoms, many patients experience neuropsychiatric disturbances such as anxiety disorders or depression more often than the normal population (Winkelmann *et al.*, 2005).

RLS symptoms can be efficiently reduced using dopaminergic drugs such as levodopa or dopamine agonists, opiods, or antiepileptics (Trenkwalder *et al.*, 2008). RLS diagnosis is mainly based on the patient's report of the typical symptoms, which also form the essential diagnostic criteria (Allen *et al.*, 2003). Additional diagnostic workup, including clinical examination, MRI, neurography, laboratory assessments, suggested immobilization test (SIT), and polysomnography (PSG) are mainly performed to rule out mimics such as neuropathic pain or sleep disturbances of other origin or to differentiate symptomatic forms of the disorder, that is, caused by chronic renal failure, iron deficiency, or peripheral neuropathy, from the much more common idiopathic form (Vignatelli *et al.*, 2006). An objective diagnostic test to prove a suspected diagnosis of RLS is not available to date; however, PSG or SIT forms the current electrophysiological gold standard (Hening, 2004).

II. From Parkinson's Disease to RLS

Although RLS does not represent a neurodegenerative disorder such as Parkinson's disease (PD), both disorders share some common features. Most evidently, the key symptoms of both disorders are excellently responsive to dopaminergic therapy. Moreover, both disorders go along with profound alterations of brain iron homeostasis and RLS occurs much more often in PD patients (about 20%) than in the normal population (Garcia-Borreguero *et al.*, 2003). Subsequently, imaging of the pre- and postsynaptic dopaminergic system has been evaluated as a diagnostic tool for RLS similar to the approach in PD, but failed to show consistent results (Hilker *et al.*, 2006).

By this time, TCS of the substantia nigra (SN) has come up as a potentially valuable method for the diagnosis and differential diagnosis of PD. Sonographic SN hyperechogenicity allows not only differentiation of PD from closely related disorders, but also shows correlations with even subclinical impairment of dopaminergic function and with increased regional iron content. Thus, it presents a distinct diagnostic procedure, which was potentially able to validly, quickly, and inexpensively visualize pathological processes of the dopaminergic system and brain iron homeostasis, which had been regarded undetectable by routine diagnostic workups so far (Berg and Becker, 2002).

Triggered by findings of reduced SN iron content in the SN of patients with RLS (Connor *et al.*, 2003; Earley, *et al.*, 2006) it was hypothesized that it might also be possible to detect abnormalities of SN iron homeostasis in RLS using TCS, in this case a reduced SN echogenicity. The first report supporting this hypothesis was published in 2005 (Schmidauer *et al.*, 2005). Since then, it has been tried to relate TCS findings of the SN and other brainstem structures in RLS to pathophysiological processes and to evaluate the value of TCS in the diagnostic approach to RLS and related disorders (see below). Still, as current as January 2010, evidence for both the diagnostic and pathophysiological approach to RLS using TCS is scarce.

III. Sonography of the Substantia Nigra in RLS

The first conceptual study on TCS in RLS in 2005 included only a rather small number of subjects, but did, however, show highly significant results. Most RLS patients showed a major reduction of the echogenicity in the anatomic area of the SN, not only compared with PD patients, but also with healthy controls (Fig. 1).

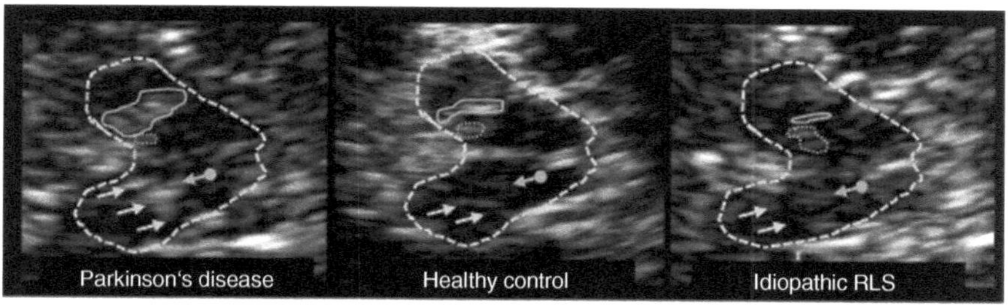

FIG. 1. Substantia nigra echogenicity in RLS. Spaced line: mesenencephalic brainstem; full line: substantia nigra ipsilateral, multiple arrows: substantia nigra contralateral; dotted line: red nucleus ipsilateral, pointed arrow: red nucleus contralateral. In idiopathic RLS, substantia nigra echogenicity is low, the area of SN echogenicity is very small, and contralaterally the SN can often not be delineated. Note also that the area of red nucleus echogenicity appears larger bilaterally in idiopathic RLS and that the brainstem midline is less echogenic compared with the healthy controls in both idiopathic RLS and Parkinson's disease. The triad of substantia nigra hypoechogenicity, red nucleus hyperechogenicity, and brainstem midline raphe hypoechogenicity allows diagnosis of idiopathic RLS with 98% diagnostic certainty (Godau *et al.*, 2008).

In 50% of the subjects, no echogenic SN area could be found at all (Schmidauer *et al.*, 2005).

These results were striking, given that until then no morphological abnormalities had been found in the brain of RLS patients (Connor *et al.*, 2003; Hening and Allen, 2003). Moreover, decreased SN echogenicity seemed to be consistent with the concept of nigral iron deficiency found in postmortem RLS brains (Connor *et al.*, 2003). However, these promising data meant a challenge for TCS as a diagnostic tool. Only when a second independent study 2 years later reproduced the findings and additionally showed excellent inter-rater correlation for the SN assessment in RLS ($r = 0.91$), the scientific community started to consider that a reliable and quantitative detection of small signals or even the absence of echogenic signals in the brainstem might be possible (Godau *et al.*, 2007).

A. Substantia Nigra Hypoechogenicity

In response to the concern that assessment of very small areas of SN echogenicity close to the limits of the spatial resolution of TCS might lead to greater diagnostic uncertainty, and considering that idiopathic RLS in contrast to PD normally shows no laterality of symptoms, it was decided to use the sum value of SN areas of echogenicity of both sides (SNsA) for the evaluation of RLS. SNsA values below the 10th percentile of the normal population were defined as pathologically decreased SN echogenicity, named SN hypoechogenicity (Godau *et al.*, 2007). The first defined limit value was $0.20\,\text{cm}^2$ for the SNsA. This calculated cut-off value was confirmed by ROC analysis, resulting in the same value for optimal differentiation of RLS patients and healthy controls. The area under the curve was 0.91 in this study, indicating SN hypoechogenicity to be an excellent marker to differentiate idiopathic RLS patients and healthy controls. SN hypoechogenicity has therefore become the sonographical key feature of RLS, which in the meantime has been replicated in four independent studies (Schmidauer *et al.*, 2005; Godau *et al.*, 2007; Godau *et al.*, 2008; Godau *et al.*, 2009).

1. *Idiopathic RLS*

An overview on the diagnostic parameters of SN hypoechogenicity for idiopathic RLS retrieved from all studies published on this subject so far is given in Table I. Cumulative data show a sensitivity of 86%, a specificity of 84%, a positive predictive value of 87%, and a negative predictive value of 82% for the

TABLE I

DIAGNOSTIC PARAMETERS FOR IDIOPATHIC RLS

	Idiopathic RLS (*n*)	Controls (*n*)	SNsA Cut-off (cm^2)	Sensitivity	Specificity	Positive predictive value	Negative predictive value	Classification accuracy
Schmidauer *et al.*, 2005	20	20	0.16*	0.90*	0.50*	0.64*	0.83*	70%
Godau *et al.*, 2007	39	49	0.20	0.82	0.90	0.88	0.77	86%
Godau *et al.*, 2009a	99	42	0.20	0.82	0.83	0.94	0.59	83%
Godau *et al.*, 2009b	73	67	0.23	0.91	0.89	0.90	0.90	90%
Cumulative	***231***	***178***	***n.a.***	***0.86***	***0.84***	***0.87***	***0.82***	***85%***

*reconstructed from published data

SNsA: sum area of SN echogenicity of both sides

diagnosis of idiopathic RLS. These values seem not too encouraging at first sight, however, the resulting classification accuracy of 85% is comparable to that of combined PSG and SIT (88%) as diagnostic gold standards, and better than the classification accuracy of each single exam (Michaud *et al.*, 2002). Given that TCS is quick to perform, cost saving, and generally available compared with PSG, TCS might indeed be a valuable additional tool in the diagnostic approach to RLS.

Until now, there are no prospective data answering the question how SN hypoechogenicity might be connected with the typical presentation of RLS symptoms and clinical progression. As far as it can be concluded from cross-sectional studies, there is no association of SN hypoechogenicity and RLS symptom severity, disease duration, or RLS treatment (Godau *et al.*, 2008). However, it was found to be related to a positive family history for RLS (Godau *et al.*, 2007; Godau *et al.*, 2008; Godau *et al.*, 2009). Moreover, idiopathic RLS patients who did not show SN hypoechogenicity have been described in one study to show untypical clinical features, that is, extended family history for RLS suggesting autosomal-dominant inheritance with anticipation, absence of additional disorders such as anxiety or depression, early need for medical intervention, and rather poor response to levodopa despite sustained response to dopamine agonists (Godau *et al.*, 2007). Taken together, these findings suggest, on the one hand, that SN hypoechogenicity may be a trait marker for RLS rather than a marker of clinical severity or treatment response, which is better achieved by PSG. On the other hand, these findings indicate that TCS might differentiate subtypes of RLS with different pathophysiological backgrounds currently subsumed under "idiopathic RLS" and might therefore be an interesting tool for study application.

2. *Symptomatic RLS*

Common causes of symptomatic RLS are iron deficiency, end-stage renal disease, peripheral neuropathy (PNP), and also neurodegenerative disorders such as PD under dopaminergic therapy and spinocerebellar ataxias (SCA). Little is known about the pathomechanism of RLS in these disorders; however, a better understanding could also contribute to the clarification of the pathophysiology underlying idiopathic RLS. Furthermore, a diagnostic marker which can differentiate idiopathic and symptomatic subtypes could be helpful in the clinical workup of patients presenting with leg discomfort.

To date, only two studies addressed the diagnostic parameters of SN echogenicity for the differentiation of idiopathic and symptomatic RLS in a normal RLS patient cohort (Godau *et al.*, 2007; Godau *et al.*, 2008). In these studies, 40 and 60% of the patients with symptomatic RLS, respectively, displayed SN hypoechogenicity, implicating that SN echogenicity may not reliably discriminate idiopathic and symptomatic RLS. This may have meant that TCS might increase the certainty of the differentiation of symptomatic RLS from closely related disorders such as neuropathic pain or other RLS mimics, since symptomatic RLS patients would more likely show SN hypoechogenicity. However, as described below, further studies on specific subtypes of symptomatic RLS failed to support this assumption. In contrast, they showed that SN hypoechogenicity is not only exclusively found in RLS, but also in a number of other disorders, such as polyneuropathy. An overview on the findings in different disorders related to a high prevalence of symptomatic RLS is given in Table II.

TABLE II
SUBSTANTIA NIGRA ECHOGENICITY IN SYMPTOMATIC RLS

	Whole cohort		Symptomatic RLS subgroup	
	N	Symptomatic RLS	SN hypoechogenicity "RLS like"	SN hyperechogenicity "PD like"
Overall RLS	49	20.4%	40.0%	0%
	132	25.0%	60.7%	6.1%*
Disorders with known nigrostriatal degeneration				
Parkinson's disease+	66	18.2%	0%	100%
Spinocerebellar ataxia type 2	6	n.a.#	0%#	67%#
Spinocerebellar ataxia type 3	15	n.a.#	0%#	40%#
Disorders without known nigrostriatal degeneration				
End stage renal disease§	3§	100%§	100%	0%
Peripheral neuropathy	68	41.2%	60.8%	3.6%
	30§	100%§	56.7%	0%
Friedreich ataxia+	57+	52.6%+	78.9%+	0%+
Pyramidal tract lesion§	6§	100%§	83.3%	0%

*both patients had Parkinson's disease as primary disorder
+preliminary, yet unpublished data from both authors' labs
#prevalence of symptomatic RLS has not been evaluated; however, no examined subject in the whole cohort showed "RLS like" SN morphology, in other studies prevalence of RLS in spinocerebellar ataxia types 2 and 3 is reported to be 28–45% (Abele *et al.*, 2001; Schols *et al.*, 1998; Reimold *et al.*, 2006).
§post hoc analysis subgroup of the "overall RLS" cohorts

3. *RLS in Peripheral Neuropathy (PNP)*

Clinical distinction of neuropathic pain and neuropathic pain with additional RLS is challenging because of the very similar presentation of symptoms, common co-occurrence of both disorders (30–54%) (Gemignani *et al.*, 2006; Nineb *et al.*, 2007), and the lack of objective markers. First data retrieved in a post hoc analysis of published data on 30 PNP patients with RLS showed that more than 55% exhibited "RLS like" SN hypoechogenicity (compare Table II) (Godau *et al.*, 2007; Godau *et al.*, 2008). Only one study published so far was directly set out to determine the value of SN hypoechogenicity to differentiate between these entities. This study failed to show clinically significant results (Godau *et al.*, 2009). The reason for the lack of difference, however, was surprising and raised new questions about the origin of SN hypoechogenicity.

Indeed, 61% of 28 PNP patients with RLS displayed SN hypoechogenicity compared with only 11% of the 67 healthy controls. Interestingly, however, no difference of SN echogenicity could be found comparing PNP patients with and without RLS ($n = 28$ vs. 40; $0.24 \pm 0.08\,\text{cm}^2$ vs. $0.25 \pm 0.06\,\text{cm}^2$), implicating that even a large number of pure PNP patients exhibit "RLS like" SN morphology. No association of SN hypoechogenicity and severity of neither neuropathic pain nor RLS symptoms could be found. Moreover, no associations could be found with disease duration, age at symptom onset, PNP etiology, or serum ferritin levels (Godau *et al.*, 2009). It remains largely unclear how PNP should affect SN morphology and how this could be connected to RLS. However, these findings might be subject to future studies in order to confirm the results in larger cohorts and to better understand the underlying pathology.

4. *RLS in Neurodegenerative Disorders*

TCS has proven to be a valuable tool in the assessment of various neurodegenerative disorders, especially SN hyperechogenicity constitutes a sensitive and specific marker in the diagnosis and differential diagnosis of idiopathic PD. RLS is a frequent comorbidity in some of these neurodegenerative disorders, such as PD (prevalence 8–50%) (Krishnan *et al.*, 2003; Braga-Neto *et al.*, 2004), spinocerebellar ataxias (up to 45%) (Schols *et al.*, 1998; Abele *et al.*, 2001), and Friedreich ataxia (~50%, own unpublished data). Investigating the characteristics of SN echogenicity in these different disorders may not only help in the understanding of SN echogenicity as a biological marker but also in the description of the mostly unclear origin of RLS in these disorders.

Idiopathic PD and spinocerebellar ataxias are known to go along with a degenerative process in the substantia nigra. Both disorders lack studies directly assessing the relation between SN echogenicity and symptomatic RLS. However, data available so far draw a rather clear picture: In two independent studies on spinocerebellar ataxia type 2 and type 3, as far as can be reconstructed from published data, patients showed "PD like" SN morphology with increased SN echogenicity in 40 and 67% of the cases, respectively, whereas no patient exhibited "RLS like" SN hypoechogenicity (Postert *et al.*, 2004; Mijajlovic *et al.*, 2008). Data on symptomatic RLS in PD show to be even clearer: two patients published so far did both show "PD like" SN hyperechogenicity (Godau *et al.*, 2008) and also 12 additional cases with PD and RLS collected by the authors all showed "PD like" SN hyperechogenicity (own unpublished data).

Contrasting these findings, in Friedreich ataxia, two independent studies from both of the authors' groups showed "RLS like" SN hypoechogenicity to be a very frequent finding, which was associated with an increased prevalence of symptomatic RLS (yet unpublished data). This seems rather surprising since Friedreich ataxia is a degenerative disorder involving increased tissue iron levels (Delatycki *et al.*, 2000), which are supposed to cause increased rather than decreased SN echogenicity (Berg *et al.*, 1999). However, this disorder has no known affection on the nigrostriatal system, suggesting that the observed abnormality in SN echogenicity may mark a distinct pathophysiological process.

5. *RLS in other Disorders*

In all studies published so far, patients with primary iron deficiency had been excluded for potential interference with SN echogenicity; therefore, no specifications can be given on SN echogenicity in this common cause of symptomatic RLS. Other disorders, such as uremia caused by end-stage renal disease or pyramidal tract lesions, have not directly been investigated; however, RLS patients with these disorders had been included in two studies (compare Table 2) (Godau *et al.*, 2007; Godau *et al.*, 2008). All except one of these totally nine patients showed "RLS like" SN hypoechogenicity, "PD like" SN morphology was not found in this group.

From these findings in neurodegenerative and non-neurodegenerative disorders, one may draw two conclusions concerning SN echogenicity in symptomatic RLS:

(1) If a neurodegenerative process affecting the SN is involved in the underlying primary disorder, symptomatic RLS patients do not show SN hypoechogenicity, but regularly exhibit "PD like" SN hyperechogenicity.
(2) In all other cases, including primary neuropathies, end-stage renal failure, pyramidal tract lesions, and degenerative disorders not primarily involving the nigrostriatal system, "RLS like" SN hypoechogenicity is a frequent finding, which due to the great variety of underlying disorders seems to be most likely related to RLS itself.

B. Causes of Substantia Nigra Hypoechogenicity

All pathophysiological data available to date implicate a multifactorial genesis of RLS, involving brain iron deficiency, alterations of the striatal and hypothalamic dopaminergic system and the opioid system, as well as disturbances in the sensory processing system on the, peripheral, spinal, thalamic, and cortical level. The integration of all these distinct disturbances in one pathophysiological concept has, however, not been achieved to date (Winkelman, 2006). Moreover, also the association of RLS pathophysiology and SN hypoechogenicity as an apparently sensitive and specific morphological marker is still unclear.

1. *The Role of Iron*

Rodent and human postmortem studies have shown that increased SN non-ferritin-bound iron content is related to increased SN echogenicity and that artificial increase of iron in the SN may cause SN hyperechogencity (Berg *et al.*, 1999; Berg *et al.*, 2006). Therefore, it sounds reasonable that SN iron deficiency, pictured in idiopathic RLS in a number of postmortem and MRI studies (Connor *et al.*, 2003; Earley *et al.*, 2006), might be a very likely cause of SN hypoechogenicity.

Data regarding the association of SN hypoechogenicity and regional iron deficiency are, however, not consistent. In favor of this hypothesis one combined MRI and TCS study showed that SN hypoechogenicity correlated with increased T2 values throughout the brain (Godau *et al.*, 2008). Increased T2 values are suggestive for decreased regional iron content, since iron would cause local field disturbances resulting in a loss of T2. Further reference comes from the finding that disorders clearly related to iron deficiency, such as primary iron deficiency, pregnancy, and end-stage renal disease, all show an increased prevalence of symptomatic RLS as well as from the finding that at least in end-stage renal disease RLS is associated with SN hypoechogenicity (see above).

Contrasting the hypothesis of an association of SN iron deficiency and SN hypoechogenicity the above-named MRI study could not verify a correlation of SN hypoechogenicity and increased T2 values specifically in the SN despite showing such an association for all other examined regions(Godau *et al.*, 2008). Furthermore, none of the clinical studies could establish a relation of SN hypoechogenicity to serum iron or ferritin content, neither in RLS patients nor in controls (see above). Finally, some disorders without known relation to iron deficiency such as peripheral neuropathy or even a known association with iron accumulation such as Friedreich ataxia display SN hypoechogenicity as a common feature, even without presence of symptomatic RLS (see above). Therefore, almost certainly also other factors are likely to contribute to SN hypoechogenicity.

2. *The Role of Fiber Integrity*

In PD, in addition to increased iron content, also the degenerative process in the dopaminergic system itself leading to a disintegration of tissue architecture has been discussed to underlie SN hyperechogenicity (Berg *et al.*, 2006). The latter can possibly be ruled out as a potential source of SN hypoechogenicity in RLS, since in two disorders with known degenerative process in the nigrostriatal system (PD and spinocerebellar ataxia) even patients with RLS exhibit "PD like" SN hyperechogenicity, implicating that the degenerative process marked by SN hyperechogenicity might mask RLS-related changes. Nevertheless, abnormalities of fiber integrity not involving degeneration may contribute to a loss of echogenicity in the SN, similar to the concept of disturbed fiber integrity as a potential cause of decreased echogenicity in the brainstem midline raphe in primary depressive disorders (Becker *et al.*, 1995; Berg *et al.*, 1999). Still, the association between alterations of SN echogenicity and RLS as well as other disorders without evidence of dopaminergic dysfunction, such as PNP or Friedreich ataxia, remains unclear. The only obvious relation between these three disorders seems to be an alteration of the primary sensory system, including peripheral nerves in all three disorders (Delatycki *et al.*, 2000; Polydefkis *et al.*, 2000), and the spinal level in RLS and Friedreich ataxia (Delatycki *et al.*, 2000; Clemens *et al.*, 2006), raising the question whether abnormalities of the primary sensory system might be related to SN hypoechogenicity. If this holds true, SN hypoechogenicity could constitute a morphological marker for a genetic vulnerability of the somatosensory system, indicated by the absence of a correlation between SN hypoechogenicity and disease progression and the proof of a correlation with a positive family history for RLS even in healthy controls (see above). However, up to now, this appears to be pure speculation.

IV. Sonography of Other Structures in RLS

In idiopathic RLS, in addition to the SN frequently other brainstem structures, that is, the raphe and the red nucleus, show sonographic abnormalities. The ventricle system, the cerebral lobes, the basal ganglia, thalamus, and the cerebellum do not seem to be altered in RLS as far as detectable by TCS (Godau *et al.*, 2008).

A. Raphe

In the majority of idiopathic RLS patients (75%), decreased echogenicity of the brainstem raphe could be found (Fig. 2) (Godau *et al.*, 2008). This finding,

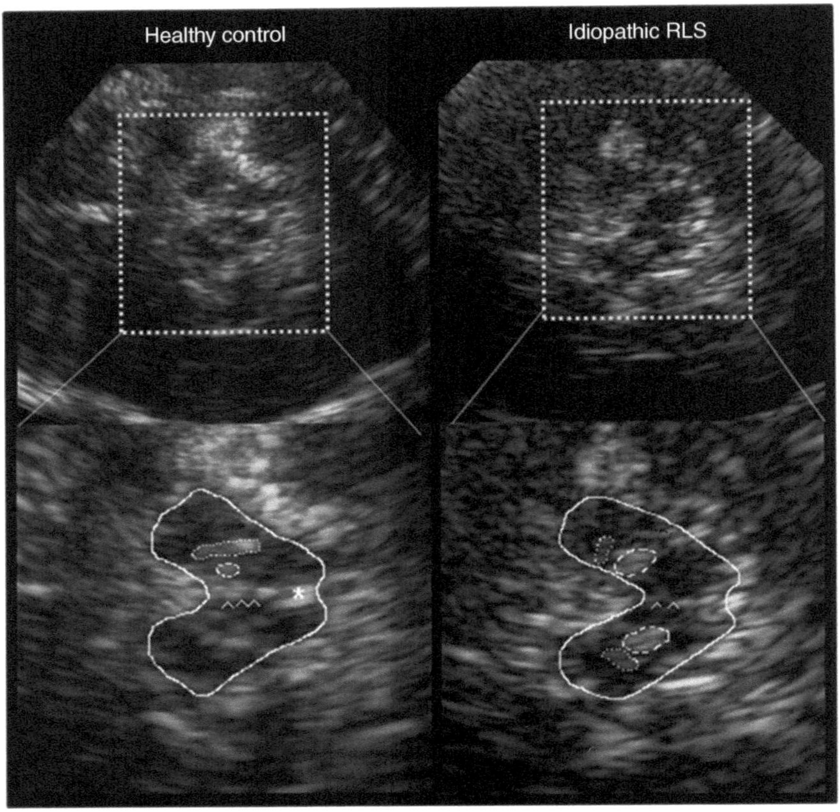

FIG. 2. Raphe and red nucleus in RLS. The upper row shows overview screens of the mesencephalic brainstem of a healthy control (left) and a patient with idiopathic RLS (right). Whereas on the left image the three main structures (SN, RN, and raphe) can be clearly differentiated, the right image shows no clear depiction of the raphe. The highly echogenic area does not belong to the SN but constitutes a prominent RN, the SN is hypoechogenic. In the lower row, the three main structures are marked in magnified sections. Full line: brainstem, dotted line: substantia nigra (SN), spaced line: red nucleus (RN), arrows: brainstem midline raphe.

however, is not specific to RLS, but has been described to be the key feature in primary depressive disorders such as major depression or depression in PD (Becker *et al.*, 1997). Similarly, also in idiopathic RLS patients, hypoechogenicity of the brainstem raphe was related to depressive symptoms (Rho = 0.45) (Godau *et al.*, 2008). Therefore, hypoechogenicity of the raphe in idiopathic RLS patients may point toward the presence of depression, which might be masked clinically by prominent RLS symptoms, but may require specific treatment in order to achieve effective therapy.

B. Red Nucleus

In addition to SN and raphe hypoechogenicity, the majority of idiopathic RLS patients (60%) display increased echogenicity of the red nucleus (RN, Fig. 2) (Godau *et al.*, 2008). RN hyperechogenicity is related to an increased prevalence of periodic limb movements (PLMs) in RLS patients and in healthy controls (Rho = 0.38 and 0.45, respectively), an association which has been described first in an fMRI study in 1997, showing that PLM occurrence is primarily related to RN activation (Bucher *et al.*, 1997). It has to be remarked that, so far, there is no standard procedure to define RN hyperechogenicity and data on quantitative RN assessment are not available to date. In the semiquantitative assessment performed in the above-named study (Godau *et al.*, 2008), RN hyperechogenicity was stated when the echogenicity of the red nucleus was in the range of the basal cisterns and/or the size of the RN appeared enlarged, compared with the normally expected "white comma." The RN was then called to be "prominent."

C. Diagnostic Value of the Assessment of RN and Raphe

SN hypoechogenicity is the key feature of idiopathic RLS. However, regarding the shortcomings in the differentiation of RLS, that is, the differentiation of symptomatic RLS and closely related disorders such as PNP, the overall diagnostic value of SN echogenicity seems to be lower for RLS than for PD. This shortcoming may be overcome by considering all three echofeatures, as shown in a recent study: If, in addition to SN hypoechogenicity, hypoechogenicity of the brainstem raphe and hyperechogenicity of the RN can be found, this triad allows the correct diagnosis of RLS in 98% of the cases; the absence of all three of the markers rules out RLS with a diagnostic certainty of 90% (Godau *et al.*, 2008). Thus, assessment of raphe and RN in addition to the SN enhances the diagnostic certainty for RLS.

V. Special Instructions for TCS in RLS

TCS for the evaluation of RLS is performed using exactly the same ultrasound systems and settings as for evaluation of other movement disorders. However, special consideration should be given to four points:

(1) Adjustment of gain and image brightness

In the absence of an ultrasound signal within the anatomic area of the SN one might feel tempted to increase gain and image brightness in order to allow a better depiction of the SN. Normally, when doing so, indeed, larger areas of SN echogenicity will occur; however, in most cases the echogenicity at the anatomical site of the SN will still be difficult to differentiate from the surrounding tissue. As an indicator for correct and reproducible adjustments one can, for example, use the contrast between the hypoechogenic brainstem, the hyperechogenic basal cisterns, and the echogenic red nucleus (for the echogenic red nucleus no cases of absent signals have been described yet). This means, if the mesencephalic brainstem can clearly be outlined in contrast to the surrounding basal cisterns and if at the same time within the brainstem a distinct echogenic signal at the anatomical site of the red nucleus can be distinguished, then gain and image brightness are correctly adjusted for SN assessments.

(2) Differentiation of substantia nigra and red nucleus

Since in RLS, the RN is often hyperechogenic, may exceed the normal expansion of the regular echogenic RN signal, and the SN at the same time is likely to be hypoechogenic or even absent, there is a good chance of mistaking the RN for the SN. The RN can be visualized in the same plane as the SN, located dorsomedially to the SN, often quite adjacent to the brainstem midline raphe. Regarding training experiences in our labs, differentiation of RN and SN is the most common pitfall in the evaluation of RLS. In unclear cases, the tissue harmonic imaging (THI) mode may help with the differentiation of both structures by improving the contrast to the surrounding tissue. Doing so, one should be aware that to date there is no report of an absent RN signal, meaning that if only one echogenic structure is seen within the mesencephalic brainstem, this in almost every case would belong to the RN with the SN signal then being absent (compare Fig. 2).

(3) Strict limitation to the butterfly-shaped mesencephalic brainstem

To our experience, when scanning higher than in the actual mesencephalic brainstem, echogenic areas may be visualized in many patients who most likely belong to lower thalamic structures but may mimic SN hyperechogenicity. Therefore, one should be determined to strictly measure SN echogenicity only when the mesencephalic brainstem with its typical butterfly-shaped structure can be visualized without sectioning the third ventricle in the same plane as a marker for the lower thalamic area.

(4) Limit values

Assessment of SN hypoechogenicity constitutes a challenge for the spatial resolution and image contrast of the ultrasound system applied. To date, only data for the Siemens Sonoline Elegra system (Siemens,

Erlangen, Germany) are available. Even for this system, different cut-off values have been determined, dependent on the group, the examiners, and the fine settings of the applied system (compare Table 1). Although this seemed not to bother the accuracy of differentiation of patients and controls, it seems much more essential to define cut-off values for SN hypoechogenicity individually for each laboratory than it had been assumed for cut-off values for SN hyperechogenicity.

VI. Summary

Assessment of mesencephalic brainstem structures in idiopathic and symptomatic RLS is an overall promising new application of TCS in movement disorders. SN hypoechogenicity, defined as a sum area of echogenicity of both sides below the 10th percentile of the normal population, is the cardinal feature of RLS, present in more than 90% of idiopathic RLS patients and in up to 60% of patients with symptomatic RLS. Hypoechogenicity of the brainstem midline raphe (depression) and hyperechogenicity of the red nucleus (PLM) as secondary features may improve the diagnostic certainty for RLS. The pathophysiological changes underlying SN hypoechogenicity are not yet known; however, regional iron deficiency as well as nondegenerative changes of fiber integrity is suspected to contribute to the lower echogenic signal at the anatomical site of the SN. Compared with SN hyperechogenicity in PD, only little is known about the value of TCS for the diagnosis of RLS and related disorders as well as about the value for the better understanding of RLS pathophysiology. More studies are essentially needed before a certified recommendation for application of TCS in the evaluation of RLS can be made.

References

Abele, M., Burk, K., Laccone, F., Dichgans, J., and Klockgether, T. (2001). Restless legs syndrome in spinocerebellar ataxia types 1, 2, and 3. *J. Neurol.* **248** (4), 311–314.

Allen, R. P., Picchietti, D., Hening, W. A., Trenkwalder, C., Walters, A. S., and Montplaisi, J. (2003). Restless legs syndrome: Diagnostic criteria, special considerations, and epidemiology. A report from the restless legs syndrome diagnosis and epidemiology workshop at the National Institutes of Health. *Sleep Med.* **4** (2), 101–119.

Becker, G., Becker, T., Struck, M,, et al. (1995). Reduced echogenicity of brainstem raphe specific to unipolar depression: A transcranial color-coded real-time sonography study. *Biol. Psychiatry* **38** (3), 180–184.

Becker, T., Becker, G., Seufert, J., et al. (1997). Parkinson's disease and depression: Evidence for an alteration of the basal limbic system detected by transcranial sonography. *J. Neurol. Neurosurg. Psychiatry* **63** (5), 590–596.

Berg, D., Becker, G. (2002). Perspectives of B-mode transcranial ultrasound. *Neuroimage.* **15** (3), 463–473.

Berg, D., Grote, C., Rausch, W. D., et al. (1999). Iron accumulation in the substantia nigra in rats visualized by ultrasound. *Ultrasound Med. Biol.* **25** (6), 901–904.

Berg, D., Hochstrasser, H., Schweitzer, K. J., and Riess, O. (2006). Disturbance of iron metabolism in Parkinson's disease – ultrasonography as a biomarker. *Neurotox. Res.* **9** (1), 1–13.

Berg, D., Supprian, T., Hofmann, E., et al. (1999). Depression in Parkinson's disease: Brainstem midline alteration on transcranial sonography and magnetic resonance imaging. *J. Neurol.* **246** (12), 1186–1193.

Braga-Neto, P., da Silva-Junior, F. P., Sueli Monte, F., de Bruin, P. F., and de Bruin, V. M. (2004). Snoring and excessive daytime sleepiness in Parkinson's disease. *J. Neurol. Sci.* **217** (1), 41–45.

Bucher, S. F., Seelos, K. C., Oertel, W. H., Reiser, M., and Trenkwalder, C. (1997). Cerebral generators involved in the pathogenesis of the restless legs syndrome. *Ann. Neurol.* **41** (5), 639–645.

Clemens, S., Rye, D., and Hochman, S. (2006). Restless legs syndrome: Revisiting the dopamine hypothesis from the spinal cord perspective. *Neurology* **67** (1), 125–130.

Connor, J. R., Boyer, P. J., Menzies, S. L., et al. (2003). Neuropathological examination suggests impaired brain iron acquisition in restless legs syndrome. *Neurology* **61** (3), 304–309.

Delatycki, M. B., Williamson, R., and Forrest, S. M. (2000). Friedreich ataxia: An overview. *J. Med. Genet.* **37** (1), 1–8.

Earley, C. J., Barker, P. B., Horska, A., and Allen, R. P. (2006). MRI-determined regional brain iron concentrations in early- and late-onset restless legs syndrome. *Sleep. Med.* **7** (5), 458–461.

Garcia-Borreguero, D., Odin, P., and Serrano, C. (2003). Restless legs syndrome and PD: A review of the evidence for a possible association. *Neurology* **61** (6 Suppl. 3), S49–55.

Gemignani, F., Brindani, F., Negrotti, A., Vitetta, F., Alfieri, S., and Marbini, A. (2006). Restless legs syndrome and polyneuropathy. *Mov. Disord.* **21** (8), 1254–1257.

Godau, J., Klose, U., Di Santo, A., Schweitzer, K., and Berg, D. (2008). Multiregional brain iron deficiency in restless legs syndrome. *Mov. Disord.* **23** (8), 1184–1187.

Godau, J., Manz, A., Wevers, A. K., Gaenslen, A., and Berg, D. (2009). Sonographic substantia nigra hypoechogenicity in polyneuropathy and restless legs syndrome. *Mov. Disord.* **24** (1), 133–137.

Godau, J., Schweitzer, K. J., Liepelt, I., Gerloff, C., and Berg, D. (2007). Substantia nigra hypoechogenicity: Definition and findings in restless legs syndrome. *Mov Disord.* **22** (2), 187–192.

Godau, J., Wevers, A. K., Gaenslen, A., et al. (2008). Sonographic abnormalities of brainstem structures in restless legs syndrome. *Sleep Med.* **9** (7), 782–789.

Hening, W. (2004). The clinical neurophysiology of the restless legs syndrome and periodic limb movements. Part I: Diagnosis, assessment, and characterization. *Clin. Neurophysiol.* **115** (9), 1965–1974.

Hening, W. A., and Allen, R. P. (2003). Restless legs syndrome (RLS): the continuing development of diagnostic standards and severity measures. *Sleep. Med.* **4** (2), 95–97.

Hilker, R., Burghaus, L., Razai, N., Jacobs, A. H., Szelies, B., and Heiss, W. D. (2006). Functional brain imaging in combined motor and sleep disorders. *J. Neurol. Sci.* **248** (1–2), 223–226.

Krishnan, P. R., Bhatia, M., and Behari, M. (2003). Restless legs syndrome in Parkinson's disease: A case-controlled study. *Mov. Disord.* **18** (2), 181–185.

Michaud, M., Paquet, J., Lavigne, G., Desautels, A., and Montplaisir, J. (2002). Sleep laboratory diagnosis of restless legs syndrome. *Eur. Neurol.* **48** (2), 108–113.

Mijajlovic, M., Dragasevic, N., Stefanova, E., Petrovic, I., Svetel, M., and Kostic, V. S. (2008). Transcranial sonography in spinocerebellar ataxia type 2. *J. Neurol.* **255** (8), 1164–1167.

Nineb, A., Rosso, C., Dumurgier, J., Nordine, T., Lefaucheur, J. P., and Creange, A. (2007). Restless legs syndrome is frequently overlooked in patients being evaluated for polyneuropathies. *Eur. J. Neurol.* **14** (7), 788–792.

Polydefkis, M., Allen, R. P., Hauer, P., Earley, C. J., Griffin, J. W., and McArthur, J. C. (2000). Subclinical sensory neuropathy in late-onset restless legs syndrome. *Neurology* **55** (8), 1115–1121.

Postert, T., Eyding, J., Berg, D., et al. (2004). Transcranial sonography in spinocerebellar ataxia type 3. *J. Neural. Transm. Suppl.* (68), 123–133.

Reimold, M., Globas, C., Gleichmann, M., et al. (2006). Spinocerebellar ataxia type 1, 2, and 3 and restless legs syndrome: Striatal dopamine D2 receptor status investigated by [11C]raclopride positron emission tomography. *Mov. Disord.* **21** (10), 1667–1673.

Schmidauer, C., Sojer, M., Seppi, K., et al. (2005). Transcranial ultrasound shows nigral hypoechogenicity in restless legs syndrome. *Ann. Neurol.* **58** (4), 630–634.

Schols, L., Haan, J., Riess, O., Amoiridis, G., and Przuntek, H. (1998). Sleep disturbance in spinocerebellar ataxias: Is the SCA3 mutation a cause of restless legs syndrome? *Neurology* **51** (6), 1603–1607.

Trenkwalder, C., Hening, W. A., Montagna, P., et al. (2008). Treatment of restless legs syndrome: An evidence-based review and implications for clinical practice. *Mov. Disord.* **23** (16), 2267–2302.

Vignatelli, L., Billiard, M., Clarenbach, P., et al. (2006). EFNS guidelines on management of restless legs syndrome and periodic limb movement disorder in sleep. *Eur. J. Neurol.* **13** (10), 1049–1065.

Walters, A. S. (1995). Toward a better definition of the restless legs syndrome. The international restless legs syndrome study group. *Mov Disord.* **10** (5), 634–642.

Winkelman, J. W. (2006). Considering the causes of RLS. *Eur. J. Neurol.* **13** (Suppl. 3), 8–14.

Winkelmann, J., Prager, M., Lieb, R., et al. (2005). "Anxietas tibiarum". Depression and anxiety disorders in patients with restless legs syndrome. *J. Neurol.* **252** (1), 67–71.

TRANSCRANIAL SONOGRAPHY IN ATAXIA

Christos Krogias,* Thomas Postert,† and Jens Eyding*

*Department of Neurology, St. Josef-Hospital, Ruhr University Bochum, Germany
†Department of Neurology, St. Vincenz-Krankenhaus, Paderborn, Germany

Transcranial sonography (TCS) identifies basal ganglia alterations in extrapyramidal movement disorders such as Parkinson's disease or dystonia. Although only rarely reported, TCS also reveals signal alterations of basal ganglia in several forms of hereditary and nonhereditary ataxia.

Here, the examination procedure and its diagnostic value for the classification of ataxia are reviewed.

Three TCS studies reported hyperechogenicity of substantia nigra (SN) as a frequent finding in spinocerebellar ataxia type 2, type 3, and type 17, indicating a vulnerability of the nigrostriatal system in SCA patients.

A new "cerebellar examination plane" was proposed, allowing better visualization of fourth ventricle enlargement and nucleus dentatus hyperechogenicity as a characteristic finding in SCA3 patients. In sporadic Creutzfeldt–Jakob disease, a blurry inhomogeneous hyperechogenic signal pattern of lentiform nucleus was identified in all of the patients in a small case series. Furthermore, distinct bilateral hyperechogenicity of pallidostriatal regions have been described as a novel diagnostic feature in the sonographic differentiation of extrapyramidal and atactic movement disorders.

TCS is a commonly available, noninvasive, and inexpensive diagnostic tool, which provides reliable information about the morphology of the brain in ataxias, even in agitated patients who do not tolerate other imaging techniques.

INTERNATIONAL REVIEW OF
NEUROBIOLOGY, VOL. 90
DOI: 10.1016/S0074-7742(10)90016-0

Further neuropathological and multimodal imaging studies are needed to elucidate the precise morphological and pathogenetic background of the detected echosignal pathology, and also to correlate these findings to the various clinical features of this disease entity.

I. Introduction

A. Terminology and Classification of Ataxias

The term "ataxia" (from Greek α- (used as a negative prefix) + -τάξις (order), meaning "lack of order") comprises a multitude of hereditary and non-hereditary diseases with a progressive cerebellar or spinal ataxia as a main clinical feature.

Up to the 1980s, the different forms of ataxia were classified according to the underlying neuropathological alterations (Greenfield, 1954), being categorized as spinal, spinocerebellar, or cerebellar degenerations. Furthermore, forms with additional degeneration of brainstem regions such as the olivopontocerebellar atrophy (OPCA) were described (Geary *et al.*, 1956). In 1983, a new classification of ataxia was proposed by Harding (Harding *et al.*, 1983), based on clinical symptomatology as well as on inheritance. These days, different forms of ataxia are categorized due to their etiology as hereditary or nonhereditary disorders (Abele and Klockgether, 2005). Hereditary forms of ataxia are classified on the basis of the underlying genetic mutation while the nonhereditary forms of ataxia are subdivided according to their underlying toxic, metabolic, (para)neoplastic, infectious, autoimmune, or vascular origin (Abele and Klockgether, 2005). An overview of hereditary and nonhereditary causes of ataxia is given in Table I.

B. TCS Experience in Several forms of Ataxia

Transcranial sonography (TCS) has become a reliable and sensitive diagnostic tool in the evaluation of extrapyramidal movement disorders, especially in the differentiation of Parkinsonian syndromes (Berg *et al.*, 2008). In this context, several studies evaluated the sonographic alterations of basal ganglia in multiple system atrophy (MSA) (Walter *et al.*, 2003; Behnke *et al.*, 2005). In MSA, the cerebellar system is frequently involved in the neurodegenerative process,

TABLE I
OVERVIEW OF HEREDITARY AND NONHEREDITARY CAUSES OF ATAXIAS

Hereditary ataxias
Autosomal-recessive ataxias
With known gene mutation
– Friedreich Ataxia (FRDA)
– Ataxia telangiectasia (AT)
– Ataxia with oculomotor apraxia (AOA)
– Abetalipoproteinemia (Bassen–Kornzweig syndrome)
– Ataxia with isolated vitamin E deficiency
– Refsum's disease
– Autosomal-recessive spastic ataxia of Charlevoix-Saguenay (ARSACS)
– Cerebrotendinose xanthomatosis
With known gene locus
– Autosomal-recessive ataxia with gene locus on chromosome 9q
– Autosomal-recessive ataxia with blindness and deafness (SCABD)
– Infantile onset spinocerecellar ataxia (IOSCA)
Unknown gene mutation or gene locus
– Early onset cerebellar ataxia (EOCA)
Autosomal-dominant ataxias
– Spinocerebellar ataxias (SCA)
– Episodic ataxias (EA)
Nonhereditary Ataxias
Degenerative cerebellar ataxias
– Multiple system atrophy, cerebellar type (MSA-C)
– Sporadic ataxias of unknown cause
Symptomatic ataxias
– Ataxia due to alcohol abuse
– Ataxia of other toxic origin
– Paraneoplastic cerebellar degeneration (PCD)
– Inflammatory or autoimmune ataxias
– Ataxias due to acquired vitamin deficiency
– Ataxias due to metabolic disturbances
– Ataxias of physical origin (fever, heatstroke...)
– Bilateral striopallidodentate calcinosis (BSPDC)
– Superficial siderosis

Adapted from Abele and Klockgether, 2005.

resulting in a distinctive ataxia, especially in the cerebellar type (MSA-C) of the disease (Klockgether, 2010). However, as the differentiation of idiopathic and atypical Parkinsonism was the major aim of these studies, patients with the Parkinsonian type (MSA-P) were primarily included. Therefore, no relevant evidence can be concluded from these studies with regard to TCS findings in ataxia. Typical sonographic findings in MSA are described and discussed in detail in Chapter IV, part 7 of this volume.

Less frequently, TCS evaluation has been performed in other movement disorders such as Huntington's disease, dystonia, and Wilson's disease (Naumann *et al.*, 1996; Postert *et al.*, 1999; Walter *et al.*, 2005). In Huntington's disease and even more in Wilson's disease, cerebellar ataxia can frequently been found (Conrad, 2005). Typical TCS findings in Wilson's disease and in Huntington's disease are described and discussed in Chapter "TCS in Brain Disorders with Trace Metal Accumulation" and in Chapter "TCS in Huntington's Disease" of this volume, respectively.

In contrast to an increasing number of studies dealing with TCS findings in different Parkinsonian syndromes, till date, there are only three small studies reporting TCS findings in spinocerebellar ataxia (Postert *et al.*, 2004; Wolters *et al.*, 2005; Mijajlovic *et al.*, 2008). These studies will be discussed in the following. Likewise, TCS findings in a case series of sporadic Creutzfeldt–Jakob disease (Müller *et al.*, 2008) as well as two cases of sonographic detection of bilateral striopallidodentate calcinosis as nonhereditary causes of ataxia have been reported (Krogias *et al.*, 2009; Ritter and Dittrich, 2010).

II. Method of Transcranial Sonography in Ataxia

In general, the methodology of transcranial sonography in ataxia corresponds to the recommendations and guidelines for the diagnostic scanning and evaluation procedure in other extrapyramidal movement disorders (Walter *et al.*, 2007a; Berg *et al.*, 2008):

In summary, TCS is performed through the transtemporal acoustic bone window while the patient is in a supine position. Ultrasound systems equipped with a 1.6- to 3.5-MHz transducer can be applied. The following parameter settings of the TCS system are recommended: penetration depth of 140–160 mm, suppression of low echosignals, dynamic range of 45–55 dB. Image brightness and time gain compensations are adapted as needed for each examination. Standardized midbrain and diencephalic examination planes are visualized in axial section. The image should be fixed and zoomed in 2- to 4-fold for further analysis. A planimetric measurement of the echogenic area of the substantia nigra (SN) is recommended: using the ultrasound system Sonoline Elegra (Siemens; Erlangen; Germany), sizes of less than 0.20 cm^2 are defined as normal, since this threshold was found to represent the upper quartile in a normal population (Berg *et al.*, 1999a). Sizes between 0.20 and 0.25 cm^2 are classified as moderately and sizes of 0.25 cm^2 and more are graded as markedly hyperechogenic, since sizes of 0.25 cm^2 and more represent upper decile in normal population (Berg *et al.*, 1999a). Almost the same thresholds have been adopted in a TCS study on patients with spinocerebellar atrophy type 2, where another ultrasound system (Esaote Technos MP, Genova, Italy) has been used (Mijajlovic *et al.*, 2008).

The brainstem raphe (BR) is evaluated by a bilateral TCS investigation. The side with the best visualization of structures is used for evaluation.

Echogenicity of BR is classified semiquantitatively on a three-point scale using the red nucleus or signal intensity of the basal cisterns as reference points: 0 = raphe structure not visible, 1 = reduced echogenicity: the echogenic line of the BR is slight and/or interrupted, 2 = normal echogenicity (echogenicity of raphe structure is not interrupted and intensity is equal to that of the red nucleus or basal cisterns). At grades 0 and 1 the BR is classified as hypoechogenic. Echogenicities of the thalami, the lentiform nuclei (LN), and the heads of the caudate nuclei (CN) are graded as hyperechogenic if they are more intense than the surrounding white matter.

For the evaluation of the ventricular system, measurement of the widths of the contralateral frontal horn is recommended (Walter *et al.*, 2007a). The measurement of the minimal width of the third ventricle is performed on a standardized diencephalic examination plane (Seidel *et al.*, 1995).

In the only TCS study published on spinocerebellar ataxia type 3 so far, the "cerebellar examination plane" was proposed for TCS examination (Postert *et al.*, 2004). To set this examination plane, the dorsal part of the ultrasound probe has to be tilted caudally in order to visualize the fourth ventricle, the cerebellar hemispheres, and the dentate nucleus. The fourth ventricle can be identified as a hypo- to anechogenic circular structure in the immediate dorsal vicinity of the mesencephalic brainstem (Fig. 1). This structure can only be displayed accurately in case of cerebellar atrophy. Size of the fourth ventricle can be measured by encircling this structure or by measuring the maximal diameter depending on which parameter can be visualized with higher accuracy. The dentate nucleus can be identified bilaterally adjacent to the hypoechogenic fourth ventricles (Fig. 2). Echogenicity of the dentate nucleus and the cerebellum are determined semiquantitatively using a three-point scale: 0=normal (isoechogenicity compared to normal white matter); 1 = mild hyperechogenicity; 2 = marked hyperechogenicity (Postert *et al.*, 2004).

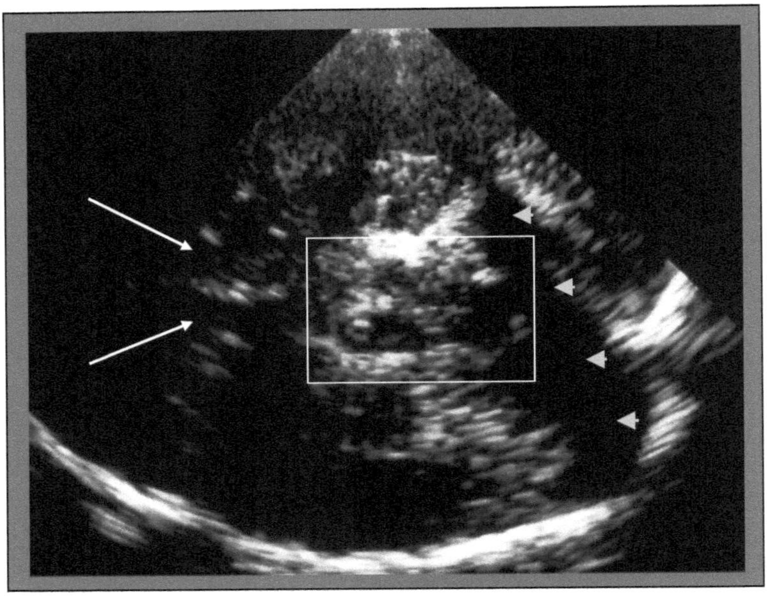

FIG. 1. TCS axial mesencephalic scanning plane in an SCA3 patient. Axial mesencephalic examination plane in an SCA3 patient: Siemens Sonoline Elegra, Erlangen, Germany; image depth 15 cm; left side of image = frontal, right side of image = occipital; white arrows = frontal horns of lateral ventricles; gray arrowheads = stripe artifact. The white rectangle contains the butterfly-shaped mesencephalic brainstem and the enlarged fourth ventricle (the zoomed image corresponding to the rectangle is shown with more details in Fig. 2).

III. TCS in Spinocerebellar Ataxia

A. TCS IN SCA2

1. *Genetics and Clinical Features of SCA2*

Spinocerebellar Ataxia type 2 (SCA2) is an autosomal-dominant inherited neurodegenerative disorder caused by an unstable expansion of the trinucleotide repeat cytosine-adenine-guanine (CAG) of the ATXN2 gene on chromosome 12q24.1 (Schöls *et al.*, 2004). Generally, a sequence of 32 or more CAG repeats results in the manifestation of the disease (Pulst *et al.*, 1996). The worldwide proportion of SCA2 patients in the totality of the autosomal-dominant inherited ataxias is about 4–40% and in Europe about 15%, while the prevalence in the Cuban province Holguin is about 100 : 100,000 (Abele and Klockgether, 2005). SCA2 is characterized by progressive cerebellar ataxia, including nystagmus, diminished velocity saccades, and, in some individuals, ophthalmoparesis. Pyramidal findings are present; deep tendon reflexes are brisk early on and are

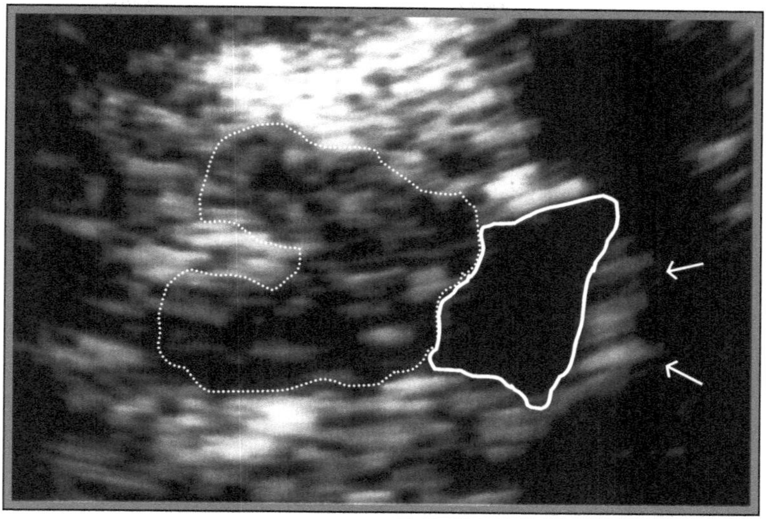

FIG. 2. TCS of the fourth ventricle in SCA3. Enlarged details in the same patient as Fig. 1. The plane of section is approximately corresponding to the white rectangle of Fig. 1. The ultrasound probe has been slightly tilted with its dorsal part into caudal direction by about 40° in order to improve visualization of the fourth ventricle. The fourth ventricle (white line) is displayed as an anechogenic structure in immediate dorsal vicinity of the butterfly-shaped mesencephalic brainstem (dotted line). The echosignals dorsal to the fourth ventricle (arrows) represent the cerebellar parenchyma bordering to the fourth ventricle. For exemplary images of the dentate nuclei, usually visualized in the same plane, confer Chapter 2 of this volume.

absent later in the course (Orozco *et al.*, 1990). Age of onset is typically in the fourth decade with a 10- to 15-year disease duration (Cancel *et al.*, 1997). The disease is more rapidly progressive when onset occurs before the age of 20 years. Interestingly, an SCA2 phenotype that includes L-dopa-responsive Parkinsonism has been reported (Furtado *et al.*, 2002; Lu *et al.*, 2004).

2. *Anatomic Pathology and Neuroimaging in SCA2*

Postmortem examinations of seven patients with SCA2 revealed a marked reduction in the number of cerebellar Purkinje cells (Orozco *et al.*, 1999). Furthermore, Purkinje cell dendrites had poor arborization as they passed through the granular layer. In the brainstem, marked neuronal loss in the inferior olive and pontocerebellar nuclei was observed. Six of seven brains also had marked loss in the substantia nigra (Orozco *et al.*, 1999). Marked demyelination was present in the posterior columns of the spinal cords and to a lesser degree in the spinocerebellar tracts. In the lumbar and sacral segments, anterior and posterior roots were partially demyelinated. Degeneration in the thalamus and reticulotegmental nucleus of the pons has also been reported (Rub *et al.*, 2003, 2004). Severe gyral atrophy, most prominent in the frontotemporal lobes, but without neuronal rarefaction has been observed additionally, as well as atrophic and gliotic cerebral white matter (Orozco *et al.*, 1989). Furthermore, a degeneration of the nigro-luyso-pallidal system mainly involving the substantia nigra has been described (Adams *et al.*, 1997).

MRI studies have shown that cerebellar cortical atrophy and olivopontocerebellar atrophy are two main patterns of morphologic changes in SCA2 (Klockgether *et al.*, 1998a). Moreover, a cruciform hyperintensity in the pons ("hot cross bun" sign) has been described as a frequent finding in patients with SCA2, reflecting atrophy and gliosis of the transverse part of the pontocerebellar fibers projecting to the middle cerebellar peduncles, the crossing part of the pontocerebellar fibers at the basis pontis, and the middle part of the reticular formation (Lee *et al.*, 2009).

3. *TCS Findings in SCA2*

There is only one report on TCS findings in SCA2, so far (Mijajlovic *et al.*, 2008). Mijajlovic *et al.* investigated echogenicities of substantia nigra (SN), caudate nucleus (CN), lentiform nucleus (LN), thalami, as well as the ventricular system in six SCA2 patients. For classification of SN echogenicity, planimetric measurement of echogenic area was performed. Using an Esaote Technos MP ultrasound system (Genova, Italy), sizes between 0.19 and 0.24 cm^2 were classified as moderately and sizes of 0.25 cm^2 and more were graded as markedly hyperechogenic.

Hyperechogenicity of substantia nigra was found in four (67%) patients (Table II). In two patients, hyperechogenicity was classified as marked. In one of them, SN hyperechogenicity was detected bilaterally, while in both patients with moderate SN hyperechogenicity these findings were detected unilaterally. In statistical comparison to findings in 30 healthy controls, the SN echogenicity was significantly larger, while no significant difference was seen in the comparison with 30 patients with Parkinson's disease (PD). No differences were observed among echogenicity of CN, LN, or thalamus.

In comparison to controls and PD patients, the widths of the third ventricle and of the frontal horn of the lateral ventricle were larger in the SCA2 group (Table II) (Mijajlovic *et al.*, 2008).

In the evaluation of this small group of SCA2 patients, no correlations between clinical features and TCS findings were reported. It was noted that the four SCA2 patients with SN hyperechogenicity did not exhibit Parkinsonism-like symptoms at the time of evaluation (Mijajlovic *et al.*, 2008). However, the authors discussed that the high prevalence of SN hyperechogenicity in SCA2 may reflect an early nigrostriatal impairment, so that this might be an

TABLE II
OVERVIEW OF PUBLISHED DATA ON TCS FINDINGS IN SCA

	SCA2 Mijajlovic et al., 2008	SCA3 Postert et al., 2004
General data		
Total number of SCA-patients	6	15
Male : female	2 : 4	10 : 5
Age in years Mean (SD)	37.0 (± 2.5)	48.5 (± 12.5)
TCS findings		
Ultrasound system	Esaote Technos MP, Genova, Italy	Siemens Sonoline Elegra, Erlangen, Germany
SN evaluation	planimetric, cm^2 Moderate: 0.19–0.24 Marked: >0.24	Planimetric, cm^2 Moderate: 0.20–0.24 Marked: >0.24
SN hyperechogenicity, marked	2/6 (33.3%)	6/15 (40.0%)
Total	4/6 (66.7%)	11/15 (73.3%)
SN hyperechogenicity bilateral	1/6 (16.7%)	5/15 (33.3%)
CN hyperechogenicity	Not specified	0/15 (0.0%)
LN hyperechogenicity	Not specified	6/15 (28.6%)
Thalamus hyperechogenicity	Not specified	Not done
Two or more basal ganglia affected	Not specified	6/15 (28.6%)
BR hypoechogenicity total (grade 0 + 1)	Not done	1/15 (6.7%)
BR not visible (grade 0)		0/15 (0.0%)
Dentatus hyperechogenicity, marked	Not done	5/15 (33.3%)
Total		7/15 (46.7%)
Cerebellum hyperechogenicity, marked	Not done	1/15 (6.7%)
Total		8/15 (53.3%)
Identification of fourth ventricle	Not done	15/15 (100%)
axial diameter (n = 9)		16.9 mm (± 2.7 mm)
circumference (n = 5)		68.6 mm (± 13.5 mm)
Width of frontal horn of side ventricle, mean (SD)	21.1 mm (± 4.7 mm)	18.5 mm (± 3.2 mm)
Width of third ventricle, mean (SD)	12.7 mm (± 2.3 mm)	4.8 mm (± 1.6 mm)

explanation for the Parkinsonism-predominant presentation of SCA2 (Furtado *et al.*, 2002; Lu *et al.*, 2004).

B. TCS IN SCA3

1. *Genetics and Clinical Features of SCA3*

Spinocerebellar ataxia type 3 (SCA3), also called Machado–Joseph disease (MJD), is an autosomal-dominant inherited neurodegenerative disorder caused by an unstable expansion of the trinucleotide repeat cytosine-adenine-guanine (CAG) of the ATXN3 gene on chromosome 14q24.3-q31 encoding for the ataxin-3 protein (Schöls *et al.*, 2004). Generally, a sequence of 55 or more CAG repeats results in the manifestation of the disease (Abele and Klockgether, 2005). Globally, SCA3 is the most frequent mutation in the whole of autosomal-dominant inherited ataxias, while in Germany the proportion is about 40% (Schöls *et al.*, 2005). The disease was first described in patients with origin from the Azores, in this Portuguese archipelago the prevalence is about 700 : 100,000 (Rosenberg, 1992; Abele and Klockgether, 2005).

SCA3 is characterized by a progressive cerebellar syndrome with ataxia and dysarthria as well as variable findings including a Parkinsonian syndrome, dystonic-rigid syndrome, or a combined syndrome of dystonia and peripheral neuropathy (Schöls *et al.*, 2004). Minor, but more specific clinical features are progressive external ophthalmoplegia, dystonia, restless legs syndrome, action-induced facial and lingual fasciculation-like movements, and bulging eyes (Klockgether, 2010). Age of onset is commonly in the third to fifth decade with a 15- to 25-year disease duration, and the longer the CAG repeat, the earlier the clinical manifestation of the disease (Klockgether *et al.*, 1998a) appears.

Based on phenotype variability, SCA3 has been subclassified into several subtypes (Fowler 1984). SCA3 type I (13% of cases) is characterized by early onset and prominent spasticity, rigidity, and bradykinesia, while atactic syndromes appear not very prominently (Lu *et al.*, 2004). SCA3 type II (with 57% of cases the most common type of the disease) is characterized by ataxia and upper motor neuron signs like spastic paraplegia (Landau *et al.*, 2000). SCA3 type III (30% of cases) is characterized by a late onset of the disease with ataxia and peripheral polyneuropathy. Interestingly, a fourth phenotype of SCA3 including L-dopa-responsive Parkinsonism has also been described (Schöls *et al.*, 2000; Lu *et al.*, 2004).

2. *Anatomic Pathology and Neuroimaging in SCA3*

Postmortem examinations in SCA3 patients revealed a marked degeneration of neurons with cell loss in the pons, substantia nigra, thalamus, anterior horn cells, and Clarke's column in the spinal cord, vestibular nucleus, many cranial motor nuclei and other brainstem nuclei (Rub *et al.*, 2002, 2003, 2004).

The cerebellum typically shows atrophic alterations, but in contrast to many other types of SCA (SCA1, SCA2, SCA6), Purkinje cells, and inferior olivary neurons are relatively spared whereas pronounced degeneration is seen in the dentate nucleus (Sequiros and Coutinho, 1993). Furthermore, intranuclear aggregation of the expanded ataxin-3 protein has been described, indicating that the disease protein ataxin-3 accumulates in ubiquitinated intranuclear inclusions selectively in neurons of affected brain regions (Paulson *et al.*, 1997).

MRI studies have shown that enlargement of the fourth ventricle, which reflects the underlying pontocerebellar atrophy, is the most commonly observed abnormal finding in SCA3 patients (Burk *et al.*, 1996; Onodera *et al.*, 1998). Voxel-based morphometric (VBM) studies revealed reduction of gray matter in the pons as well as in the vermis, documenting the pontocerebellar atrophy with predominant affection of pontine nuclei and fiber tracts in SCA3 (Lukas *et al.*, 2006). On T2-weighted and FLAIR MR images, hyperintensity of the globus pallidus interna has also been observed (Yamada *et al.*, 2005).

Moreover, spinal cord atrophy has been documented in MRI studies, while VBM of the upper cervical cord correlated with clinical signs, suggesting that spinal cord atrophy could serve as a noncerebellar neurodegenerative process in SCA3 (Lukas *et al.*, 2008).

3. *TCS Findings in SCA 3*

There is only one report on TCS findings in SCA3 patients so far (Postert *et al.*, 2004). According to the methodology used in previous TCS studies, echogenicities of substantia nigra (SN), caudate nucleus (CN), lentiform nucleus (LN), as well as the ventricular system were investigated in 15 SCA3 patients. For classification of SN echogenicity, planimetric measurement of echogenic areas was performed. Using a Sonoline Elegra ultrasound system (Siemens Medical, Erlangen, Germany), sizes between 0.20 and 0.24 cm^2 were classified as moderately and sizes of 0.25 cm^2 and more were graded as markedly hyperechogenic.

Hyperechogenicity of substantia nigra was found in 11 (73.3%) of the patients (Table II). In six patients, hyperechogenicity was classified as marked. In five of the patients with SN hyperechogenicity, these findings were detected bilaterally.

Furthermore, in six patients (28.6%), hyperechogenicity of lentiform nucleus was revealed. This observation was made bilaterally in all of these six SCA3 patients. As LN hyperechogenicity is thought to be a typical finding in patients with idiopathic dystonia (Naumann *et al.*, 1996), it was stressed that the only SCA3 patient with dystonia as an additional symptom in this series did not exhibit abnormal lenticular echogenicity.

As MRI studies revealed enlargement of the fourth ventricle to be the most characteristic finding in SCA3 (Klockgether *et al.*, 1998a), Postert *et al.* proposed a new "cerebellar examination plane" to make this finding visible also in TCS. In fact, in all SCA3 patients the fourth ventricle could be easily identified as a hypo- to anechogenic structure dorsally to the mesencephalic brainstem (Figs. 1 and 2), whereas this was not possible in healthy individuals (Postert *et al.*, 2004). Diameters and circumferences of the fourth ventricle varied between patients depending on the extent of cerebellar atrophy. Moreover, the dentate nucleus can be evaluated in this "cerebellar examination plane" (Fig. 2). In seven (46.7%)

of the 15 SCA3 patients, hyperechogenicity of DN could be assessed. This finding is in accordance to the observation of pronounced degeneration in this cerebellar area in SCA3 patients (Dürr *et al.*, 1996).

Comparing the TCS findings of SCA3 patients with those of SCA2 (Table II), a larger width of third ventricle has been reported in patients with SCA2. It would be important to test the possible significance of this observation in differentiating the two types of SCA in a larger patient series.

C. TCS in SCA17

1. *Genetics and Clinical Features of SCA17*

SCA17 is a rare type of autosomal-dominant spinocerebellar ataxia caused by a CAG/CAA expansion in the gene encoding the TATA-binding protein (TBP) (Koide *et al.*, 1999). Generally, a sequence of more than 42 repeats results in the manifestation of the disease (Gao *et al.*, 2008). Fewer than 100 families with SCA17 have been reported. In the Japanese population, the prevalence of SCA17 is estimated 0.47 : 1,000,000 and SCA17 accounts for roughly 0.3% of autosomal-dominant SCA (Maruyama *et al.* 2002). In northeast England, the minimum prevalence of SCA17 is approximately 0.16 : 1,000,000 (Craig *et al.*, 2005).

SCA17 is characterized by progressive dementia, cerebellar ataxia, involuntary movements including dystonia, and chorea (Hagenah *et al.*, 2004; Toyoshima *et al.*, 2004). Pyramidal as well as extrapyramidal signs like rigidity are common. Of note is the high proportion of individuals with psychiatric problems and chorea (Rolfs *et al.*, 2003). Although the disease course is variable, ataxia and psychiatric abnormalities are frequently the initial findings followed by involuntary movement, Parkinsonism, dementia, and pyramidal signs. Age of onset is typically in the fourth decade (Koide *et al.*, 1999). The clinical features correlate with the length of the CAA/CAG repeat expansion (Gao *et al.*, 2008).

2. *Anatomic Pathology and Neuroimaging in SCA17*

Neuropathological examinations in SCA17 patients demonstrated atrophy of cerebellar, cortical, and subcortical regions (Bruni *et al.*, 2004). Histologically, neuronal loss is observed in the striatum and Purkinje cell layer. Loss of cerebral cortical neurons is seen in some individuals (Rolfs *et al.*, 2003).

MRI shows prominent cerebellar atrophy as well as variable atrophy of cortical and subcortical regions (Lasek *et al.*, 2006). In a voxel-based morphometric (VBM) study, a correlation between the clinical cerebellar, extrapyramidal, and psychiatric scores and degeneration in specific areas could be documented.

Regarding motor dysfunction, atrophy of the gray matter involved mainly the cerebellum and other motor networks, in particular, the basal ganglia were found (Lasek *et al.*, 2006). In contrast, correlations with psychiatric scores revealed gray matter degeneration patterns in the frontal and temporal lobes, the cuneus and cingulum. A highly significant correlation between the clinical Mini-Mental State Examination scores and atrophy of the nucleus accumbens was documented (Lasek *et al.*, 2006).

3. *TCS Findings in SCA17*

There is only one report on TCS findings in SCA17 so far (Wolters *et al.*, 2005). Wolters *et al.* investigated basal ganglia echogenicities and the ventricular system in eight patients with SCA17. The authors describe a prevalent visualization of the fourth ventricle in "cerebellar examination plane," reflecting the prominent cerebellar atrophy in this disease. Furthermore, a hypoechogenicity of brainstem raphe was documented frequently, and this finding correlated with disease severity. This is of particular interest, since psychiatric features like depression are reported frequently in SCA17 (Rolfs *et al.*, 2003). Additionally, a correlation between hyperechogenicity of caudate nucleus (CN) and degree of dementia was documented (Wolters *et al.*, 2005). Interestingly, in SCA17 patients with additional extrapyramidal syndromes, substantia nigra was not hyperechogenic in general. Thus, the authors discussed that this finding may reflect a nigrostriatal impairment different from that in patients with Parkinson's disease.

IV. TCS in other Neurological Disorders with Ataxia

A. TCS in Sporadic Creutzfeldt–Jakob Disease

Creutzfeld–Jakob disease (CJD) is a rapidly progressive, fatal neurodegenerative disease caused by accumulation of human prion protein, typically presenting with psychiatric and neurological symptoms like rapidly progressive dementia, myoclonus, and ataxia (Sharma *et al.*, 2009). CJD can be classified into sporadic, hereditary, iatrogenic, and variant cases, while sporadic CJD (sCJD) represents the most common form of the disease with a worldwide incidence of one case per million per year (Sharma *et al.*, 2009). Periodic sharp wave complexes on electroencephalography and detection of 14.3.3 protein in cerebrospinal fluid are two characteristic diagnostic findings (Zerr *et al.*, 1998). In MRI imaging diffusion-weighted images (DWIs) and fluid-attenuate inversion recovery (FLAIR) images reveal characteristic abnormalities involving deep gray matter (striatum and/or thalamus) in about 70% of the cases (Young *et al.*, 2005).

Recently, TCS findings in a case series of three patients with sCJD were reported (Müller *et al.*, 2008). Using a Sonoline Antares ultrasound system (Siemens, Erlangen, Germany), the authors investigated two patients, with "probable" one patient with "definite" diagnosis. In all of the three cases, a blurry inhomogeneous hyperechogenic pattern was detected in large parts of the lentiform nucleus, while the thalamus exhibited a normal hypoechogenic signal. In two of the cases, this finding was depicted bilaterally, while in one of the "probable" sCJD cases LN hyperechogenicity was detected unilaterally. Copper accumulation, as well as the pathoanatomically described vacuolization of brain parenchyma, is discussed as responsible factors for the echogenic ultrasound signal. On the other hand, it remains unclear why the thalami exhibited a normal hypoechogenic signal in TCS investigations, particularly since MRI examinations of the same patients demonstrated signal alteration within thalamic structures. Regarding echogenicity of substantia nigra or brainstem raphe, no pathologic details are given. As MRI pathology usually is observed in late stages of the disease, it remains to be seen if TCS may help to establish the diagnosis earlier.

B. TCS in Bilateral Striopallidodentate Calcinosis

Bilateral striopallidodentate calcinosis (BSPDC) is an idiopathic or secondary calcification of the basal ganglia and of the cerebellum (Manyam *et al.* 2001). Formerly, the term "Fahr's disease" was used, but due to an inconsistent definition, the different subforms of this disease are subsumed as BSPDC (Manyam, 2005). Clinically, this disease can be associated with extrapyramidal movement disorders, with seizures, neuropsychological and psychiatric disturbances, and frequently with ataxia (Lopez-Villegas *et al.*, 1996; Manyam *et al.*, 2001). One case illustration reports the first sonographic detection of BSPDC (Krogias *et al.*, 2009). In a 65-year-old patient presenting with progressive ataxia and additional extrapyramidal signs, TCS examination revealed massive bilateral hyperechogenicity of pallidostriatal areas (Fig. 3). In a subsequent cranial computer tomography, symmetric striopallidododentate calcinosis was detected. The authors pointed out that this finding should be considered in TCS examinations as an additional diagnostic sign for the differentiation of extrapyramidal and atactic movement disorders. Recently, the potential of TCS in detecting bilateral hyperechogenicity in patients with BSPDC was verified in a second case report by another group (Ritter and Dittrich, 2010).

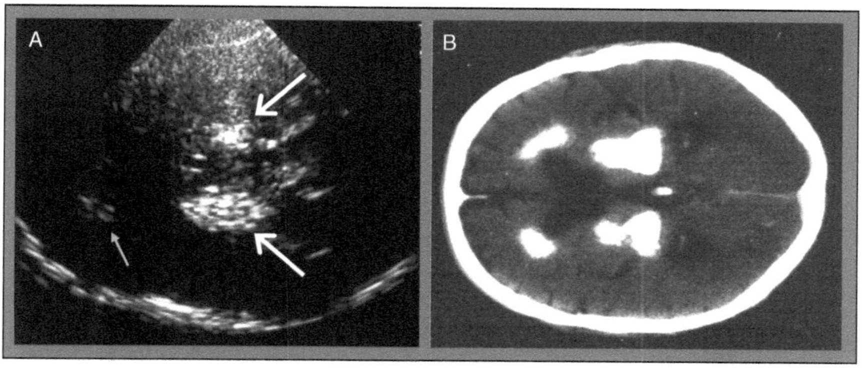

FIG. 3. Sonographic detection of bilateral striopallidodentate calcinosis. (A) Axial diencephalic examination plane in BSPDC patient (ultrasound system: Siemens Sonoline Elegra; parameter settings: examination depth 15 cm). Large white arrows indicate bilateral hyperechogenicity in pallidostriatal regions; small gray arrows, hyperechogenicity frontolateral to the frontal horn of side ventricle. (B) CT scan corresponding to the sonographic image shown in (A). CT shows excessive symmetric calcinosis of pallidostriatal regions.

V. Conclusions and Future Perspectives

Two TCS studies revealed comparable frequencies of hyperechogenicities as well as a similar pattern of basal ganglia lesions in SCA2 and SCA3. SN hyperechogenicity was a frequent finding in both reports, indicating a vulnerability of nigrostriatal system in SCA patients.

Although evidence of TCS alterations in ataxia is limited, with the proposed cerebellar examination plane the enlargement of the fourth ventricle and nucleus dentatus hyperechogenicity could be visualized as a characteristic finding in SCA3 as well as in SCA17 patients. Furthermore, bilateral distinct hyperechogenicity of pallidostriatal regions have been described as a novel diagnostic feature in the sonographic differentiation of extrapyramidal and atactic movement disorders.

TCS is a commonly available, noninvasive, and inexpensive diagnostic tool, which provides reliable information about the morphology of the brain in ataxia, even in agitated patients who do not tolerate other imaging techniques.

Further neuropathological and multimodal imaging studies are needed to elucidate the precise morphological and pathogenetic underlying cause of the detected echosignal pathology and to correlate these findings to the various clinical features of this disease entity.

References

Abele, M., and Klockgether, T. (2005). Ataxien. In Bewegungsstörungen (A. Ceballos-Baumann, and B. Conrad, Eds.), 2nd ed., pp. 287–307. Georg Thieme, Stuttgart.

Adams, C., Starkman, S., and Pulst, S. M. (1997). Clinical and molecular analysis of a pedigree of southern Italian ancestry with spinocerebellar ataxia type 2. *Neurology* **49**, 1163–1166.

Berg, D., Becker, G., Zeiler, B., Tucha, O., Hofmann, E., Preier, M., Benz, P., Jost, W., Reiners, K., and Lange, K. W. (1999). Vulnerability of the nigrostriatal system as detected by transcranial ultrasound. *Neurology* **53**, 1026–1031.

Berg, D., Godau, J., and Walter, U. (2008). Transcranial sonography in movement disorders. *Lancet Neurol.* **7**, 1044–1055.

Behnke, S., Berg, D., Naumann, M., and Becker, G. (2005). Differentiation of Parkinson's disease and atypical parkinsonian syndromes by transcranial ultrasound. *J. Neurol. Neurosurg. Psychiatry* **76**, 423–425.

Bruni, A. C., Takahashi-Fujigasaki, J., Maltecca, F., Foncin, J. F., Servadio, A., Casari, G., D'Adamo, P., Maletta, R., Curcio, S. A., De Michele, G., Filla, A., El Hachimi, K. H., and Duyckaerts, C. (2004). Behavioral disorder, dementia, ataxia, and rigidity in a large family with TATA box-binding protein mutation. *Arch. Neurol.* **61**, 1314–1320.

Burk, K., Abele, M., Fetter, M., Dichgans, J., Skalej, M., Laccone, F., Didierjean, O., Brice, A., and Klockgether, T. (1996). Autosomal dominant cerebellar ataxia type I clinical features and MRI in families with SCA1, SCA2 and SCA3. *Brain* **119**, 1497–1505.

Conrad, B. (2005). Wilson-Krankheit. In Bewegungsstörungen A. Ceballos-Baumann, and B. Conrad, Eds.), 2nd ed., pp. 287–307. Georg Thieme, Stuttgart.

Craig, K., Keers, S. M., Walls, T. J., Curtis, A., and Chinnery, P. F. (2005). Minimum prevalence of spinocerebellar ataxia 17 in the north east of England. *J. Neurol. Sci.* **239**, 105–109.

Dürr, A., Stevanin, G., Cancel, G., Duyckaerts, C., Abbas, N., and Didierjean, O. (1996). Spinocerebellar ataxia 3 and Machado-Joseph disease: Clinical, molecular, and neuropathological features. *Ann. Neurol.* **39**, 490–499.

Fowler, H. L. (1984). Machado-Joseph-Azorean disease. A ten-year study. *Arch. Neurol.* **41**, 921–925.

Furtado, S., Farrer, M., Tsuboi, Y., Klimek, M. L., de la Fuente-Fernandez, R., Hussey, J., Lockhart, P., Calne, D. B., Suchowersky, O., Stoessl, A. J., and Wszolek, Z. K. (2002). SCA-2 presenting as parkinsonism in an Alberta family: Clinical, genetic, and PET findings. *Neurology* **59**, 1625–1627.

Gao, R., Matsuura, T., Coolbaugh, M., Zühlke, C., Nakamura, K., Rasmussen, A., Siciliano, M. J., Ashizawa, T., and Lin, X. (2008). Instability of expanded CAG/CAA repeats in spinocerebellar ataxia type 17. *Eur. J. Hum. Genet.* **16**, 215–222.

Greenfield, J. G. (1954). The Spino-cerebellar Degenerations. Charles C. Thomas, Springfield.

Geary Jr, J. R., Earle, K. M., and Rose, A. S., 1956. Olivopontocerebellar atrophy. *Neurology* **6**, 218–224.

Hagenah, J. M., Zuhlke, C., Hellenbroich, Y., Heide, W., Klein, C. (2004). Focal dystonia as a presenting sign of spinocerebellar ataxia 17. *Mov. Disord.* **19**, 217–220.

Harding, A. E. (1983). Classification of the hereditary ataxias and paraplegias. *Lancet* **1**, 1151–1155.

Klockgether, T., Skalej, M., Wedekind, D., Luft, A. R., Welte, D., Schulz, J. B., Abele, M., Bürk, K., Laccone, F., Brice, A., and Dichgans, J. (1998). Autosomal dominant cerebellar ataxia type I. MRI-based volumetry of posterior fossa structures and basal ganglia in spinocerebellar ataxia types 1, 2, and 3. *Brain* **121**, 1687–1693.

Klockgether, K. (2010). Sporadic ataxia with adult onset: Classification and diagnostic criteria. *Lancet Neurol.* **9**, 94–104.

Koide, R., Kobayashi, S., Shimohata, T., Ikeuchi, T., Maruyama, M., Saito, M., Yamada, M., Takahashi, H., and Tsuji, S. (1999). A neurological disease caused by an expanded CAG trinucleotide repeat in the TATA-binding protein gene: A new polyglutamine disease? *Hum. Mol. Genet.* **8**, 2047–2053.

Krogias, C., Meves, S., Schoellhammer, M., Gold, R., and Andrich, J. (2009). Sonographic detection of bilateral striopallidodentate calcinosis. *J. Neurol.* **256**, 266–267.

Landau, W. M., Merims, D., Abeliovich, D., Zlotogora, J., and Gadoth, N. (1996). Machado-Joseph disease: Correlation between the clinical features, the CAG repeat length and homozygosity for the mutation. *Eur J Hum Genet.* **4**, 3–7.

Lasek, K., Lencer, R., Gaser, C., Hagenah, J., Walter, U., Wolters, A., Kock, N., Steinlechner, S., Nagel, M., Zuhlke, C., Nitschke, M. F., Brockmann, K., Klein, C., Rolfs, A., and Binkofski, F. (2006). Morphological basis for the spectrum of clinical deficits in spinocerebellar ataxia 17 (SCA17). *Brain* **129**, 2341–2352.

Lee, Y. C., Liu, C. S., Wu, H. M., Wang, P. S., Chang, M. H., and Soong, B. W. (2009). The "hot cross bun" sign in the patients with spinocerebellar ataxia. *Eur. J. Neurol.* **16**, 513–516.

Lopez-Villegas, D., Kulisevsky, J., Deus, J., Junque, C., Pujol, J., Guardia, E., and Grau, J. M. (1996). Neuropsychological alterations in patients with computed tomography-detected basal ganglia calcification. *Arch. Neurol.* **53**, 251–256.

Lu, C. S., Wu Chou, Y. H., Kuo, P. C., Chang, H. C., and Weng, Y. H. (2004). The parkinsonian phenotype of spinocerebellar ataxia type 2. *Arch. Neurol.* **61**, 35–38.

Lukas, C., Schöls, L., Bellenberg, B., Rüb, U., Przuntek, U., Schmid, G., Köster, O., and Suchan, B. (2006). Dissociation of grey and white matter reduction in spinocerebellar ataxia type 3 and 6: A voxel-based morphometry study. *Neurosci. Lett.* **408**, 230–235.

Lukas, C., Hahn, H. K., Bellenberg, B., Hellwig, K., Globas, C., Schimrigk, S. K., Köster, O., and Schöls, L. (2008). Spinal cord atrophy in spinocerebellar ataxia type 3 and 6: Impact on clinical disability. *J Neurol.* **255**, 1244–1249.

Manyam, B. V., Walters, A. S., and Narla, K. R. (2001). Bilateral striopallidodentate calcinosis clinical characteristics of patients seen in a registry. *Mov. Disord.* **16**, 258–264.

Manyam, B. V. (2005). What is and what is not "Fahr's disease". *Parkinsonism Relat. Disord.* **11**, 73–80.

Maruyama, H., Izumi, Y., Morino, H., Oda, M., Toji, H., Nakamura, S., and Kawakami, H. (2002). Difference in disease-free survival curve and regional distribution according to subtype of spinocerebellar ataxia: A study of 1,286 Japanese patients. *Am. J. Med. Genet.* **114**, 578–583.

Mijajlovic, M., Dragasevic, N., Stefanova, E., Petrovic, I., Svetel, M., and Kostic, V. S. (2008). Transcranial sonography in spinocerebellar ataxia type 2. *J. Neurol.* **255**, 1164–1167.

Müller, M., Treumann, T., Hewer, E., and Stoeck, K. (2008). Transcranial ultrasound of the basal ganglia in sporadic Creutzfeld–Jakob disease. *J. Neuroimaging* **18**, 154–157.

Naumann, M., Becker, G., Toyka, K. V., Supprian, T., and Reiners, K. (1996). Lenticular nucleus lesion in idiopathic dystonia detected by transcranial sonography. *Neurology* **47**, 1284–1290.

Onodera, O., Idezuka, J., Igarashi, S., Takiyama, Y., Endo, K., Takano, H., Oyake, M., Tanaka, H., Inuzuka, T., Hayashi, T., Yuasa, T., Ito, J., Miyatake, T., and Tsuji, S. (1998). Progressive atrophy of cerebellum and brainstem as a function of age and the size of the expanded CAG repeats in the MJD1 gene in Machado-Joseph disease. *Ann. Neurol.* **43**, 288–296.

Orozco, G., Estrada, R., Perry, T. L., Arana, J., Fernandez, R., Gonzalez-Quevedo, A., Galarraga, J., and Hansen, S. (1989). Dominantly inherited olivopontocerebellar atrophy from eastern Cuba. Clinical, neuropathological, and biochemical findings. *J. Neurol. Sci.* **93**, 37–50.

Orozco Diaz, G., Nodarse Fleites, A., Cordovés Sagaz, R., and Auburger, G. (1990). Autosomal dominant cerebellar ataxia: Clinical analysis of 263 patients from a homogeneous population in Holguin, Cuba. *Neurology* **40**, 1369–1375.

Paulson, H. L., Perez, M. K., Trottier, Y., Trojanowski, J. Q., Subramony, S. H., Das, S. S., Vig, P., Mandel, J. L., Fischbeck, K. H., and Pittman, R. N. (1997). Intranuclear inclusions of expanded polyglutamine protein in spinocerebellar ataxia type 3. *Neuron* **19**, 333–344.

Postert, T., Lack, B., Kuhn, W., Jergas, M., Andrich, J., Braun, B., Przuntek, H., Sprengelmeyer, R., Agelink, M., and Büttner, T. (1999). Basal ganglia alterations and brain atrophy in Huntington's disease depicted by transcranial real time sonography. *J. Neurol. Neurosurg. Psychiatry* **67**, 457–462.

Postert, T., Eyding, J., Berg, D., Przuntek, H., Becker, G., Finger, M., and Schöls, L. (2004). Transcranial sonography in spinocerebellar ataxia type 3. *J. Neural Transm.* **68** (Suppl.), 123–133.

Pulst, S. M., Nechiporuk, A., Nechiporuk, T., Gispert, S., Chen, X. N., Lopes-Cendes, I., Pearlman, S., Starkman, S., Orozco-Diaz, G., Lunkes, A., DeJong, P., Rouleau, G. A., Auburger, G., Korenberg, J. R., Figueroa, C., and Sahba, S. (1996). Moderate expansion of a normally biallelic trinucleotide repeat in spinocerebellar ataxia type 2. *Nat. Genet.* **14**, 269–276.

Ritter, M. A., and Dittrich, R. (2010). Fahr's disease – Diagnosis of the bilateral striopallidodentate calcinosis by cCT and transcranial sonography (TCS). *Klin. Neurophysiol.* **41**, 35–36.

Rolfs, A., Koeppen, A. H., Bauer, I., Bauer, P., Buhlmann, S., Topka, H., Schöls, L., and Riess, O. (2003). Clinical features and neuropathology of autosomal dominant spinocerebellar ataxia (SCA17). *Ann. Neurol.* **54**, 367–375.

Rosenberg, R. N. (1992). Machado-Joseph disease: An autosomal dominant motor system degeneration. *Mov. Disord.* **7**, 193–203.

Rub, U., de Vos, R. A., Schultz, C., Brunt, E. R., Paulson, H., and Braak, H. (2002). Spinocerebellar ataxia type 3 (Machado–Joseph disease): Severe destruction of the lateral reticular nucleus. *Brain* **125**, 2115–2124.

Rub, U., Del Turco, D., Del Tredici, K., de Vos, R. A., Brunt, E. R., Reifenberger, G., Seifried, C., Schultz, C., Auburger, G., and Braak, H., 2003. Thalamic involvement in a spinocerebellar ataxia type 2 (SCA2) and a spinocerebellar ataxia type 3 (SCA3) patient, and its clinical relevance. *Brain* **126**, 2257–2272.

Rub, U., Burk, K., Schöls, L., Brunt, E. R., de Vos, R. A., Diaz, G. O., Gierga, K., Ghebremedhin, E., Schultz, C., Del Turco, D., Mittelbronn, M., Auburger, G., Deller, T., and Braak, H. (2004). Damage to the reticulotegmental nucleus of the pons in spinocerebellar ataxia type 1, 2, and 3. *Neurology* **63**, 1258–1263.

Schöls, L., Peters, S., Szymanski, S., Krüger, R., Lange, S., Hardt, C., Riess, O., and Przuntek, H. (2000). Extrapyramidal motor signs in degenerative ataxias. *Arch. Neurol.* **57**, 1495–2000.

Schöls, L., Bauer, P., Schmidt, T., Schulte, T., and Riess, O. (2004). Autosomal dominant cerebellar ataxias: Clinical features, genetics, and pathogenesis. *Lancet Neurol.* **3**, 291–304.

Schöls, L., Amoiridis, G., Langkafel, M., Büttner, T., Przuntek, H., Riess, O., Vieira-Saecker, A. M., and Epplen, J. T. (1995). Machado-Joseph disease mutations as the genetic basis of most spinocerebellar ataxias in Germany. *J. Neurol. Neurosurg. Psychiatry* **59**, 449–450.

Seidel, G., Kaps, M., Gerriets, T., and Hutzelmann, A. (1995). Evaluation of the ventricular system in adults by transcranial duplex sonography. *J Neuroimaging* **5**, 105–108.

Sequeiros, J., and Coutinho, P. (1993). Epidemiology and clinical aspects of Machado-Joseph disease. *Adv. Neurol.* **61**, 139–153.

Sharma, S., Mukherjee, M., Kedage, V., Muttigi, M. S., Rao, A., and Rao, S. (2009). Sporadic Creutzfeldt-Jakob disease – A review. *Int. J. Neurosci.* **119**, 1981–1994.

Toyoshima, Y., Yamada, M., Onodera, O., Shimohata, M., Inenaga, C., Fujita, N., Morita, M., Tsuji, S., and Takahashi, H. (2004). SCA 17 homozygote showing Huntington's disease-like phenotype. *Ann. Neurol.* **55**, 281–286.

Walter, U., Niehaus, L., Probst, T., Benecke, R., Meyer, B. U., and Dressler, D. (2003). Brain parenchyma sonography discriminates Parkinson's disease and atypical parkinsonian syndromes. *Neurology* **60**, 74–77.

Walter, U., Krolikowski, K., Tarnacka, B., Benecke, R., Czlonkowska, A., and Dressler, D. (2005). Sonographic detection of basal ganglia lesions in asymptomatic and symptomatic Wilson disease. *Neurology* **64**, 1726–1732.

Walter, U., Behnke, S., Eyding, J., Niehaus, L., Postert, T., Seidel, G., and Berg, D. (2007). Transcranial brain parenchyma sonography in movement disorders: State of the art. *Ultrasound Med. Biol.* **33**, 15–25.

Wolters, A., Walter, U., Benecke, R., and Rolfs, A. (2005). Characterization of autosomal dominant spinocerebellar ataxias with transcranial magnetic stimulation and transcranial brain parenchyma sonography. *Klin. Neurophysiol.* **36**, 9–13.

Yamada, S., Nishimiya, J., Nakajima, T., and Taketazu, F. (2005). Linear high intensity area along the medial margin of the internal segment of the globus pallidus in Machado-Joseph disease patients. *J. Neurol. Neurosurg. Psychiatry* **76**, 573–575.

Young, G. S., Geschwind, M. D., Fischbein, N. J., Martindale, J. L., Henry, R. G., Liu, S., Lu, Y., Wong, S., Liu, H., Miller, B. L., and Dillon, W. P. (2005). Diffusion-weighted and fluid-attenuated inversion recovery imaging in Creutzfeldt-Jakob disease: High sensitivity and specificity for diagnosis. *AJNR Am. J. Neuroradiol.* **26**, 1551–1562.

Zerr, I., Bodemer, M., Gefeller, O., Otto, M., Poser, S., Wiltfang, J., Windl, O., Kretschmer, H. A., and Weber, T. (1998). Detection of 14-3-3 protein in the cerebrospinal fluid supports the diagnosis of Creutzfeldt-Jakob disease. *Ann. Neurol.* **43**, 32–40.

TRANSCRANIAL SONOGRAPHY IN HUNTINGTON'S DISEASE

Christos Krogias,* Jens Eyding,* and Thomas Postert†

*Department of Neurology, St. Josef-Hospital, Ruhr University Bochum, Germany
†Department of Neurology, St. Vincenz-Krankenhaus, Paderborn, Germany

Transcranial sonography (TCS) has become a reliable and sensitive diagnostic tool in the evaluation of extrapyramidal movement disorders, especially in the differentiation of Parkinsonian syndromes. Although only rarely reported, TCS reveals also signal alterations of basal ganglia in Huntington's disease (HD). Distinct findings are related to all of the three symptom domains of the clinical triad of the disease.

The TCS finding of substantia nigra hyperechogenicity was related to higher clinical disease severity. A poorer cognitive performance correlated with larger width of third ventricle. Moreover, widths of frontal horns of lateral ventricles measured with TCS corresponded closely to diameters estimated by CT imaging. Depressive symptoms were found to be associated with abnormal echogenicity of mesencephalic raphe structures. Furthermore, a larger number of CAG repeats in the *huntingtin* gene correlated with presence of SN hyperechogenicity.

INTERNATIONAL REVIEW OF
NEUROBIOLOGY, VOL. 90
DOI: 10.1016/S0074-7742(10)90017-2

This review provides information about the examination procedure and its diagnostic value in HD. Possible morphological and pathophysiological mechanisms leading to changes in the reflection of ultrasound waves are discussed in the context of established neuroimaging modalities.

I. Introduction

A. Genetics and Pathophysiology

Huntington's disease (HD) is an autosomal-dominant inherited neurodegenerative disorder caused by an unstable expansion of the trinucleotide repeat cytosine-adenine-guanine (CAG) of the IT-15 gene on chromosome 4p16.3 (Huntington's Disease Collaborative Research Group, 1993). The CAG repeat is translated into an enlarged extension of the polyglutamine tract in the cytoplasmatic protein huntingtin (HTT) (Walker 2007). Generally, a sequence of 36 or more glutamines results in an altered form of this protein, called mutant huntingtin (mHTT). The presence and accumulation of mHTT result in gradual damage to certain types of neurons.

The underlying biochemical process of the observed neural decay remains unclear, but it is apparently related to a particular vulnerability of striatal spiny neurons (Davies *et al.*, 1997). Such "medium spiny neurons" comprise the neurotransmitters gamma-aminobutyric acid (GABA) and enkephalin or substance P (MacMillan and Quarrell, 1996). Furthermore, striatal D1 and D2 dopamine as well as *N*-methyl-D-aspartate (NMDA) receptor loss could be identified in symptomatic and even in presymptomatic mutation carriers (Weeks *et al.*, 1996; Young *et al.*, 1988). Different cerebral regions have differing amounts and reliance on these type of neurons, and are affected accordingly (Walker 2007). Thus, a progressive neurodegeneration of basal ganglia, particularly of neostriatum, can be observed even prior to the outbreak of clinical symptoms (Aylward *et al.*, 2004; Saft *et al.*, 2009). However, neuronal loss has been identified also in other subcortical and cortical structures (Fennema-Notestine *et al.*, 2004; Rosas *et al.*, 2008).

The genetic confirmation of an abnormal expansion of more than 36 CAG repeats is not automatically considered the definite diagnosis of the disease, since it may be obtained decades before the symptoms begin (Saft *et al.*, 2009). Usually, symptoms of HD begin between 35 and 44 years of age (Walker 2007). Generally, the number of CAG repeats correlates with the severity of the eventual clinical picture and, in addition, it accounts for 60% of the variation of the age of the onset of symptoms (Brinkmann *et al.*, 1997). The remaining variation is attributed to environment and to modifier genes (Wexler *et al.*, 2004). 36–40 CAG repeats result in a reduced penetrance, with a much later onset and slower progression of symptoms. Recently, a case of a 75-year-old man with clinical features of HD and with only 34 CAG repeats has been reported (Andrich *et al.*, 2008). In case of very large CAG expansion, HD has full penetrance and can occur already under the age of 20, when it is then referred to as juvenile or Westphal variant of HD (Nance *et al.*, 2001).

B. Clinical Manifestation

The typical clinical manifestation of Huntington's disease is described as a triad of motor, cognitive, and emotional abnormalities (Walker, 2007). Typically, onset of symptoms is in middle age, but the disorder can manifest at any time between infancy and senescence (Walker, 2007).

Even though choreatic involuntary movements are the hallmark of the disease, this feature is only part of a generalized motor disturbance comprising oculomotor signs, dysarthria, hyperreflexia, difficulties with alternating movements, rigidity, bradykinesia, and dystonia (Sánchez-Pernaute *et al.*, 2000). Furthermore, disturbances of the autonomic nervous system have been described (Andrich *et al.*, 2002).

The juvenile or Westphal variant accounts for about 7–10% of HD carriers, and it is attended by a predominant hypokinetic-rigid syndrome (Nance *et al.*, 2001).

In addition, this variant shows a more severe and rapid course of the disease, resulting in a drastic reduction of life expectancy (Nance *et al.*, 2001).

While diagnosis and treatment of HD focus on the movement disorder, the clinical phenotype is associated also with psychiatric symptoms like depression, irritability, aggression, apathy, anxiety disorders, as well as psychotic and obsessive symptoms (Paulsen *et al.*, 2001).

Depression is one of the most frequent psychiatric features in patients with HD (Slaughter *et al.*, 2001). Even in the neurologically presymptomatic-stage depressive symptoms exist in 30–58% of the mutation carriers (Gargiulo *et al.*, 2009). Rates of suicides vary between 3 and 7.3% (Di Maio *et al.*, 1993).

The psychiatric disturbances often appear years before the outbreak of neurological symptoms (Duff *et al.*, 2007).

Due to the primary affection of subcortical regions, the pattern of cognitive deficits in HD has been characterized as a subcortical dementia suggesting dysfunction of the frontal–subcortical neuronal circuitry (Montoya *et al.*, 2006).

Cognitive dysfunction is an intrinsic feature of the disease, and there is evidence for an early cognitive impairment in HD patients even prior to the onset of motor disturbances (Ho *et al.*, 2003). In asymptomatic mutation carriers, subtle cognitive deficits have been found concerning psychomotor, attentional, and executive functions, as well as deficits in semantic verbal fluency and visual working memory

(Kirkwood *et al.*, 2000; Lemiere *et al.*, 2002). In the further course of the disease, patients show a progressive impairment of attention, planning, and executive function as well as immediate memory (Ho *et al.*, 2003). Furthermore, patients exhibit an overall deficit in recognizing emotion, with particularly severe deficits regarding recognition of expressions of disgust from both voices and faces (Sprengelmeyer *et al.*, 1996). At later stages of the disease, memory deficits become more manifest and the cognitive decline leads to subcortical dementia being characterized by a slowing of information processing, decreased motivation, depression, apathy, and personality changes, although language is relatively spared (Paulsen *et al.*, 2001). Additionally, different visuospatial abilities have been shown to be impaired in patients with HD (Brouwers *et al.*, 1984; Montoya *et al.*, 2006).

C. Anatomic Pathology and Neuroimaging

In HD patients, the mutant huntingtin dominantly affects specific brain structures. The most prominent neurodegeneration takes place in a part in the striatum, composed of the caudate nucleus and putamen (Walker, 2007). Striatal spiny neurons are most vulnerable, particularly the ones with projections toward the external globus pallidus, with interneurons and spiny cells projecting to the internal pallidum being less affected (DeLong, 2000). Other areas affected include the substantia nigra, layers 3, 5, and 6 of the cerebral cortex, the hippocampus, Purkinje cells in the cerebellum, lateral tuberal nuclei of the hypothalamus, and parts of the thalamus (Walker, 2007). These areas are affected according to their structure and the types of neurons they contain. Due to the neuronal degeneration, these structures become reduced in size. Thus, different bicaudate diameter indices have been described as the earliest signs in structural neuroimaging of the disease (Stober *et al.*, 1984). Most commonly, the ratio of the greatest distance of the frontal horns and the shortest distance between the heads of the caudate nuclei (bicaudate diameter) is used for the quantification of caudate atrophy (Weindl and Conrad, 2005). Furthermore, the ratio of the width of both lateral ventricles at the level of the heads of the caudate nuclei to the distance between the outer tables of the skull at the same level is used as the bicaudate index (Fig. 1) (Ho *et al.*, 1995). Besides, the sum of the maximum distance of the frontal horns and the intercaudate distance is another method used in clinical practice to monitor brain atrophy in HD patients quantitatively (Huckmann *et al.*, 1975). All these features can be assessed by computed tomography (CT) as well as by magnetic resonance imaging (MRI) (Saft *et al.*, 2009). Voxel-based morphometric MRI studies revealed a reduced volume of basal ganglia, particularly of striatal regions (Kassubek *et al.*, 2004). Studying extrapyramidal regions, it could be shown that white matter volume is lower than

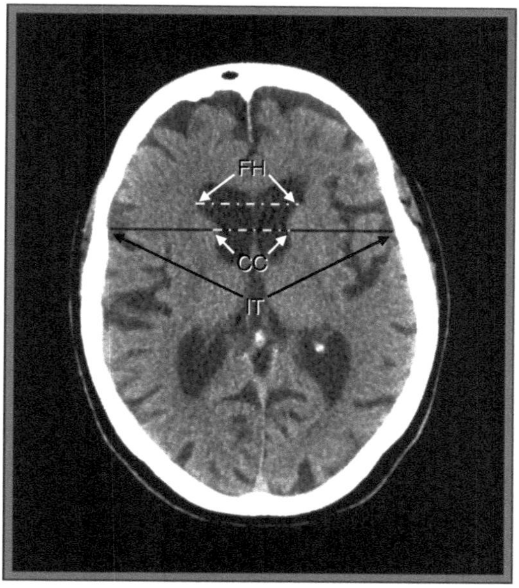

FIG. 1. Diameters used for quantification of brain atrophy in HD. Computer tomography of an HD patient with axial projection at the level of the heads of the caudate nuclei at with the frontal horn width (FH), the intercaudate distance (CC), and calvarial inner table (IT) width can be measured. The sum of FH and CC form the "Huckmann-number".

normal, while the cortical gray matter is relatively spared (Rosas *et al.*, 2008). Cortical degeneration seems to be regionally specific (Nopoulos *et al.*, 2007).

Clinical manifestations of disease do not simply depend on the extent of tissue loss, but represent a complex balance among neuronal dysfunction, tissue repair, and circuitry reorganization. Thus, studies dealing with functional neuroimaging modalities such as positron emission tomography (PET), single photon emission computer tomography (SPECT), and functional MRI (fMRI) have shown to be more sensitive to the earliest changes in HD than are structural imaging approaches (Paulsen, 2009). A hypometabolism of striatal regions with reduced glucose utilization could be determined by 18-fluorodeoxyglucose (FDG)-PET imaging even in early HD patients without detectable caudate atrophy on structural CT imaging (Kuhl *et al.*, 2005). Not only glucose, but also oxygen utilization is reduced in striatum and frontal cortex (Leeders *et al.*, 1986). Regarding pre- and postsynaptic dopaminergic systems, HD patients show a normal striatal ^{18}F-dopa uptake while ^{11}C-methylspiperone binding is reduced, suggesting a reduced density of postsynaptic D2 dopamine receptors (Leeders *et al.*, 1986). These findings go well with the described pathology of HD where fall out of spiny striatal neurons is seen, while the nigrostriatal pathway is spared (Brooks and Frackowiak, 1989). In a longitudinal study, serial ^{11}C-raclopride PET demonstrates a linear progression of striatal loss of

D2 dopamine receptors in HD patients over 3 years (Pavese *et al.*, 2003). Statistical parametric mapping also revealed a progressive loss of temporal and frontal D2 binding (Pavese *et al.*, 2003). Compared with healthy controls, HD gene carriers showed a dysfunctional decreased activation within the left dorsal anterior insula during processing of disgusted facial expressions in functional MRI (fMRI) imaging (Hennenlotter A *et al*, 2004).

II. Method of Transcranial Sonography in Huntington's Disease

Generally, the methodology of TCS of brain parenchyma in HD corresponds to the recommendations and guidelines for the diagnostic scanning and evaluating procedure in other movement disorders (Berg *et al.*, 2008; Walter *et al.*, 2007a):

In summary, TCS is performed through the transtemporal acoustic bone window while the patient is in a supine position. Ultrasound systems equipped with a 1.6- to 3.5-MHz transducer can be applied. The following parameter settings of the TCS system are recommended: penetration depth of 140–160 mm, suppression of low echosignals, dynamic range of 45–55 dB. Image brightness and time gain compensation are adapted as needed for each examination. Standardized midbrain and diencephalic scanning planes are visualized in the axial section. The image should be fixed and zoomed in 2- to 4-fold for further analysis. In the sole published TCS study in HD so far, only a semiquantitative analysis was performed (Postert *et al.*, 1999). Nevertheless, in case of hyperechogenicity of the substantia nigra (SN) a quantitative planimetric measurement can be recommended. Using the ultrasound system Sonoline Elegra (Siemens; Erlangen; Germany), sizes of less than 0.20 cm^2 are defined as normal, since this threshold was found to represent the upper quartile in normal population (Berg *et al.*, 1999a). Sizes between 0.20 and 0.25 cm^2 are classified as moderately and sizes of 0.25 cm^2 and above are graded as markedly hyperechogenic, since sizes of 0.25 cm^2 and above represent the upper decile in normal population (Berg *et al.*, 1999a). The brainstem raphe (BR) is evaluated by bilateral TCS investigation. The side with the best visualization of BR is used for evaluation. Echogenicity of BR is classified semiquantitatively on a three-point scale using the red nucleus or signal intensity of the basal cisterns as reference points: 0 = raphe structure not visible, 1 = reduced echogenicity: the echogenic line of the BR is slight and/or interrupted, 2 = normal echogenicity (echogenicity of raphe structure is not interrupted and intensity is equal to that of the red nucleus or basal cisterns). At grades 0 and 1 the BR is classified as hypoechogenic. The thalamus, the lentiform nucleus (LN), and the head of the caudate nucleus (CN) is graded as hyperechogenic if the echogenicity of the referring structure is more intense than that of the surrounding white matter.

The assessment of the ventricular system in HD patients shows a peculiarity: Generally, for the evaluation of the ventricular system, the measurement of the widths of the contralateral frontal horn is recommended (Seidel *et al.*, 1995; Walter *et al.*, 2007a). Frontal horn diameters and intercaudate distance measured by CT or MRI images have great relevance in the assessment of HD (see chapter I, part C of this review). Due to the tilted diencephalic scanning plane in TCS evaluation, the measurement of the width of the contralateral frontal horn comes along with a deviation from the corresponding width in an untilted axial CT scan. Therefore, in the only published TCS study in HD to date, the measurement of frontal horn diameter was performed on the frontal horn being located ipsilaterally to the ultrasound probe (Postert *et al.*, 1999). Precisely, the maximum width of the frontal horn of the lateral ventricle and in parallel line the distance between the septum pellucidum and the head of CN were measured (Fig. 2). With this approach, a significant correlation between measurements of diameters in TCS and cranial CT images could be achieved (Postert *et al.*, 1999). Nonetheless, in future studies an additional measurement of the widths of the contralateral frontal horns as described in the guidelines should be recommended, so that comparability with other TCS

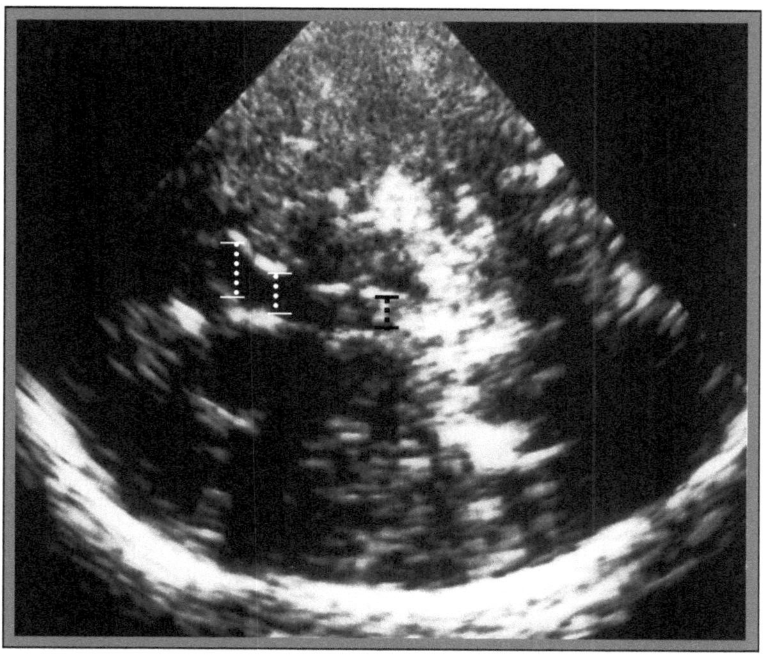

FIG. 2. Sonographic evaluation of the ventricular system in HD. Axial deincephalic TCS examination plane showing enlargement of trontal horns and of third ventricle in a HD patient. White dotted lines mark the large and mall diameter of the ipsilateral frontal horn, black dotted line marks the diameter of the third ventricle.

studies on movement disorders is facilitated. The measurement of the minimal width of the third ventricle requires no specific consideration and is performed on a standardized diencephalic examination plane.

III. Frequency of Abnormal TCS Findings in HD

In contrast to an increasing number of studies dealing with TCS findings in different Parkinsonian syndromes, there is only one manuscript reporting of TCS findings in HD so far (Postert *et al.*, 1999). Postert *et al.* investigated the echogenicities of substantia nigra (SN), caudate nucleus (CN), lentiform nucleus (LN), thalamus, as well as the ventricular system in 45 HD patients. Eighteen (40%) out of the 45 HD patients showed abnormal hyperechogenicity in at least one basal ganglia region. SN hyperechogenicity was found in 12 (26.7%) patients. In 10 (83.3%) of these cases, this finding was detected bilaterally. The number of CAG repeats correlated with the presence of SN hyperechogenicity (Postert *et al.*, 1999). The correlation of SN hyperechogenicity to clinical motor features is discussed in section IV of this review. Six (13.3%) patients showed hyperechogenicity of the CN and three (6.7%) of the LN. Echogenicity of thalamus region was documented consistently as normal. To determine the widths of the ventricular system using an established approach, in 33 of the patients a cerebral CT scan was performed. A significant correlation between measurements of ventricular diameters in TCS and CT images could be demonstrated. Additionally, a cranial MRI analysis was performed in 21 of the HD patients. Hyperechogenic lesions of the CN in TCS were associated with an increased signal intensity in T2-weighted MR images, suggesting that this finding was caused by gliosis or elevated concentration of trace metals (Postert *et al.*, 1999). For TCS analysis in this study, the Ultramark 3000 and Ultramark 9 (Advanced Technology Laboratories, Washington, USA) ultrasound systems were used.

In a subsequent TCS study, we further evaluated findings of deep brain structures in HD (Krogias *et al.*, 2010 submitted). Using a Sonoline Elegra ultrasound system (Siemens; Erlangen, Germany), for the first time echogenicity of the brainstem raphe (BR) was analyzed in addition to the echogenicity of substantia nigra (SN) and basal ganglia in 39 HD patients. Preliminary results of this study have been published before, reporting TCS findings of 32 HD patients (Krogias *et al.*, 2008). Eight patients (25.0%) showed SN hyperechogenicity. In four (50%) of them, this finding was depicted bilaterally. Four (12.5%) patients exhibited hyperechogenicity of the CN and two patients (6.3%) of the LN (Fig. 3). No signal alterations were seen in thalamic regions. In general, the previously published frequency of abnormal basal ganglia findings in HD patients could be replicated by another examiner using another ultrasound system. Although data

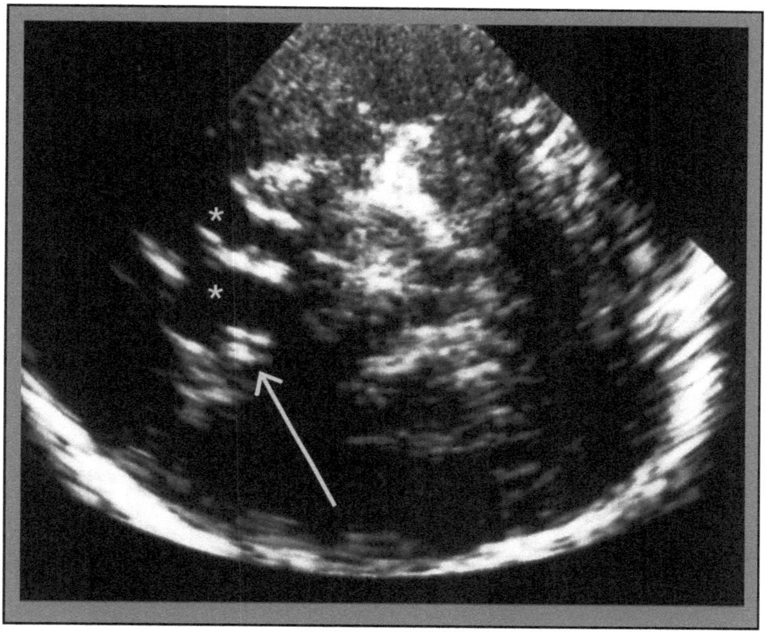

FIG. 3. TCS image of caudate nucleus hyperechogenicity in HD. White stars mark the frontal horns. Compared to surrounding brain parenchyma, echogenicity of the contralateral caudate nucleus is increased (arrow).

of the completed study revealed a frequency of SN hyperechogenicity being slightly higher compared to the earlier study (Postert *et al.*, 1999), this conclusion does not change (Krogias *et al.*, 2010 submitted). BR echogenicity was analyzed for the first time. As an interesting novel finding, 17 (53.2%) of the HD patients exhibited a reduced BR signal. A possible explanation of this finding and its correlation to clinical features is discussed in detail in chapter VI of this review.

Regarding the evaluation of the ventricular system, the recent study documented larger widths of third ventricle in HD patients (range 4.2–15.8 mm) compared with the earlier study (Postert *et al.*, 1999), where the values ranged from 3.1 to 7.5 mm. While a satisfactory explanation of this difference cannot be given, this difference may partly be attributed to the different ultrasound systems applied in the two studies. Established reference values refer only to the Elegra TCS system (normal range <7 mm for subjects aged 20–60 years; <10 mm for subjects aged more than 60 years) (Walter *et al.*, 2007a). If these reference values also applied for the Ultramark TCS system, the width of third ventricle in almost all HD patients in the study of Postert *et al.* would have to be classified as normal. Since the widths of frontal horns of lateral ventricles were not specified in the more recent TCS study, comparison of the values is not applicable. The association between the TCS measures of the ventricular system and clinical features in patients with HD is particularly discussed in section V of this review.

An overview of published TCS findings in HD is given in Table I. One TCS study was performed on a subgroup of HD patients with hypokinetic rigid symptoms

TABLE I
Overview of Published Data on TCS Findings in HD

	Postert et al., 1999	Krogias et al., 2008 (abstract)	Summarized
General data			
Total number of HD patients	49	34	83
Male : female	31 : 18	15 : 19	46 : 37
Age in years Mean (range)	50.0 (27–75)	47.8 (28–69)	n/a
CAG repeats Mean (range)	45.2 (39–58)	45.5 (39–63)	n/a
HD patients with sufficient bone window	45/49 (91.8%)	32/34 (94.1%)	77/83 (92.8%)
TCS findings			
Ultrasound system	Ultramark 3000 and Ultramark 9 (Advanced Technology Laboratories, Washington, USA)	Siemens Sonoline Elegra, Erlangen, Germany	n/a
SN evaluation	Semiquantitatively	Planimetric	n/a
SN hyperechogenicity total	12/45 (26.7%)	8/32 (25.0%)	20/77 (26.0%)
SN hyperechogenicity bilateral	10/45 (22.2%)	4/32 (12.5%)	14/77 (18.2%)
CN hyperechogenicity	6/45 (13.3%)	4/32 (12.5%)	10/77 (13.0%)
LN hyperechogenicity	3/45 (6.7%)	2/32 (6.3%)	5/77 (6.5%)
Thalamus hyperechogenicity	0/45 (0.0%)	0/32 (0.0%)	0/77 (0.0%)
Two or more basal ganglia affected	2/45 (4.4%)	2/32 (6.3%)	4/77 (5.2%)
BR hypoechogenicity total (grade 0 + 1)	Not done	17/32 (53.1%)	n/a
BR not visible (grade 0)	Not done	3/32 (9.4%)	n/a
Width of side ventricle	Small diameter: range = 7.8– 16.6 mm Large diameter: range = 14.8–26.0 mm	Not done	n/a
Width of third ventricle	Mean = not specified Range = 3.1–7.5 mm	Mean = 8.2 mm Range = 4.2–15.8 mm	n/a

n/a, = not applicable

(Lambeck *et al.*, 2008), so these findings are not itemized in Table I. In this subgroup of HD patients, the authors report a very high frequency of hyperechogenicity in different basal ganglia: hyperechogenic SN was detected in 14 (93.3%) of 15 cases, while hyperechogenicity of CN and LN was detected in 12 (80.0%) and in 8 (53.3%) of the 15 cases, respectively (Lambeck *et al.*, 2008). For TCS analysis in this study, the IU22 (Philips, Netherlands) ultrasound system was used. No data about TCS findings in presymptomatic mutation carriers have been published so far.

IV. TCS Findings Related to Motor Features in HD

A. Clinical Assessment of Motor Features in HD

Usually, the clinical status of HD patients is evaluated by the standardized modified version of the Unified Huntington's Disease Rating Scale (UHDRS) (Huntington Study Group, 1996). Among others, this scale comprises ratings of motor performance, independent functioning, and total functional capacity. In the first part of the UHDRS, the motor score (ms) is calculated as the sum of the scores of items 1–15. In these items, oculomotor performance, tongue movements, involuntary and alternating movements, as well as gait performance are rated. The sum ranges from 0 (healthy) to 124 points (most severely affected). The sum of the scores of items 70–74 composes the total functional capacity (tfc), ranging from 0 (most severely affected) to 13 points (healthy). The independent functioning, and therewith the overall severity of the disease, is graded on the independence scale ranging from 0–35% (dependent) over 40–75% (limited self-care) to 80–100% (self-care maintained).

B. Correlation between Motor Features and TCS Findings in HD

In both patient series (Krogias *et al.*, 2008; Postert *et al.*, 1999), no significant correlation between motor function scores and hyperechogenic lesions in any of the deep brain structures studied could be found. For further subgroup analysis, Postert *et al.* dichotomized the patient cohort according to the score on the independence scale (75% or less, corresponding to partial or complete independence, compared with 80% or more, corresponding to self-care maintained). Frequency of SN hyperechogenicity was significantly higher in the clinically severely affected patients than in the less affected patients ($P < 0.01$; Fisher's exact test) (Postert *et al.*, 1999). Such an analysis was not performed in our recent study. Even though statistical significance was not reached in the comparison with the control group, both TCS studies revealed an increased frequency of CN hyperechogenicity in HD patients. In addition, CN hyperechogenicity correlated with an increased signal intensity in T2-weighted MRI ($r = 0.378$, $P < 0.05$, Spearman's rank correlation) (Postert *et al.*, 1999). Moreover, it was surprising that disease severity in HD did not correlate with signal alteration of CN but with hyperechogenicity of SN, bearing in mind that PET studies indicated primary striatal pathology but an intact nigrostriatal pathway (Brooks and Frackowiak, 1989). The fact that in both TCS studies abnormalities could be visualized more often in the SN than in the CN might be due to the more difficult discrimination of the atrophic caudate nucleus from the highly echogenic border of the frontal horn on TCS. SN hyperechogenicity is a highly characteristic finding in idiopathic Parkinson's disease (PD) (Berg *et al.*, 2008, Walter *et al.*, 2003). It is assumed that this echosignal alteration is based on increased amounts of iron bound to proteins other than ferritin (Berg *et al.*, 2002). Furthermore, it has been shown that individuals without the diagnosis of idiopathic PD but with SN hyperechogenicity develop more frequently akinetic-rigid extrapyramidal symptoms (Berg *et al.*, 2001b, 2001c). Although choreatic hyperkinesia is the most conspicuous feature of HD, rigidity and bradykinesia can also be part of a complex generalized motor disturbance in HD patients (Sánchez-Pernaute *et al.*, 2000). These motor features are most significant in functional terms (Quarrel and Harper, 1991). This might be an explanation for the observed correlation between the SN hyperechogenicity and clinical status in HD patients (Postert *et al.*, 1999). Most of the HD patients included in the two discussed TCS studies were on antichoreatic and/or neuroleptic medication. Thus, possible exptrapyramidal side effects of the medication might bias an adequate evaluation of this hypothesis, particularly since appropriate PET studies have been performed on neuroleptic-free mutation carriers (Sánchez-Pernaute *et al.*, 2000). In this regard, an interesting observation has been recently reported in an abstract form: in HD patients with hypokinetic-rigid symptoms, hyperechogenicity of the SN was detected in 14 (93.3%) of 15 cases (Lambeck *et al.*, 2008). In 7 of these 14 subjects, hyperechogenicity was detected bilaterally. Also in our experience, the juvenile or Westphal variant of HD, which is attended by a predominant hypokinetic-rigid syndrome (Nance *et al.*, 2001), is associated with a higher frequency of SN hyperechogenicity. An exemplary image of SN hyperechogenicity in a 25-year-old HD patient with onset of hypokinetic motor symptoms at the age of 15 is shown in Fig. 4.

The precise histopathological and pathophysiological cause of basal ganglia hyperechogenicities in HD remains speculative. Gliotic lesions are known to account for high signal intensity on T2-weighted MRI (Grafton *et al.*, 1991). Hence, this may support the hypothesis that hyperechogenicity depicted by TCS reflects neurodegeneration. On the other hand, signal alteration on MRI can also be caused by accumulation of several metals (Hayflick *et al.*, 2006) This is of particular interest, since neuroradiological examinations revealed increased

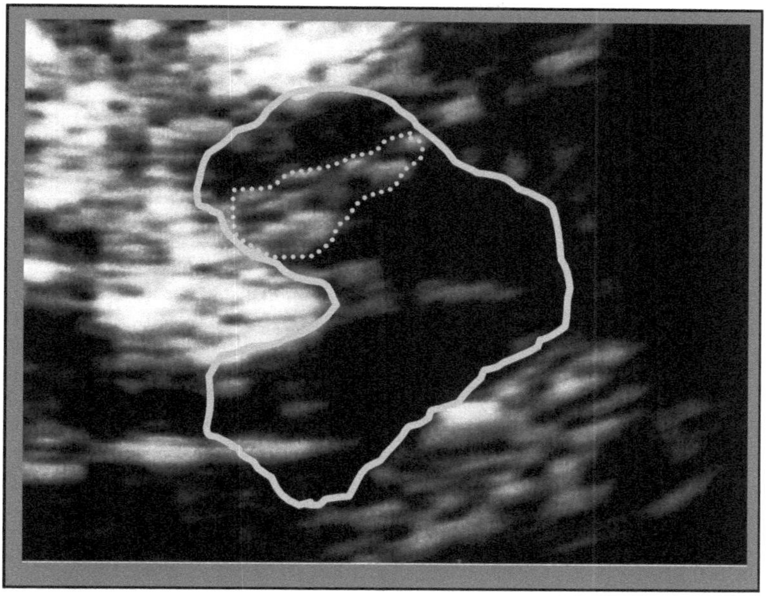

FIG. 4. TCS image of substantia nigra hyperechogenicity in a patient with Westphal-variant of HD. Mesencephalic axial examination plane in a 25-year-old patient with Westphal variant of HD showing onset of hypokinetic motor symptoms at the age of 15. The butterfly-shaped midbrain is encircled for better visualisation. Note the ipsilateral hyperechogenicity of the substantia nigra (SN). Echogenic area of the SN was encircled for computerized planimetric measurement.

ferritin–iron levels in the basal ganglia of HD patients (Bartzokis *et al.*, 2000). This condition may result in changes of iron–protein binding, which might cause echogenic alterations similar to SN hyperechogenicity in PD (Walter, 2007a). Moreover, molecular findings indicate that *huntingtin* plays an essential role in cellular iron homeostasis (Hilditch-Maguire *et al.*, 2000). Additionally, postmortem studies showed that not only iron but also copper levels are increased in Huntington's disease (Bartzokis *et al.*, 2000).

Copper accumulation is supposed to cause hyperechogenicity primarily of LN in patients with Wilson's disease (Walter *et al.*, 2005). LN hyperechogenicity is a frequent finding also in patients with cervical or upper-limb dystonia (Naumann *et al.*, 1996). Neuropathological examinations in idiopathic dystonia depicted significantly increased copper levels in globus pallidum and putamen compared with control brain samples (Becker *et al.*, 1999). Interestingly, increased concentration of copper has been found in the SN of HD patients (Dexter *et al.*, 1992). This fact supports the "copper hypothesis" as a possible underlying cause of observed echo-signal alterations in HD patients. Further neuropathological and multimodal imaging studies are needed to elucidate the pathogenesis of hyperechogenicities in HD.

V. TCS Findings Related to Cognitive Features in HD

A. Clinical Assessment of Cognitive Features in HD

The cognitive assessment of HD patients typically comprises a battery of neuropsychological tests (verbal fluency, symbol digit, color naming, word reading, interference). These tests deal with attention, execution function, memory, visuospatial skills, and intelligence, and they are part of the cognitive assessment on the UHDRS (Huntington Study Group, 1996). The sum of the scores of these five items add up to the cognitive score (the higher the cognitive performance the higher the score).

Further commonly used psychometric tests are the Wechsler Memory Scale (WMS) (Hahn-Barma *et al.*, 1998) and the "Hopkins Verbal Learning Test—revised" (HVLTR) (Duff *et al.*, 2007; Saft *et al.*, 2009). In later stages of the disease, the Mini-Mental Status Test (MMST) is used as a rough screening procedure (Rosenblatt *et al.*, 2006).

B. Correlation of Cognitive Features and TCS Findings in HD

Cognitive changes are one of the first symptoms in HD (Duff *et al.*, 2007; Saft *et al.*, 2009). In structural brain imaging studies, striatal and cortical atrophy are the most common findings to correlate with cognitive deficits in attention, working memory, and executive functions (Montoya *et al.*, 2006). The sonographic evaluation of ventricular system including ventricular enlargement has been described previously (Seidel *et al.*, 1995). Although Postert *et al.* could demonstrate a good agreement of TCS and CT measurements of frontal horn diameters in HD patients, no correlation between frontal horn widths and neuropsychological findings was found (Postert *et al.*, 1999). In a more recent TCS study, however, the measured width of the third ventricle was inversely correlated with the cognitive performance of HD patients (Krogias *et al.*, 2008). This could be explained by a better correlation of third-ventricle width with general brain atrophy. Voxel-based morphometric studies suggested a correlation between cognitive dysfunction and cerebral atrophy (Peinemann *et al.*, 2005). On the other hand, since the third ventricle is the median cleft between the thalami, also a regional accentuation of cerebral atrophy could be discussed as the underlying cause, particularly since morphometric MRI studies demonstrated that thalamic atrophy co-varies with cognitive performance in HD patients (Kassubek *et al.*, 2005).

VI. TCS Findings Related to Psychiatric Features in HD

A. Clinical Assessment of Psychiatric Features in HD

While diagnosis and treatment of HD focus predominantly on the motor features, the clinical phenotype is also associated with psychiatric symptoms like depression, irritability, aggression, apathy, anxiety disorders, as well as psychotic and obsessive symptoms (Paulsen *et al.*, 2001; Phillips *et al.*, 2008). The psychiatric disturbances often appear years before the outbreak of neurological symptoms (Julien *et al.*, 2007).

The complete UHDRS also comprises items assessing psychiatric disturbances like depression and psychotic symptoms (Huntington Study Group, 1996). Depression is one of the predominant psychiatric features, frequently resulting in suicide attempts or even completed suicide. Therefore, it is favorable to use additional depression scales (Di Maio *et al.*, 1993). One of the commonly used scales in the evaluation of depressive symptoms is the Hamilton Depression Rating Scale (HDRS) (Hamilton, 1967). It has to be considered that motor symptoms might have an influence on the rating of items quantifying somatic features and psychomotor functions (items 7–9, 13, 16), so that these parameters should be excluded from further analysis. Despite a low validity in the evaluation of HD patients, the Beck's Depression Inventory (BDI) is still one of the most frequently used standardized tests for the self-assessment of depression in HD (Beck *et al.*, 1961; De Souza *et al.*, 2009).

B. Correlation between Psychiatric Features and TCS Findings in HD

Postert *et al.* reported no statistical correlation between depression or psychosis and basal ganglia alterations (Postert *et al.*, 1999). However, echogenicity of brainstem raphe (BR) was not evaluated in this study. Alteration of mesencephalic raphe on TCS has been observed in major depression and in depressed patients with Parkinson's disease (Becker *et al.*, 1995; 2001; Berg *et al.*, 1999b; Walter *et al.*, 2007b). Up to now, only one study of echogenicity of BR in HD has been presented in an abstract form (Krogias *et al.*, 2008). In this study, a significant correlation between abnormal reduced echosignal of BR and depressive symptoms in HD patients was found (Fig. 5) (Krogias *et al.*, 2008). The precise morphological and pathophysiological interpretation of BR hypoechogenicity is still speculative. It is assumed that this finding might reflect a structural disruption of mesencephalic serotonergic raphe structures, especially since this finding was correlated to BR signal alteration on MRI studies (Becker *et al.*, 2001). Positron emission

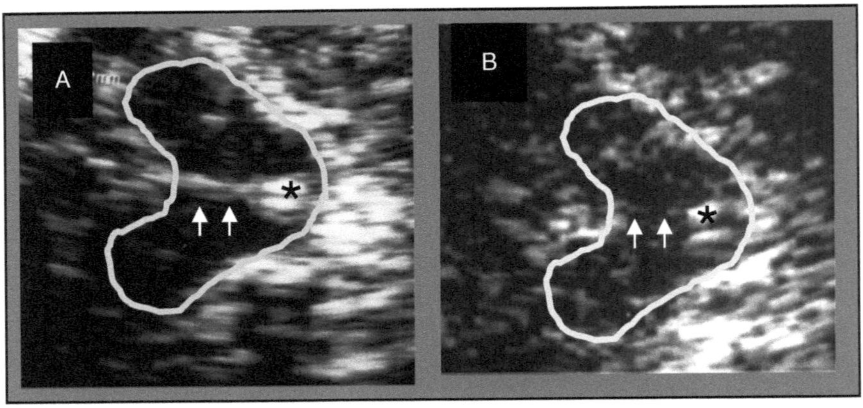

FIG. 5. TCS images with normal and pathologic findings of brainstem raphe structures in HD. Corresponding mesencephalic axial examination planes in two HD-patients (butterfly-shaped midbrain encircled for better visualisation). Black star marks the aqueduct. Arrowheads indicate the brainstem raphe. (A) Normal echogenicity: the sagittal running hyperechogenic band is not interrupted, Grade 2, normal finding. (B) Raphe structure not visible: no hyperechogenic band can be identified, pathologic finding, grade 0.

tomography (PET) studies analyzing the neurobiological basis of affective disturbances in HD demonstrated a widespread reduction of activity within the frontal and parietal lobes (Paradiso *et al.*, 2008). Mesencephalic raphe nuclei form complex serotonergic interconnections with forebrain structures, making an association between disrupted serotonergic projections at midbrain level and frontal hypoactivity conceivable (Becker *et al.*, 2001). In one histopathological study, however, neuronal loss within the dorsal raphe nucleus was not prominent in HD (Zweig *et al.*, 1992). On the other hand, studies have shown that major depression in mutation carriers can be a consequence of being genetically tested for HD (Licklederer *et al.*, 2008). Furthermore, high rates of affective disturbances have also been reported in HD family members not carrying the genetic mutation, leading to the hypothesis that depressive symptoms arise in response to emotional stressors, such as being at risk, or the burden of growing up in a family with affected members (Julien *et al.*, 2007).

Recently, BR hypoechogenicity has been also reported in patients with adjustment disorder with depressed mood (Walter *et al.*, 2007b). From this point of view, BR hypoechogenicity in HD patients might be a finding reflecting a predisposition for depressive reaction, independent from the neurodegenerative process of the disease.

So far, no studies have been performed using TCS to prospectively identify individuals being at risk for depression in any population. In presymptomatic HD mutation carriers, it would be of great interest to evaluate whether BR hypoechogenicity goes along with an occurrence of depressive symptoms prior to motor symptoms. Interestingly, of the 32 HD patients reported in the abstract, all of the eight HD patients without previous depression exhibited a regular BR echogenicity (Krogias *et al.*, 2008).

VII. Conclusions and Future Perspectives

In conclusion, two TCS studies revealed comparable frequencies of hyperechogenicities as well as a similar pattern of basal ganglia lesions in HD. A correlation with all of the three features of the typical triad of symptoms of the disease could be documented. Clinical severity and the number of CAG repeats within the *huntingtin* gene have been shown to be correlated with increased SN echogenicity. Depressive symptoms were associated with a decreased echosignal of mesencephalic raphe structures. Cognitive decline correlated with larger width of third ventricle, which may be caused by thalamus atrophy. Moreover, depicted enlargement of frontal horns of lateral ventricles corresponded closely to diameters measured on CT. TCS is a commonly available, noninvasive, and inexpensive diagnostic tool, which provides reliable information about the morphology of deep brain structures in HD, even in agitated patients who do not tolerate other imaging techniques.

Further neuropathological and multimodal imaging studies are needed to elucidate the precise morphological and pathogenetic underlying cause of the detected echosignal pathology.

Future studies should investigate possible differences of basal ganglia echogenicities in different clinical subtypes of HD, particularly in the juvenile or Westphal variant of the disease. Further research should also focus on examinations of presymptomatic mutation carriers to find out whether TCS is useful for indicating increased risk for depressive symptoms, already before they become clinically obvious.

References

Andrich, J., Schmitz, T., Saft, C., Postert, T., Kraus, P., Epplen, J. T., Przuntek, H., and Agelink, M. W. (2002). Autonomic nervous system function in Huntington's disease. *J. Neurol. Neurosurg. Psychiatry.* **72**, 726–731.

Andrich, J., Arning, L., Wieczorek, S., Kraus, P. H., Gold, R., and Saft, C. (2008). Huntington's disease as caused by 34 CAG repeats. *Mov. Disord.* **23**, 879–881.

Aylward, E. H., Sparks, B. F., Field, K. M., Yallapragada, V., Shpritz, B. D., Rosenblatt, A., Brandt, J., Gourley, L. M., Liang, K., Zhou, H., Margolis, R. L., and Ross, C. A. (2004). Onset and rate of striatal atrophy in preclinical Huntington disease. *Neurology.* **63**, 66–72.

Bartzokis, G., and Tischler, T. A. (2000). MRI evaluation of basal ganglia ferritin iron and neurotoxicity in Alzheimer's and Huntington's disease. *Cell Mol. Biol.* **46**, 821–833.

Beck, A. T., Ward, C., Mendelson, M., Mock, J., and Erbaugh, J. (1961). An inventory for measuring depression. *Arch. Gen. Psychiatry.* **4**, 561–571.

Becker, G., Becker, T., Struck, M., Lindner, A., Burzer, K., Retz, W., Bogdahn, U., and Beckmann, H. (1995). Reduced echogenicity of brainstem raphe specific to unipolar depression: A transcranial color-coded real-time sonography study. *Biol. Psychiatry.* **38**, 180–184.

Becker, G., Berg, D., Lesch, K. P., and Becker, T. (2001). Basal limbic system alteration in major depression: A hypothesis supported by transcranial sonography and MRI findings. *Int. J. Neuropsychopharmacol.* **4**, 21–31.

Becker, G., Berg, D., Rausch, W. D., Lange, H. K., Riederer, P., and Reiners, K. (1999). Increased tissue copper and manganese content in the lentiform nucleus in primary adult-onset dystonia. *Ann. Neurol.* **46**, 260–263.

Berg, D., Becker, G., Zeiler, B., Tucha, O., Hofmann, E., Preier, M. et al. (1999a). Vulnerability of the nigrostriatal system as detected by transcranial ultrasound. *Neurology.* **53**, 1026–1031.

Berg, D., Supprian, T., Hofmann, E., Zeiler, B., Jäger, A., Lange, K. W., Reiners, K., Becker, T., and Becker, G. (1999b). Depression in Parkinson's disease: Brainstem midline alteration on transcranial sonography and magnetic resonance imaging. *J. Neurol.* **246**, 1186–1193.

Berg, D., Siefker, C., Ruprecht-Dorfler, P., and Becker, G. (2001b). Relationship of substantia nigra echogenicity and motor function in elderly subjects. *Neurology.* **56**, 13–17.

Berg, D., Jabs, B., Merschdorf, U., Beckmann, H., and Becker, G. (2001c). Echogenicity of substantia nigra determined by transcranial ultrasound correlates with severity of parkinsonian symptoms induced by neuroleptic therapy. *Biol. Psychiatry.* **50**, 463–467.

Berg, D., Roggendorf, W., Schroder, U., Klein, R., Tatschner, T., Benz, P., et al. (2002). Echogenicity of the substantia nigra: Association with increased iron content and marker for susceptibility to nigrostriatal injury. *Arch. Neurol.* **56**, 999–1005.

Berg, D., Godau, J., and Walter, U. (2008). Transcranial sonography in movement disorders. *Lancet Neurol.* **7**, 1044–1055.

Brinkmann, R. R., Mezei, M. M., Theilmann, J., et al. (1997). The likelihood of being affected with Huntington's disease by a particular age, for a specific CAG size. *Am. J. Hum. Genet.* **60**, 1202–1210.

Brooks, D. J., and Frackowiak, R.S.J. (1989). PET and movement disorders. *J. Neurol. Neurosurg. Psychiatry.* (Suppl), 68–77.

Brouwers, P., Cox, C., Martin, A., et al. (1984). Differential perceptual–spatial impairment in Huntington's and Alzheimer's disease. *Arch. Neurol.* **41**, 1073–1076.

Davies, S. W., Turmaine, M., Cozens, B. A., et al. (1997). Formation of neuronal intranuclear inclusions underlies the neurological dysfunction in mice transgenic for the HD mutation. *Cell.* **90**, 537–548.

DeLong, M. R. (2000). The basal ganglia. In Principles of Neural Science (E. R. Kandel, J. H. Schwartz, and T. M. Jessel, Eds.), pp. 853–867. Fourth Edition. McGraw-Hill, New York.

De Souza, J., Jones, L. A., and Rickards, H. (2009). Validation of self-report depression rating scales in Huntington's disease. *Mov. Disord.* Nov 11 Epub ahead of print

Di Maio, L., Squitieri, F., Napolitano, G., Campanella, G., Trofatter, J. A., and Conneally, P. M. (1993). Suicide risk in Huntington's disease. *J. Med. Genet.* **48**, 231–233.

Duff, K., Paulsen, J. S., Beglinger, L. J. et al. (2007). Psychiatric symptoms in Huntington's disease before diagnosis: The predict-HD study. *Biol. Psychiatry.* **62**, 1341–1346.

Fennema-Notestine, C., Archibald, S. L., Jacobson, M. W., Corey-Bloom, J., Paulsen, J. S., Peavy, G. M., Gamst, A. C., Hamilton, J. M., Salmon, D. P., and Jernigan, T. L. (2004). In vivo evidence of cerebellar atrophy and cerebral white matter loss in Huntington disease. *Neurology.* **63**, 989–995.

Gargiulo, M., Lejeune, S., Tanguy, M. L., Lahlou-Laforet, K., Faudet, A., Cohen, D., Feingold, J., and Durr, A. (2009). Long-term outcome of presymptomatic testing in Huntington disease. *Eur. J. Hum. Gen.* **17**, 165–171.

Grafton, S. T., Sumi, S. M., Stimac, G. K., et al. (1991). Comparison of post-mortem magnetic resonance imaging and neuropathological findings in the cerebral white matter. *Arch. Neurol.* **48**, 293–298.

Hahn-Barma, V., Deweer, B., Dürr, A., Dodé, C., Feingold, J., Pillon, B., Agid, Y., Brice, A., and Dubois, B. (1998). Are cognitive changes the first symptoms of Huntington's disease? A study of gene carriers. *J. Neurol. Neurosurg. Psychiatry.* **64**, 172–177.

Hamilton, M. (1967). Development of a rating scale fpr primary depressive illness. *Br. J. of. Soc. Clin. Psychology.* **6**, 278–296.

Hayflick, S. J., Harman, M., Coryell, J., Gitschier, J., and Rowley, H. (2006). Brain MRI in neurodegeneration with brain iron accumulation with and without PANK2 mutations. *Am. J. Neuroradiol.* **27**, 1230–1233.

Hennenlotter, A., Schroeder, U., Erhard, P., Haslinger, B., Stahl, Weindl,A., von Einsiedel, H. G., Lange, K. W., and Ceballos-Baumann, A. O. (2004). Neural correlates associated with impaired disgust processing in pre-symptomatic Huntington's disease. *Brain.* **127**, 1446–1453.

Hilditch-Maguire, P., Trettel, F., Passani, L. A., Auerbach, A., Persichetti, F., and MacDonald, M. E. (2000). Huntingtin: An iron-regulated protein essential for normal nuclear and perinuclear organelles. *Hum. Mol. Genet.* **9**, 2789–2797.

Ho, A. K., Sahakian, B. J., Brown, R. G., et al. (2003). Profile of cognitive progression in early Huntington's disease. *Neurology.* **61**, 1702–1706.

Ho, V. B., Chuang, S., Rovira, M. J., and Koo, B. (1995). Juvenile Huntington Disease: CT and MR features. *Am. J. Neuroradiol.* **16**, 1405–1412.

Huckmann, M. S., Fox, J., and Topel, J. (1975). The validity of criteria for the evaluation of cerebral atrophy by computed tomography. *Radiology.* **116**, 85–92.

Huntington's Disease Collaborative Research Group. (1993). A novel gene containing a trinucleotide repeat that is expanded and unstable on Huntington's disease chromosomes. *Cell.* **72**, 971–983.

Huntington Study Group. (1996). Unified Huntington's disease rating scale: Reliability and consistency. *Mov. Disord.* **11**, 136–142.

Julien, C. L., Thompson, J. C., Wild, S., Yardumian, P., Snowden, J. S., Turner, G., and Craufurd, D. (2007). Psychiatric disorders in preclinical Huntington's disease. *J. Neurol. Neurosurg. Psychiatry.* **78**, 939–943.

Kassubek, J., Juengling, F., Kioschies, T., Henkel, K., Karitzky, J., Kramer, B., Ecker, D., Andrich, J., Saft, C., Kraus, P., Aschoff, A., Ludolph, A., and Landwehrmeyer, G. (2004). Topography of cerebral atrophy in early Huntington's disease: A voxel based morphometric MRI study. *J. Neurol. Neurosurg. Psychiatry.* **75**, 213–220.

Kassubek, J., Juengling, F. D., Ecker, D., et al. (2005). Thalamic atrophy in Huntington's disease co-varies with cognitive performance: A morphometric MRI analysis. *Cereb. Cortex.* **15**, 846–853.

Kirkwood, S. C., Siemers, E., Hodes, M. E., et al. (2000). Subtle changes among presymptomatic carriers of the Huntington's disease gene. *J. Neurol. Neurosurg. Psychiatry.* **69**, 773–779.

Krogias, C., Saft, C., Strassburger, K., Gold, R., Norra, C., Juckel, G., and Ninphius, D. (2008). Hirnparenchym-Sonographie bei depressiven und nicht-depressiven Huntington-Patienten. *Nervenarzt* **79** (Suppl. 4), 372.

Krogias, C., Strassburger, K., Eyding, J., Gold, R., Norra, C., Juckel, G., Saft, C., and Ninphius, D. (2010). Depression in Huntington's disease correlates with alterations of brainstem raphe depicted by transcranial sonography. *J Psychiatry Neurosci.* (submitted).

Kuhl, D. E., Phelps, M. E., Markham, C. H., et al. (1982). Cerebral metabolism and atrophy in Huntington's disease determined by 18FDG and computed tomographic scan. *Ann. Neurol.* **12**, 425–434.

Lambeck, J., Niesen, W., and Zucker, B. (2008). Substantia nigra hyperechogenicity in Huntington's disease. *Akt. Neurologie.* **35** (Suppl. 1): S152, abstract.

Leeders, K. L., Frackowiak, R.S.J., Quinn, N., and Marsden, C. D. (1986). Brain Energy and Dopaminergic Function in Huntington's disease measured in vivo using Positron Emission Tomography. *Movement Disorders.* **1**, 69–77.

Lemiere, J., Decruyenaere, M., Evers-Kiebooms, G., et al. (2002). Longitudinal study evaluating neuropsychological changes in so-called asymptomatic carriers of the Huntington's disease mutation after 1 year. *Acta Neurol. Scand.* **106**, 131–141.

Lickléderer, C., Wolff, G., and Barth, J. (2008). Mental health and quality of life after genetic testing for Huntington disease: A long-term effect study in Germany. *Am. J. Med. Genet.* **146A**, 2078–2085.

MacMillan, J. C., and Quarrell, O.W.J. (1996). The neurobiology of Huntington's disease. In Huntington's disease (P. S. Harper, Ed.), pp. 317–357. Second Edition. WP Saunders, London.

Montoya, A., Price, B. H., Menear, M., and Lepade, M. (2006). Brain imaging and cognitive dysfunctions in Huntington's disease. *J. Psychiatry Neursci.* **31**, 21–29.

Nance, M. A., and Myers, R. H. (2001). Juvenile onset Huntington's Disease—Clinical and research perspectives. *Ment. Retard. Dev. Disabil. Res. Rev.* **7**, 153–157.

Naumann, M., Becker, G., Toyka, K. V., Supprian, T., and Rainers, K. (1996). Lenticular nucleus lesion in idiopathic dystonia detected by transcranial sonography. *Neurology.* **47**, 1284–1290.

Nopoulos, P., Magnotta, V. A., Mikos, A., Paulson, H., Andreasen, N. C., and Paulsen, J. S. (2007). Morphology of the cerebral cortex in preclinical Huntington's disease. *Am. J. Psychiatry.* **164**, 1428–1434.

Paradiso, S., Turner, B. M., Paulsen, J. S., Jorge, R., Ponto, L. L., and Robinson, R. G. (2008). Neural bases of dysphoria in early Huntington's disease. *Psychiatry Res.* **162**, 73–87.

Paulsen, J. S., Ready, R. E., Hamilton, J. M., Mega, M. S., and Cummings, J. L. (2001). Neuropsychiatric aspects of Huntington's disease. *J. Neurol. Neurosurg. Psychiatry.* **71**, 310–314.

Paulsen, J. S. (2009). Functional imaging in Huntington's disease. *Exp. Neurol.* **216**, 272–277.

Peinemann, A., Schuller, S., Pohl, C., Jahn, T., Windl, A., and Kassubek, J. (2005). Executive dysfunction in early stages of Huntington's disease is associated with striatal and insular atrophy: A neuropsychological and voxel-based morphometric study. *J. Neurol. Sci.* **239**, 11–19.

Phillips, W., Shannon, K. M., and Barker, R. A. (2008). The current clinical management of Huntington's disease. *Mov. Disord.* **23**, 1491–1504.

Postert, T., Lack, B., Kuhn, W., Jergas, M., Andrich, J., Braun, B., Przuntek, H., Sprengelmeyer, R., Agelink, M., and Büttner, T. (1999). Basal ganglia alterations and brain atrophy in Huntington's disease depicted by transcranial real time sonography. *J. Neurol. Neurosurg. Psychiatry.* **67**, 457–462.

Quarrel, O., and Harper, P. S. (1991). The clinical neurology of Huntington's disease. In Huntington's disease (P. Harper, Ed.), pp. 37–80. WB Saunders, London.

Rosas, H. D., Koroshetz, W. J., Chen, Y. I., Skeuse, C., Vangel, M., Cudkowicz, M. E., Caplan, K., Marek, K., Seidman, L. J., Makris, N., Jenkins, B. G., and Goldstein, J. M. (2003). Evidence for more widespread cerebral pathology in early HD: An MRI-based morphometric analysis. *Neurology.* **60**, 1615–1620.

Rosas, H. D., Salat, D. H., Lee, S. Y., Zaleta, A. K., Pappu, V., Fischl, B., Greve, D., Hevelone, N., and Hersch, S. M. (2008). Cerebral cortex and the clinical expression of Huntington's disease: Complexity and heterogeneity. *Brain.* **131**, 1057–1068.

Rosenblatt, A., Liang, K. Y., Zhou, H., Abbott, M. H., Gourley, L. M., Margolis, R. L., Brandt, J., and Ross, C. A. (2006). The association of CAG repeat length with clinical progression in Huntington's disease. *Neurology.* **66**, 1016–1020.

Saft, C., Kosinski, C. M., and Landwehrmeyer, G. B. (2009). Progress in premanifest and manifest diagnostics in Huntington's disease. *Akt. Neurol.* **36**, 506–523.

Sánchez-Pernaute, R., Künig, G., del Barrio Alba, A., de Yébenes, J. G., Vontobel, P., and Leenders, K. L. (2000). Bradykinesia in early Huntington's disease. *Neurology.* **54**, 119–125.

Seidel, G., Kaps, M., Gerriets, T., and Hutzelmann, A. (1995). Evaluation of the ventricular system in adults by transcranial duplex sonography. *J. Neuroimaging.* **5**, 105–108.

Slaughter, J. R., Martens, M. P., and Slaughter, K. A. (2001). Depression and Huntington's disease: Prevalence, clinical manifestations, etiology and treatment. *CNS Spectr.* **6**, 306–326.

Sprengelmeyer, R., Young, A. W., Calder, A. J., et al. (1996). Loss of disgust. Perception of faces and emotions in Huntington's disease. *Brain.* **119**, 1647–1665.

Stober, T., Wussow, W., and Schimrigk, K. (1996). Bicaudate diameter—The most specific and simple CT parameter in the diagnosis of Huntington's disease. *Arch. Neurol.* **53**, 1293–1296.

Walker, F. O. (2007). Huntington's disease. *Lancet.* **369**, 218–228.

Walter, U., Niehaus, L., Probst, T., Benecke, R., Meyer, B. U., and Dressler, D. (2003). Brain parenchyma sonography discriminates Parkinson's disease and atypical parkinsonian syndromes. *Neurology.* **60**, 74–77.

Walter, U., Dressler, D., Wolters, A., Probst, T., Grossmann, A., and Benecke, R. (2004). Sonographic discrimination of corticobasal degeneration vs progressive supranuclear palsy. *Neurology.* **63**, 504–509.

Walter, U., Krolikowski, K., Tarnacke, B., Benecke, R., Czlonkowska, A., and Dressler, D. (2005). Sonographic detection of basal ganglia lesions in asymptomatic and symptomatic Wilson disease. *Neurology.* **64**, 1726–1732.

Walter, U., Behnke, S., Eyding, J., Niehaus, L., Postert, T., Seidel, G., and Berg, D. (2007a). Transcranial brain parenchyma sonography in movement disorders: State of the art. *Ultrasound Med. Biol.* **33**, 15–25.

Walter, U., Prudente-Morrisey, L., Herpertz, S. C., Benecke, R., and Hoeppner, J. (2007b). Relationship of brainstem raphe echogenicity and clinical findings in depressive states. *Psychiatry Res.* **155**, 67–73.

Weeks, R. A., Picini, P., Harding, A. E., et al. (1996). Striatal D1 and D2 dopamine receptor loss in asymptomatic mutation carriers of Huntington's disease. *Ann. Neurol.* **40**, 49–54.

Weindl, A., and Conrad, B. (2005). Chorea. In Bewegungsstörungen (A. Ceballos-Baumann, and B. Conrad, Eds.), pp. 197–211. Second Edition. Georg Thieme Verlag, Stuttgart, New York.

Wexler, N. S., Lorimer, J., Porter, J., et al. (2004). Venezuelan kindreds reveal that genetic and environmental factors modulate Huntington's disease age of onset. *PNAS (USA).* **101**, 3498–3503.

Young, A. B., Greenamyre, J. T., Hollingsworth, Z., et al. (1988). NMDA receptor losses in putamen form patients with Huntington's disease. *Science.* **241**, 981–983.

Zweig, R. M., Ross, C. A., Hedreen, J. C., Peyser, C., Cardillo, J. E., Folstein, S. E., and Price, D. L. (1992). Locus coeruleus involvement in Huntington's disease. *Arch. Neurol.* **49**, 152–156.

TRANSCRANIAL SONOGRAPHY IN DEPRESSION

Milija D. Mijajlovic

Institute of Neurology, Clinical Center of Serbia; School of Medicine University of Belgrade, Serbia

Transcranial sonography, which displays tissue echogenicity of the brain through the intact skull, reveals reduced echogenicity of the brainstem raphe (BR) as a characteristic finding in unipolar depression and in depression associated with Parkinson's or Wilson's disease, but in general not in healthy adults, bipolar affective disorders, schizophrenia, multiple sclerosis with depression, or Parkinson's disease without concomitant depression. The reason for the reduced echogenicity of the BR is not entirely clear. Magnetic resonance imaging investigations, however, provide evidence that the signal alteration may be caused by a modification of tissue cell density, the interstitial matrix composition, or an alteration of fiber tracts integrity. An involvement of the basal limbic system in the pathogenesis of unipolar depression and depression associated with certain neurodegenerative diseases is assumed.

I. Transcranial Sonography in Depression—First Findings and Implications

The first paper published on structural alterations in neurodegenerative and psychiatric disorders applying transcranial sonography (TCS) was on reduced

INTERNATIONAL REVIEW OF NEUROBIOLOGY, VOL. 90
DOI: 10.1016/S0074-7742(10)90018-4

echogenicity of the mesencephalic midline raphe in depression by Georg Becker and co-workers in 1994 (Becker *et al.*, 1994). In comparison with healthy controls, patients suffering from unipolar depression showed a reduced echogenicity, defined as interruption or invisibility of the normally continuous midline of the mesencephalic brainstem (Becker *et al.*, 1994, 1995). This structural abnormality, which was reported to occur in unipolar depressed patients, seemed to be unrelated to the severity of the current illness and was absent in patients with bipolar depressive disorders or schizophrenia (Becker *et al.*, 1995). The same structural abnormality was reported in other neurodegenerative disorders associated with depression in subsequent investigations, for example, in Parkinson's disease (PD) (Becker *et al.*, 1997; Berg *et al.*, 1999), idiopathic dystonia (Naumann *et al.*, 1996), and Wilson's disease (Walter *et al.*, 2005). Interestingly, in depressed multiple sclerosis patients, raphe echogenicity was in general normal (Berg *et al.*, 2000a, 2000b). This led to the hypothesis that disorders affecting the basal limbic system can present with the clinical phenotype of depression and reduced echogenicity of the midline raphe on TCS while disorders affecting primarily projection areas of the basal limbic system—that is, subcortical lesions in multiple sclerosis and cortical/subcortical lesions in stroke—may also have a clinical phenotype of depression but normal echogenicity of the brainstem raphe (Becker *et al.*, 2001)—see below.

II. Method of Transcranial Sonography in Depressive Disorders

Modern clinical TCS systems display deep echogenic brain structures with high image resolution of up to 0.7 × 1 mm, which is even higher than that of magnetic resonance imaging (MRI) as used in clinical conditions (Walter *et al.*, 2008). Meanwhile, consensus guidelines for a standardized TCS procedure of midbrain structures, basal ganglia, and ventricles have been established (Berg *et al.*, 2006, 2008; Walter *et al.*, 2007a) allowing comparison of TCS findings between different research groups. Applying the standard technique (see also chapter on methods), the BR is usually seen as a highly echogenic continuous line in the middle of the mesencephalic brainstem with an echogenicity that is identical to that of the red nucleus (RN). (Fig. 1A) Echogenicity of the BR is rated semiquantitatively, using either a three-point (grade 1: raphe invisible; grade 2: slightly echogenic or interrupted BR; grade 3: hyperechogenic, identical or comparable to that of the RN or basal cisterns) or, preferably, using a two-point (grade 0: invisible, hypoechogenic or interrupted BR; grade 1: hyperechogenic BR as a continuous line) grading system (Berg *et al.*, 2006, 2008, Walter *et al.*, 2007a). In the first two studies on raphe echogenicity, even a four-point grading system was used. It is important to scan the subject investigated from both sides, as the bone window may vary allowing

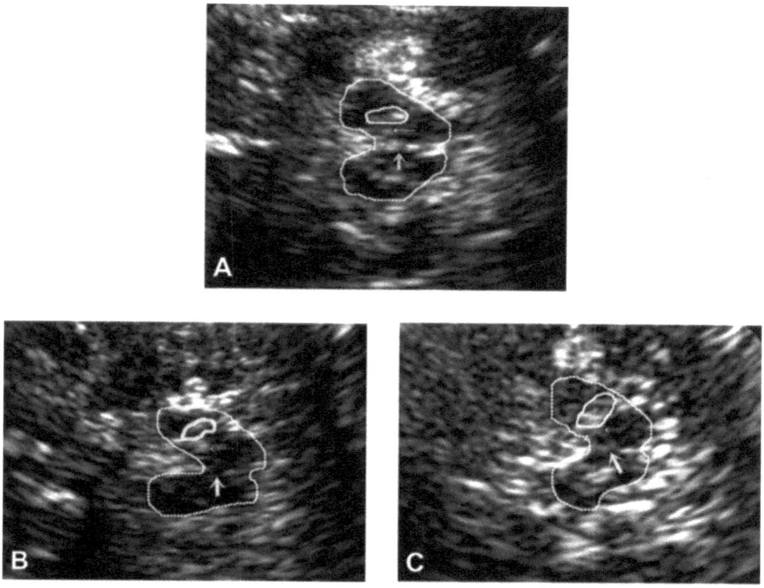

FIG. 1. TCS images of corresponding midbrain axial sections in three subjects. The butterfly-shaped midbrain was encircled for better visualization (dotted line). Thick arrows indicate brainstem raphe; thin arrows: red nucleus; substantia nigra is encircled with full line. (A) **Mesencephalic brainstem of a healthy individual** with normal, only slightly visible substantia nigra, and normal, highly echogenic brainstem raphe, which has the same echogenicity as the red nucleus; (B) **Mesencephalic brainstem of a patient with unipolar depression** with normal, substantia nigra, visible only as very narrow line and abnormal, reduced echogenicity of the brainstem raphe; (C) **Mesencephalic brainstem of a PD patient with depression** displaying marked bilateral hyperechogenicity of the substantia nigra and abnormal, reduced echogenicity of the brainstem raphe.

sufficient visualization of the raphe only if both sides are considered. Therefore, if the raphe can be depicted as continuous line from one side, it is rated as a normal—that is, hyperechogenic, noninterrupted midline.

III. TCS Findings in Unipolar Depression and Depression Associated with Neurodegenerative Diseases and other Etiologies

A. TCS FINDINGS IN UNIPOLAR DEPRESSION

Raphe hypoechogenicity has been shown to be a common finding in 50–70% of patients with unipolar depression (Becker *et al.*, 1994; Walter *et al.*, 2007b) and is associated with responsivity to serotonin reuptake inhibitors (SSRIs) (Walter *et al.*, 2007c). In a pioneer study, echogenicity of the BR was examined by TCS in 20 patients with unipolar depression and 20 healthy adult controls. The echogenicity of the BR in the present study was compared with that of the RN and semiquantitatively rated on a four-point scale. A marked reduction of raphe echogenicity in depressed patients was found. The mean scores of raphe echogenicity in patients with major depression and healthy adults were 1.3 and 2.8, respectively (Becker *et al.*, 1994). This finding was confirmed 1 year later in a larger number of patients in a study that compared echogenicity of the BR between patients with unipolar depression, bipolar disorder, schizophrenia and healthy controls. Each group comprised 40 subjects and the echogenicity of the BR was rated on a four-point scale as in the previous study. The mean echogenicity score of the BR in patients with unipolar depression was significantly reduced: 1.4 compared with 2.8 in healthy controls. A significant difference in raphe echogenicity, but in the opposite direction, was also found between bipolar disorder (mean BR echogenicity score: 3.1) and healthy controls. No significant difference was found between schizophrenic patients (mean BR echogenicity score: 2.5) and healthy controls.

Importantly, no correlation was found between raphe echogenicity and age, sex, or disease severity (Becker *et al.*, 1995).

Several further studies confirmed the finding of reduced raphe echogenicity in unipolar depression. In the study of Walter and co-workers, the frequency of patients with reduced echogenicity of the BR was higher in unipolar depression compared with healthy individuals (53 *vs* 9%) and in depressed PD patients compared with nondepressed (38 *vs* 16%). The frequency of reduced echogenicity of the BR was the highest in patients with unipolar depression (Walter *et al.*, 2007b).

TCS findings of another study (Walter *et al.*, 2007c) showed that reduced echogenicity of the pontomesencephalic BR is frequent in depressive states (54% had reduced BR echogenicity), but only rarely found in healthy subjects without any history of psychiatric disorder (8% had hypoechogenicity of the BR). BR echogenicity could not discriminate between a major depressive disorder and adjustment disorder with depressed mood, as hypoechogenicity of the BR was found with similar frequency in both types of depression (53 *vs* 60%). This study also gave first evidence that BR echogenicity may be related to the response to specific antidepressive therapy, as BR echogenicity scores were significantly lower in SSRI responders than in SSRI nonresponders. Reduced BR echogenicity indicated SSRI responsivity with a positive predictive value of 88%. Also in this study, no impact of age, gender, or antidepressant medication on BR echogenicity was found. These results indicate that reduced BR echogenicity is not related to the diagnostic category of a depressive state but may give implications on treatment outcome.

In contrast to the first study, in which no relation of BR echogenicity and severity of depression was found, a recent study did reveal an association of BR hypoechogenicity and more severe depressive symptoms. In this study, reduced raphe echogenicity was found in 47% of 17 patients with major depressive disorder but only in 15% of 40 healthy controls. In patients with suicidal ideations, this finding was even more pronounced (86% of 14 suicidal patients had reduced BR echogenicity) with the highest frequency (72%) of a completely invisible raphe on TCS. Although the sample of patients was relatively small, data from this study showed for the first time that reduced echogenicity of the BR is frequent in patients with suicidal ideations. Normal raphe echogenicity in patients with major depression was associated with less severe depressive symptoms and rarely with the presence of suicidal ideations (Budisic *et al.*, 2009). In this study, an Aloka Alpha 10 ultrasound machine was used, while in the first studies a Siemens Sonoline Elegra machine was applied. Resolution of ultrasound machines have improved by large since the first studies. Therefore, the recently discovered correlation of BR hypoechogenicity and severity of depression may have become possible by the advanced ultrasound techniques.

B. TCS Findings in PD and Depression

Depression is a frequent comorbid condition in PD. The mean prevalence reported in various studies is 40%, ranging from 4 to 70% (Cummings, 1992). A review of all studies on depression in PD published from 1992 to 1998, comprising a total of 5911 PD patients, found that 31% of them were, at any given time, depressed (Brooks *et al.*, 2001). In large register studies, each lasting more than 15 years, the risk for depressed patients to develop PD was 2.2- to 3.1-fold compared with nondepressed controls (Leentjens *et al.*, 2003; Schuurman *et al.*, 2002). These studies suggest that either depression is an early symptom of PD or that depression is a risk factor for PD (Liebermann 2006).

Midbrain raphe hypoechogenicity is frequently found in depressed PD patients. Compared with nondepressed PD patients in whom raphe hypoechogenicity was detected in only 6–27%, 35–85% of depressed PD patients have been reported to show raphe hypoechogenicity (Becker *et al.*, 1997, 2001; Berg *et al.*, 1999; Walter *et al.* 2007b, 2007e)

Becker *et al.* were the first to evaluate brainstem changes in depressed PD patients using TCS (Becker *et al.*, 1997). Thirty PD patients (13 with associated depression) were compared with 30 healthy individuals. BR raphe was rated on a three-point scale. In the group of PD patients with depression, BR echogenicity was distinctly reduced compared with nondepressed PD patients and healthy controls (more than 90% of depressed PD patients had BR hypoechogenicity). The distribution of raphe echogenicity scores in controls was very similar to that found in nondepressed patients with PD (about 40% in both groups had a BR grade of 1 or 2). Differences in raphe echogenicity between depressed and nondepressed patients with PD and between depressed patients with PD and controls were significant. Raphe echogenicity correlated negatively with the degree of motor impairment in PD patients with depression.

Additionally, a significant reduction of mesencephalic BR echogenicity has been found in PD patients with overactive bladder symptoms compared to PD patients without these symptoms (Walter *et al.*, 2006). These results suggested that a brain region, which may be afferent to the BR, could primarily be affected at least in this subpopulation of PD patients (Walter *et al.*, 2006).

The value of TCS in the differential diagnosis of PD (with or without concomitant depression) versus isolated depression is most obvious in the very early stages of the disease, in which clinical features alone do often not allow diagnosing a patient to have either PD or depression (Gaenslen *et al.*, 2008). But even in the later stages of PD, in which hypomimia and bradykinesia may mimic depressed mood, TCS finding may be helpful.

In subjects who are only depressed, normal echogenicity of the substantia nigra (SN) is found in the majority of cases, while the midline raphe is hypoechogenic, interrupted or not visible (Fig. 1B). In PD patients, normal raphe echogenicity can be expected besides hyperechogenicity of the SN (Berg *et al.*, 1999; Walter *et al.*, 2007b, 2010). In patients suffering from PD and concomitant depression, both pathological findings of SN hyperechogenicity and an interrupted, hypoechogenic, or missing midline raphe is found in general (Fig. 1C).

The fact that TCS displays a fairly specific feature in patients with PD makes this method unique in the differential diagnosis of depression, PD, and the concomitant occurrence of both disorders. In PD, characteristically increased echogenicity of the SN can already be visualized in the very first stages of the disease and most likely even before motor symptoms allow the clinical diagnosis (see also chapter on early and premotor diagnosis of PD). The co-occurrence of SN hyperechogenicity and depression may therefore give implications on the etiology and possibly later development of symptoms.

SN hyperechogenicity has been shown to correlate with motor asymmetry and reduced verbal fluency in patients with depressive disorders. This relationship was even stronger in younger patients (<50 years) and independent from age in patients who had reduced BR echogenicity (Hoeppner *et al.*, 2009). Since both liability for developing PD and frequency of PD-like TCS findings were found to be increased in depression, patients with depressive disorders might be an important population to screen for sonographic and clinical signs of early PD. In previous studies (Walter *et al.*, 2007b, 2007c), coexistence of SN hyperechogenicity and reduced BR echogenicity in non-Parkinsonian depressed patients was associated also with right-to-left asymmetry on the unified PD rating scale and in PD patients with a history of depressive disorder prior to PD onset. Therefore, the co-occurrence of SN hyperechogenicity and raphe hypoechogencity in subjects with depression or with very mild motor slowing could indicate a subgroup at risk for the development of PD. Follow-up of these patients is needed to evaluate whether SN hyperechogenicity in depressed patients indeed alludes to a higher risk of conversion to PD than normal SN echogenicity in depressive subjects.

C. TCS in Huntington's Disease and Depression

There is only one published study that investigated TCS findings in depressed patients with Hunington's disease (Postert *et al.*, 1999). It included 49 patients with manifest Huntington's disease and 39 healthy subjects. Depression was rated on the four points modified version of the unified Huntington's disease rating scale (Huntington Study Group, 1996). Considering clinical subtypes, there was no significant statistical correlation between cognitive impairment, depression, or psychosis and any basal ganglia alterations including raphe echogenicity in this study. Further studies on larger number of patients are needed in order to conclude if depressed patients with Huntington's disease have the same TCS changes as depressed patients with unipolar depression or depressed PD patients.

D. TCS and Depression in Wilson's Disease

In the study of Walter and co-workers (Walter *et al.*, 2005), 21 patients with Wilson's disease (WD) with or without neurologic symptoms were investigated. Echogenicity of the BR was rated semiquantitatively. Disease severity was assessed using a WD rating scale (WDRS) with the items dysarthria, akinesia, ataxia, tremor, and dystonia; the median WDRS score was used for further analysis. Reduced echogenicity of the BR was detected in three symptomatic patients who were clinically most severely affected according to the WDRS scores. In all other patients with WD, normal echogenicity of the BR was found. Similar to unipolar depression and depression in PD, depression in WD has also been correlated to dysfunction of the presynaptic serotonergic system (Eggers *et al.*, 2003). Neuropsychological testing in patients with WD revealed affective changes, most pronounced in neurologically severely affected patients, and also symptoms of depression in those patients with WD who had Parkinsonian features (Seniow *et al.*, 2002, 2003).

E. TCS in Multiple Sclerosis and Depression

Depression is a frequent finding in multiple sclerosis (MS), with a prevalence of about 40% (Minden *et al.*, 1990). Disconnection between subcortical and cortical areas associated with the limbic system has been proposed as a pathogenic mechanism of depression in MS (Sabatini *et al.*, 1996).

In the study of Berg and co-workers (Berg *et al.*, 2000a), echogenicity of the BR was rated in 78 patients, fulfilling the criteria for definite multiple sclerosis (Poser *et al.*, 1983). BR was not visible in 10, slightly echogenic in 43, and of normal echogenicity in 25 patients. There was no difference in midline echogenicity in depressed compared with nondepressed MS patients. In another study of this group, no significant correlation of the widths of third and lateral ventricles on TCS and severity of depression as measured with the Hamilton Depression Rating Scale, the Montgomery-Åsberg Depression Rating Scale, and the Beck Depression Inventory was found (Berg *et al.*, 2000b). The latest study of Walter *et al.* confirmed that TCS findings of midbrain structures, including raphe echogenicity, basal ganglia, and ventricles, did not correlate with the occurrence of fatigue or depression in MS patients (Walter *et al.*, 2007d)

These data suggest that the pattern of lesions as visualized by different neuroimaging methods is different in depression in MS compared with unipolar depression or depression associated with PD—see below.

IV. The Concept of the Basal Limbic System in Depression

Strong evidence suggests an involvement of the basal limbic system in the pathogenesis of mood disorders. Anatomically, the midline, seen as a continuous echogenic structure on TCS, corresponds to an accumulation of several nuclei and fiber tracts, including ascending and descending fiber tracts of the medial forebrain bundle, dorsal longitudinal fascicle, mammilotegmental tract, and fasciculus retroflexus. The midbrain midline also includes the dorsal raphe nucleus as well as tracts connecting numerous brainstem and cerebellar nuclei (e.g., interpeduncular nuclei, central superior nucleus) with diencephalic and telencephalic brain areas. Its projections reach almost all brain areas and it is the major monoamine serotonin brain source (Carpenter *et al.*, 1993; Crossman 2005; Nieuwenhuys *et al.*, 2008). The finding of an impaired structure of this midline on TCS in unipolar depression and depression associated with neurodegenerative diseases was both astonishing and elucidating: For the first time it was shown that a morphological alteration of the BR seems to be involved in the pathogenesis of different forms of depressive disorders. Changes in raphe echogenicity reflect changes in tissue impedance and point toward an alteration of the brainstem micro-architecture, which could be due to a shift in tissue cell density, a change in interstitial matrix composition, or an alteration of fiber tract integrity (Berg *et al.*, 1999; Becker *et al.*, 2001). The exact cause of the hypoechogenic signal of the BR raphe has not been completely elucidated, yet

V. MRI Studies Supporting TCS Findings and the Concept of the Basal Limbic System

Although there are several reports of MRI signal alteration of the BR in depression, a characteristic neuroimaging pattern of BR abnormality has not yet been found (Steele *et al.*, 2005; Hoeppner *et al.*, 2009). Still, the findings published so far give important clues on alterations of the BR possibly underlying the etiology of depression involving the basal limbic system.

Increased intensity of the midline has been reported for unipolar depressed patients than for bipolar patients and controls in a retrospective study using T2-weighted MRI (Becker *et al.*, 1998). A difference between patients with major depression and control subjects with regard to T2-relaxation times was found in a region of interest located along the midline of the pons, while no difference was found between patients with bipolar disorder and control subjects. The above-mentioned suggestion that alterations of T2-relaxation times might indicate subtle tissue changes (Supprian *et al.*, 2004) are in line with the results of pathoanatomic and PET studies demonstrating morphological and functional alteration of the dorsal raphe nucleus in major depression, with decreased serotonin type 1A receptor binding and fewer neurons expressing serotonin transporter mRNA compared with findings in healthy subjects (Arango *et al.*, 2001; Meltzer *et al.*, 2004). The relationship of BR echogenicity and SSRI responsivity, which was found in the study of Walter (Walter *et al.*, 2007c), further supports the idea that reduced BR echogenicity reflects an alteration of the serotonergic system.

Increased T2-relaxation time in the pontine brainstem in patients with major depression could be in line with previous reports of brainstem pathology in these patients (Becker *et al.*, 2001). This observation might indicate a subtle tissue alteration, which cannot be identified by visual inspection of the images. T2-relaxation time depends on physical tissue characteristics and is influenced by the hydration status or iron content. Differences in T2-relaxation time of specific brain areas between patients with major depression and healthy controls may indicate different tissue composition caused by histological changes, which could, for instance, be caused by neurodegeneration. For example, it has been speculated that the finding of reduced echogenicity of the midline and increased T2-

weighted signal intensity may be consistent with disruption of myelinated fibers running along the brainstem, which include the monoaminergic tracts (Becker *et al.*, 1997, 2001).

Furthermore, other MRI studies of depressed PD patients reported a shift of signal in the midbrain using a semiquantitative assessment of T2-relaxation time (Berg *et al.*, 1999). Rating of signal intensity and T2 relaxometry of the ponto-mesencephalic midline structures revealed a significant difference between depressed and nondepressed PD patients. This corresponded to a significant reduction in mesencephalic midline echogenicity of depressed PD patients on TCS images. No correlation was found between raphe signal intensity, T2-relaxation times, or TCS echogenicity and the severity of motor symptoms or depression. The findings of changes in signal intensity and T2-relaxation time of the pontomesencephalic midline structures on MRI in PD patients with depression therefore confirm previous TCS findings, which also did not show any relation to the severity of depression or motor symptoms (Berg *et al.*, 1999).

In contrast to previous reports, no difference in the echogenicity of the BR of unipolar depressed patients was found in the study of Steele, the first published study, that investigated possible structural changes of the BR in unipolar depression using diffusion tensor magnetic resonance imaging (DTI) (Steele *et al.*, 2005). Fifteen unipolar depressed patients and fifteen controls were studied using TCS and DTI. The mean BR echogenicity scores were calculated as 2.07 and 2.57 for the patient and control groups, respectively. This difference was not significant but a possible trend in the direction of previous studies, which showed decreased BR echogenicity in unipolar depression, was found. No significant differences in fractional anisotrophy (FA) values between patient and control groups were identified in the midbrain using DTI. Therefore, this study on a fairly small number of patients did not confirm the finding of structural changes of the BR in unipolar depressive patients using DTI (Steele *et al.*, 2005). Whether patient number or diagnostic categories may have influenced these results cannot be stated with certainty. Interestingly, another study did show decreased FA values and increased mean diffusivity (MD), which was calculated based on DTI of the BR in a cohort of 16 depressed PD patients who had also a hypoechogenic BR, compared with unchanged FA values and MD in 16 healthy controls (Mijajlovic, submitted). The latter study therefore seems to support the hypothesis that morphological changes are associated with neuronal loss and possibly with demyelization and decreased directionality of aminergic fiber tracts running along the BR. As already hypothesized based on the TCS findings, these fiber tract alterations may cause changes of acoustic impedance, which then may present as hypoechogenic signal on TCS (Mijajlovic *et al.*, in preparation).

The concept of an affection of the basal limbic system in some forms of depression is underscored by MRI findings in other forms of depression. Whereas in unipolar depression and in PD and depression, generally no supratentoriell structural alteration can be found by routine MRI techniques, it could be shown for depression in MS that normal BR echogenicity is associated with supratentoriell structural abnormalities. These alterations were, among others, found in the projection areas of the basal limbic system on MRI, that is, in temporal, frontal, and parietal regions (Berg *et al.*, 2000a), while no significant differences in MRI T2-relaxation times and relative signal intensities were found in the pontomesencephalic midline compared with nondepressed MS patients. (Berg *et al.*, 2000a).

VI. Postmortem Studies

There are few postmortem studies of the brainstem of patients who suffered from depressive disorders. One small study reported that patients with unipolar depression have a distinct disruption of the mesencephalic fiber tracts, which may be consistent with both the ultrasound and MRI findings. It has been suggested that such an abnormality could result in a reduction of brain monoamines and could therefore be addressed with antidepressants (Becker *et al.*, 2001).

Baumann and co-workers performed a morphometric postmortem study on neuronal morphology in all subnuclei of the dorsal raphe nucleus in the brains of six patients with major depressive disorder and 12 healthy subjects. The number of neurons of the ventrolateral subnucleus of the dorsal raphe was reduced by 31% in patients with major depressive disorder compared with nonpsychiatric control subjects. Ventrally located subnuclei of the rostral dorsal raphe (ventrolateral, ventral, interfascicular) also showed a smaller number of neurons. Results of this study indicated that patients with a major depressive disorder have a circumscribed numerical neuronal deficiency in the dorsal raphe (Baumann *et al.*, 2002).

These studies underscore the assumption that a structural alteration of the BR is involved in depression, which may be depicted by TCS and other specific neuroimaging methods like T2-relaxation time.

VII. Conclusions and Further Perspectives

TCS is a relatively new neuroimaging method, which can be used as a helpful supplementary tool in the diagnosis of unipolar depression and depression associated with certain neurodegenerative diseases like PD, dystonic syndromes, or WD. The method is easy to apply, even in nonresting patients with different movement disorders, it is fast to perform, cheap, and has no side effects. Its application helps in the differentiation of PD and depression as a cause of hypokinesia. Hypokinetic and hypomimic patient with depression and BR hypoechogenicity and with a normal signal of SN points toward the diagnosis of depression as a reason for motor slowness. On the other hand, SN hyperechogenicity and BR hypoechogenicity could either mean PD and concomitant depression or isolated depression and a susceptibility to develop PD. In unclear cases (nonresponsiveness to antidepressive agents and dopaminergic treatment), functional neuroimaging methods (i.e., DATScan) should be performed.

The finding of a reduced BR echogenicity has led to a revival of the model of basal limbic system affection in unipolar depression and depression associated with some forms of neurodegenerative diseases. Further investigations may help to better understand the etiology of different forms of depression, responsiveness to therapeutic strategies, and underlying pathology.

References

Arango, V., Underwood, M. D., Boldrini, M., Tamir, H., Kassir, S. A., Hsiung, S., et al. (2001). Serotonin 1A receptors, serotonin transporter binding and serotonin transporter mRNA expression in the brainstem of depressed suicide victims. *Neuropsychopharmacology* **25**, 892–903.

Baumann, B., Bielau, H., Krell, D., Agelink, M. W., Diekmann, S., Wurthmann, C., et al. (2002). Circumscribed numerical deficit of dorsal raphe neurons in mood disorders. *Psychol. Med.* **32**, 93–103.

Becker, G., Struck, M., Bogdahn, U., and Becker, T. (1994). Echogenicity of the brainstem raphe in patients with major depression. *Psychiatry Res.* **55**, 75–84.

Becker, G., Becker, T., Struck, M., Lindner, A., Burzer, K., Retz, W., et al. (1995). Reduced echogenicity of brainstem raphe specific to unipolar depression: A transcranial color-coded real-time sonography study. *Biol. Psychiatry* **38**, 180–184.

Becker, T., Becker, G., Seufert, J., Hofmann, E., Lange, K. W., Naumann, M., et al. (1997). Parkinson's disease and depression: Evidence for an alteration of the basal limbic system detected by transcranial sonography. *J. Neurol. Neurosurg. Psychiatry* **63**, 590–596.

Becker, T., Becker, D., Berg, D., et al. (1998). Pathological findings in neuropsychiatric diseases. In: Echoenhancers and Transcranial Color Duplex Sonography (U. Bogdahn, G. Becker, F. Schlacketzki, Eds.), pp. 359–373. Blackwell Science, London.

Becker, G., Berg, D., Lesch, K. P., and Becker, T. (2001). Basal limbic system alteration in major depression: A hypothesis supported by transcranial sonography and MRI findings. *Int. J. Neuropsychopharmacol.* **4**, 21–31.

Berg, D., Supprian, T., Hofmann, E., Zeiler, B., Jager, A., Lange, K. W., et al. (1999). Depression in Parkinson's disease: Brainstem midline alteration on transcranial sonography and magnetic resonance imaging. *J. Neurol.* **246**, 1186–1193.

Berg, D., Supprian, T., Thomae, J., Warmuth-Metz, M., Horowski, A., Zeiler, B., et al. (2000a). Lesion pattern in patients with multiple sclerosis and depression. *Mult. Scler.* **6**, 156–162.

Berg, D., Mäurer, M., Warmuth-Metz, M., Rieckmann, P., and Becker, G. (2000b). The correlation between ventricular diameter measured by transcranial sonography and clinical disability and cognitive dysfunction in patients with multiple sclerosis. *Arch. Neurol.* **57**, 1289–1292

Berg, D., Behnke, S., and Walter, U. (2006). Application of transcranial sonography in extrapyramidal disorders: Updated recommendations. *Ultraschall. Med.* **27**, 12–19.

Berg, D., Godau, J., and Walter, U. (2008). Transcranial sonography in movement disorders. *Lancet Neurol.* **7**, 1044–1055.

Brooks, D. J., and Doder, M. (2001). Depression in Parkinson's disease. *Curr. Opin. Neurol.* **14**, 465–470.

Budisic, M., Karlovic, D., Trkanjec, Z., Lovrencic-Huzjan, A., Vukovic, V., Bosnjak, J., et al. (2009). Brainstem raphe lesion in patients with major depressive disorder and in patients with suicidal ideation recorded on transcranial sonography. *Eur. Arch. Psychiatry Clin. Neurosci.* DOI 10.1007/S00406-009-0043.

Carpenter, M. B., and Sutin, J. (1993). Human Neuroantomy. Williams and Wilkins, Baltimore.

Crossman, A. R. (2005). Brain stem. In: Gray's Anatomy: The Anatomical Basis of Clinical Practice (S. Standring, Ed.), pp. 327–351. Churchill Livingston, Edinburgh.

Cummings, J. L. (1992). Depression and Parkinson's disease: A review. *Am. J. Psychiatry* **149**, 443–454.

Eggers, B., Hermann, W., Barthel, H., Sabri, O., Wagner, A., and Hesse, S. (2003). The degree of depression in Hamilton rating scale is correlated with the density of presynaptic serotonin transporters in 23 patients with Wilson's disease. *J. Neurol.* **250**, 576–580.

Gaenslen, A., Unmuth, B., Godau, J., Liepelt, I., Di Santo, A., Schweitzer, K. J., et al. (2008). The specificity and sensitivity of transcranial ultrasound in the differential diagnosis of Parkinson's disease: A prospective blinded study. *Lancet Neurol.* **7**, 417–424.

Hoeppner, J., Prudente-Morrissey, L., Herpertz, S. C., Benecke, R., and Walter, U. (2009). Substantia nigra hyperechogenicity in depressive subjects relates to motor asymmetry and impaired word fluency. *Eur. Arch. Psychiatry Clin. Neurosci.* **259**, 92–97.

Huntington Study Group. (1996). Unified Huntingtons disease rating scale: Reliability and consistency. *Mov. Disord.* **11**, 136–142.

Leentjens, A. F., Van den Akker, M., Metsemakers, J. F., Lousberg, R., and Verhey, F. R. (2003). Higher incidence of depression preceding the onset of Parkinson's disease: A register study. *Mov. Disord.* **18**, 414–418.

Lieberman, A. (2006). Depression in Parkinson's disease–a review. *Acta Neurol. Scand.* **113**, 1–8.

Meltzer, C., Price, C., Mathis, A., Butters, A., Ziolko, K., Moses-Kolko, E., et al. (2004). Serotonin 1A receptor binding and treatment response in late-life depression. *Neuropsychopharmacology* **29**, 2258–2265.

Minden, S. L., and Schiffer, R. B. (1990). Affective disorders in multiple sclerosis. Review and recommendations for clinical research. *Arch. Neurol.* **47**, 98–104.

Naumann, M., Becker, G., Toyka, K. V., Naumann, M., Becker, G., Toyka, K. V., Supprian, T., and Reiners, K. (1996). Lenticular nucleus lesion in idiopathic dystonia detected by transcranial sonography. *Neurology* **47**, 1284–1290.

Nieuwenhuys, R., Voogd, J., and Van Huijzen, C. (2008). The Human Central Nervous System. pp. 889–916. Springer, Berlin

Poser, C. M., Paty, D. W., Scheinberg, L., McDonald, W. I., Davis, F. A., Ebers, G. C., et al. (1983). New diagnostic criteria for multiple sclerosis: Guidelines for research protocols. *Ann. Neurol.* **13**, 226–231.

Postert, T., Lack, B., Kuhn, W., Jergas, M., Andrich, J., Braun, B., et al. (1999). Basal ganglia alterations and brain atrophy in Huntington's disease depicted by transcranial real time sonography. *J. Neurol. Neurosurg. Psychiatry* **67**, 457–462.

Sabatini, U. et al. (1996). Involvement of the limbic system in multiple sclerosis patients with depressive disorders. *Biol. Psychiatry* **39**, 970–975.

Schuurman, A. G., van den Akker, M., Ensinck, K. T., Metsemakers, J. F., Knottnerus, J. A., Leentjens, A. F., et al. (2002). Increased risk of Parkinson's disease after depression: A retrospective cohort study. *Neurology* **58**, 1501–1504.

Seniow, J., Bak, T., Gajda, J., Poniatowska, R., and Czlonkowska, A. (2002). Cognitive functioning in neurologically symptomatic and asymptomatic forms of Wilson's disease. *Mov. Disord.* **17**, 1077–1083.

Seniow, J., Mroziak, B., Czlonkowska, A., and Jedryka-Goral, A. (2003). Self-rated emotional functioning of patients with neurological or asymptomatic form of Wilson's disease. *Clin. Neuropsychol.* **17**, 367–373.

Steele, J. D., Bastin, M. E., Wardlaw, J. M., and Ebmeier, K. P. (2005). Possible structural abnormality of the brainstem in unipolar depressive illness: A transcranial ultrasound and diffusion tensor magnetic resonance imaging study. *J. Neurol. Neurosurg. Psychiatry* **76**, 1510–1515.

Supprian, T., Reiche, W., Schmitz, B., Grunwald, I., Backens, M., Hofmann, E., et al. (2004). MRI ofthe brainstem in patients with major depression, bipolar affective disorder and normal controls. *Psychiatry Res.* **131**, 269–276.

Walter, U., Krolikowski, K., Tarnacka, B., Benecke, R., Czlonkowska, A., and Dressler, D. (2005). Sonographic detection of basal ganglia lesions in asymptomatic and symptomatic Wilson disease. *Neurology* **64**, 1726–1732.

Walter, U., Dressler, D., Wolters, A., Wittstock, M., and Benecke, R. (2006). Overactive bladder in Parkinson's disease: Alteration of brainstem raphe detected by transcranial sonography. *Eur. J. Neurol.* **13**, 1291–1297.

Walter, U., Behnke, S., Eyding, J., Niehaus, L., Postert, T., Seidel, G., et al. (2007a). Transcranial brain parenchyma sonography in movement disorders: State of the art. *Ultrasound Med. Biol.* **33**, 15–25.

Walter, U., Hoeppner, J., Prudente-Morrissey, L., Horowski, S., Herpertz, S. C., and Benecke, R. (2007b). Parkinson's disease-like midbrain sonography abnormalities are frequent in depressive disorders. *Brain* **130**, 1799–1807.

Walter, U., Prudente-Morrissey, L., Herpertz, S. C., Benecke, R., and Hoeppner, J. (2007c). Relationship of brainstem raphe echogenicity and clinical findings in depressive states. *Psychiatry Res.* **155**, 67–73.

Walter, U., Horowski, S., Benecke, R., and Zettl, U. (2007d). Transcranial brain sonography findings related to neuropsychological impairment in multiple sclerosis. *J. Neurol.* **254** (Suppl. 2), 49–52.

Walter, U., Dressler, D., Wolters, A., Wittstock, M., and Benecke, R. (2007e). Transcranial brain sonography findings in clinical subgroups of idiopathic Parkinson's disease. *Mov. Disord.* **22**, 48–54.

Walter, U., Kanowski, M., Kaufmann, J., Grossmann, A., Benecke, R., and Niehaus, L. (2008). Contemporary ultrasound systems allow high-resolution transcranial imaging of small echogenic deep intracranial structures similarly as MRI: A phantom study. *Neuroimage* **40**, 551–558.

Walter, U., Skoloudík, D., and Berg, D. (2010). Transcranial sonography findings related to nonmotor features of Parkinson's disease. *J. Neurol. Sci.* **289** (1–2), 123–127.

PART IV

FUTURE APPLICATIONS AND CONCLUSION

TRANSCRANIAL SONOGRAPHY-ASSISTED STEREOTAXY AND FOLLOW-UP OF DEEP BRAIN IMPLANTS IN PATIENTS WITH MOVEMENT DISORDERS

Uwe Walter

Department of Neurology, University of Rostock, Rostock, Germany

Transcranial sonography (TCS) of the brain parenchyma meanwhile allows a high-resolution imaging of deep brain structures in the majority of adults. Two pilot studies have demonstrated that the intraoperative visualization with TCS and the TCS-assisted insertion of deep-brain stimulation (DBS) electrodes into the subthalamic nucleus and the globus pallidus interna are feasible and safe provided there is exact knowledge on the extent of electrode TCS imaging artifacts. Moreover, TCS can be recommended for the postoperative monitoring of DBS electrode position. Dislocation of a DBS electrode can be easily detected. It is likely that the obvious advantages of TCS will promote its increasing use for the intra- and postoperative monitoring of deep brain implants.

I. Introduction

Transcranial B-mode sonography (TCS) of the brain parenchyma and the intracranial ventricular system have been shown to be reliable and clinically useful in children already in the 1980s and 1990s of the past century (Babcock, 1995; Machado *et al.*, 1991). For example, the guidance of programming a shunt valve system in the treatment of a fluctuating child hydrocephalus has been shown to be well possible with TCS (Mursch *et al.*, 1996). Moreover, a high-resolution imaging of the infant brain was demonstrated to be possible

INTERNATIONAL REVIEW OF
NEUROBIOLOGY, VOL. 90
DOI: 10.1016/S0074-7742(10)90019-6

with TCS already in the 1980s (Schöning *et al.*, 1988). This allowed early an astonishingly broad range of TCS applications in infants, for example, to diagnose periventricular leukomalacia and hypoxic/ischemic injury (Babcock, 1995).

In adults, the TCS imaging conditions are much more difficult than in children because of the thickening of temporal bones with increasing age. Nevertheless, due to the technological advances of the past decades, a high-resolution imaging of deep brain structures is meanwhile possible even in the majority of adults (Berg *et al.*, 2008; Walter *et al.*, 2007). This opens a new field of TCS application, the intraoperative assistance of deep brain implant placement and the postoperative monitoring of brain implant position in patients with movement disorders. The present chapter reviews the current literature and the experience of our lab in the application of TCS for the localization of deep brain stimulation (DBS) electrodes in patients with movement disorders.

II. Advances in TCS Image Resolution

In recent years, TCS image resolution was refined by improved transducer technologies and advances in signal processing (Claudon *et al.*, 2002; Clement, 2005; Crosgrove, 2006). As a result, present-day TCS systems can achieve a higher image resolution in comparison not only to former-generation systems, but currently also to MRI under clinical conditions (Walter *et al.*, 2008). This holds true if the requirements for optimal TCS are fulfilled, that is, sufficient acoustic bone window, increased echogenicity of target structure, and its localization around midline within the skull. A sophisticated clinical high-end TCS system was shown to gain an in-plane image resolution of intracranial structures in the focal zone of about 0.7 × 1.1 mm (Walter *et al.*, 2008). With the type of phased-array probes usually applied for TCS in adults, using a center frequency of 2.0–3.5 MHz, the focal zone of maximum resolution is at a distance of 5–7 (4–8) cm from the contact plane of the probe. This means that the best-quality images of intracranial structures are obtained in deep brain areas near the midline. This feature has promoted TCS applications mainly in the diagnostic evaluation of deep brain structures such as midbrain, basal ganglia, ventricles, corpus callosum, and subcortical areas (Berg *et al.*, 2008; Walter *et al.*, 2007). For the same reason, also the imaging of DBS electrodes is easily feasible (Moringlane *et al.*, 2005; Walter *et al.*, 2009).

III. Technical and Safety Issues Concerning TCS of Brain Implants

When intracranial devices containing metal parts such as DBS electrodes are visualized with TCS, one has to bear in mind that these structures cause several imaging artifacts due to their high echogenicity. On the one hand, due to poorer lateral image resolution compared with axial image resolution, the DBS electrode appears more extended in the lateral direction than in the axial direction. On the other hand, reverberation artifacts are generated behind the DBS electrode due to its high echogenicity (Fig. 1). We have measured the extent of the DBS electrode imaging artifacts using a human skull phantom. With the TCS system

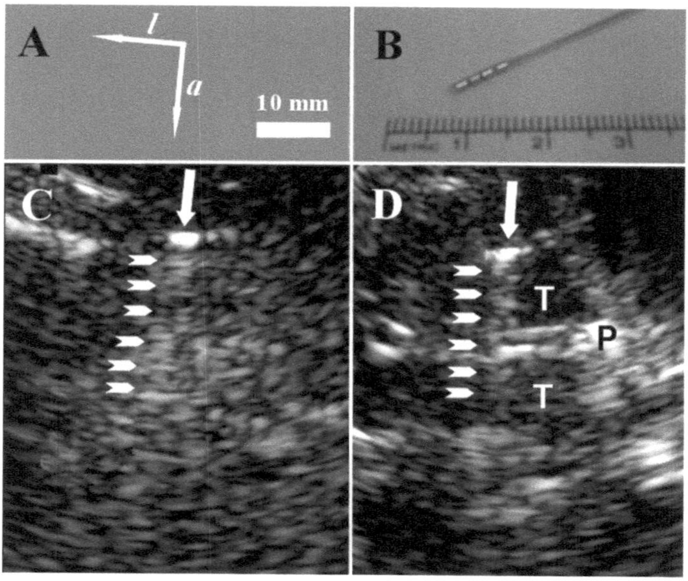

FIG. 1. Transcranial sonography (TCS) of an intracranial deep brain stimulation (DBS) electrode (transverse scan through cranium and electrode). (A) The arrows denote the so-called axial (*a*) and lateral (*l*) direction of ultrasound propagation and image resolution, as referring to the TCS images shown in (C) and (D). The bar denotes a length of 10 mm referring to the TCS images shown in (C) and (D). (B) Photograph of the DBS electrode with a diameter of 1.27 mm displayed as axial (transverse) scan in the TCS images shown in (C) and (D). (C) TCS appearance of the DBS electrode (arrow) placed within a human skull phantom filled with ultrasound gel. Note that, due to poorer lateral compared to axial image resolution, the echosignal of the DBS electrode transection appears to be lengthy instead of round. Moreover, reverberation artifacts of the DBS electrode can be seen (arrow heads). (D) TCS appearance of axial transection through a DBS electrode (arrow) of same type as that shown in (B) and (C), located in the globus pallidus interna of a patient with idiopathic dystonia. Note that the shape and reverberation artifacts (arrow heads) are identical to that seen in the skull phantom as shown in (C). T denotes thalamus; P, pineal gland.

Acuson Antares (Siemens; Erlangen, Germany), in the lateral direction of insonation, usually applied to monitor DBS electrode depth intra- and postoperatively, the highly echogenic imaging artifact of the metal part of the DBS lead used for globus pallidus interna (GPI) stimulation in dystonia was found to exceed the 1-mm rubber tip by minimum 0.1 mm (range, 0.1–1.5 mm, depending on image brightness). In the axial direction of insonation, the imaging artifact exceeding the real boundary of the DBS lead was smaller (range, 0.3–0.6 mm; resulting seeming DBS lead diameter, 1.9–2.5 mm, depending on image brightness; real diameter, 1.27 mm) (Walter *et al.*, 2009). It should be stressed that, before any application of TCS for intraoperatively guiding the positioning of DBS lead in patients, the sizes of imaging artifacts need to be estimated separately for each different ultrasound system and each different DBS lead type to account for differences of imaging technologies and lead shape.

We also have investigated whether the insonation of intracranially located DBS electrodes might be associated with a heating of the electrode. Using a skull phantom, we found a constant temperature (22.3 °C) of the intracranial DBS lead when exposed to TCS or transcranial color-coded sonography (TCCS) for 30 min each with ultrasound frequencies of 2.0, 2.5, or 3.1 MHz (ultrasound intensity: mechanical index 1.4) (Walter *et al.*, 2009). Therefore, it is unlikely that a heating of DBS electrodes occurs during TCS application, considering also the effective heat transfer within the brain due to the intense blood perfusion of the brain.

IV. TCS for Intraoperative Optimization of Implant Placement

While a considerable number of studies have dealt with the intraoperative application of sonography in patients with open cranium (e.g., Comeau *et al.*, 2000; Letteboer *et al.*, 2005; Lindner *et al.*, 2006; Rygh *et al.*, 2008), applications of intraoperative TCS have been rarely described. White *et al.* (2009) recently reported on preliminary findings using a novel intraoperative brain-shift monitor using shear-mode transcranial ultrasound. Despite the obvious advantages of ultrasound in an intraoperative setting compared with other neuroimaging methods (Walter, 2009), such as high temporal resolution, portability, and non-ionizing mode of radiation, the application of commercially available TCS systems for intraoperative monitoring of DBS electrode placement has been reported only in two reports so far.

The first report describes findings obtained with a former-generation TCS system (Sonoline Elegra, Siemens; Erlangen, Germany) during implantation of DBS electrodes into the targeted subthalamic nucleus (STN) in patients with Parkinson's disease (Moringlane *et al.*, 2005). The authors found that the 0.8-mm-thick electrode was easy to visualize. The position of the imaging artifact of the tip of the DBS electrode appeared to be within the substantia nigra. Additionally, the segment of the laterally running posterior cerebral artery at the corresponding level could also be displayed. The authors found the appearing correct position of the DBS electrode tip on TCS at a place just touching the echosignals of the substantia nigra. Limitations of this pilot study were the poorer lateral image resolution of the TCS system applied compared with contemporary TCS systems (Walter *et al.*, 2008), and the missing estimation of imaging artifacts of the DBS electrode, which caused some uncertainty with regard to the exact electrode tip position.

In the second study, a contemporary TCS system (Acuson Antares, Siemens; Erlangen, Germany) was applied to intraoperatively monitor the placement of DBS electrode into the GPI in patients with idiopathic dystonia (Walter *et al.*, 2009). Fig. 2 shows the intraoperative setting applied this study. In this study, not

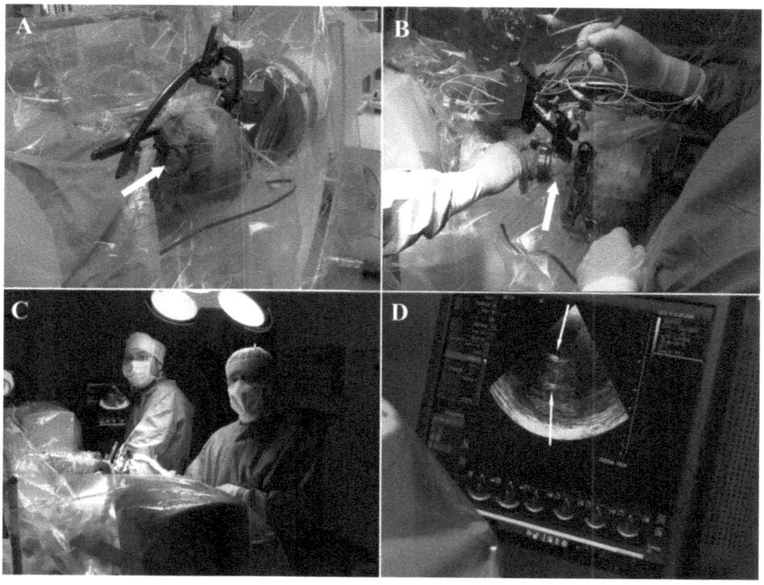

FIG. 2. Operation theatre setting of transcranial sonography (TCS)-assisted stereotaxy. (A) Under monitored anesthesia care, the patients' head is fixed in the Leksell stereotactic head frame and wrapped with sterile layers of foil. Underneath the foil, ultrasound contact gel was placed at the temporal plane of the patient. The arrow indicates the access through the frame for the ultrasound transducer. (B) Under sterile conditions, the ultrasound transducer (arrow) is placed at the temporal plane of the patients' head for intraoperative TCS. Through a 10-mm burr hole in the cranium the microelectrodes were inserted using guide tubes. During left-side electrode implantation, the sonographer stands on the left to the surgeon. (C) During subsequent right-side electrode implantation the sonographer stands on the right to the surgeon. (D) Display of the TCS system showing a semi-axial scan through the brain with the bilateral deep brain stimulation electrodes (arrows).

only the visualization of the final DBS electrodes was possible but also the simultaneous visualization of two to five closely located microelectrodes used for detection of the optimal trajectory of the final electrode (Fig. 3). The main advantage of the intraoperative TCS monitoring was that the distance of the DBS electrode tip to the artery at the anatomic target (penetrating branch of the posterior communicating artery) could exactly be assessed (Fig. 3D). This was possible since the extent of the imaging artifact of the electrode had been estimated in advance for the referring TCS system and implant (see Section 3

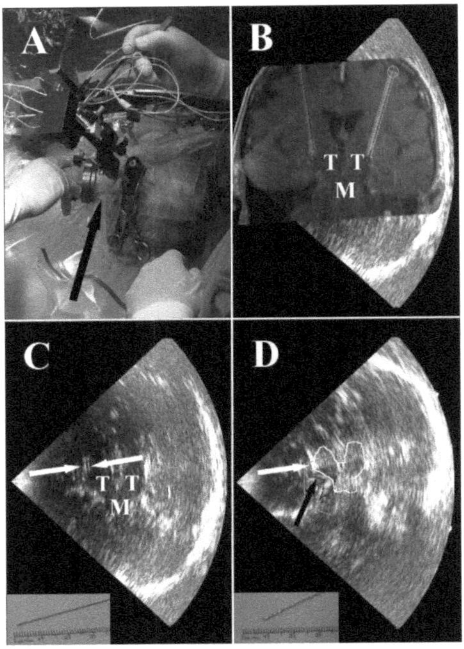

FIG. 3. Transcranial sonography (TCS) images of deep brain stimulation (DBS) electrodes obtained during stereotactic implantation. (A) Position of the TCS transducer placed at the temporal plane of the patient (arrow). (B) Fusion image of preoperative magnetic resonance imaging (MRI) scan and intraoperative TCS scan. The stereotactic trajectories are highlighted on the MRI image. T indicates thalamus; M, midbrain. (C) Shown is the same TCS image as in (B). Two closely located microelectrodes (arrows) of 0.5 mm diameter are clearly visualized in the target region. Inserted panel in the left lower corner: photograph of a microelectrode. (D) Intraoperative TCS image showing the tip of the final DBS lead (white arrow) and its distance to the perforating branch (black arrow) of the posterior communicating artery. The midbrain and thalami are surrounded for better recognition. Considering the previously determined imaging artifact of the DBS lead tip, monitoring of its distance to the perforating artery near the anatomical target allowed the intraoperative decision to slightly further insert the DBS lead for improvement of targeting accuracy. Inserted panel in the left lower corner: photograph of a DBS lead.

of this chapter). This enabled us to intraoperatively decide to insert the final DBS electrode somewhat deeper than it would have been done using only the preoperatively planned navigation data (Walter *et al.*, 2009). Simultaneous visualization of the artery at the anatomic target prevented hemorrhages at the target site. The outcome of the patients was excellent.

The possibility of intraoperatively optimizing the electrode position is of high interest since discrepancies of up to 4 mm (average 2 mm) between the initial target, selected on preoperative MRI, and the final DBS lead location are caused mainly by caudal brain shift that occurs once the cranium is open (Khan *et al.*, 2008; White *et al.*, 2009). Intraoperative MRI may overcome targeting inaccuracy but is expensive and not widely available (Hall and Truwit, 2008). The higher mobility and temporal resolution achieved with TCS may promote an increasing use for the intraoperative guidance of deep brain implant placement.

V. TCS for Postoperative Position Control of Deep Brain Implants

Postoperatively, it is even easier to localize DBS electrodes on TCS since the patient and the investigator are in a much more comfortable setting. Especially, there is less constriction in finding the optimal temporal acoustic bone in order to achieve high-quality brain images. Typical aspects of DBS electrodes targeting the pars ventralis intermedius (VIM) of the thalamus, the GPI, and the STN are shown in Fig. 4. It is recommendable to define some landmarks that can be used as reference points for estimating the exact position of the DBS electrode tip. Typical measures are the shortest distance of the electrode tip from the midline and/or the outer boundary of the third ventricle (VIM, GPI, STN), the distance of the electrode tip from the pineal gland (VIM, GPI) and the position of the electrode in relation to the substantia nigra (STN).

In our experience, the detection of slight differences of 2 mm and more between right- and left-sided electrodes with respect to their distance to midline, and also in their rostrocaudal position, is possible with TCS. Gross electrode dislocation can easily be diagnosed with TCS. Fig. 5 shows images of a patient in whom DBS electrode dislocation was diagnosed first with TCS and subsequently confirmed by CCT. After successful re-implantation, the post-operative TCS proved the correct position of the DBS electrodes in the STN (Fig. 4).

In conclusion, TCS can be recommended for the postoperative monitoring of DBS electrode position. Dislocation of a DBS electrode can be easily detected.

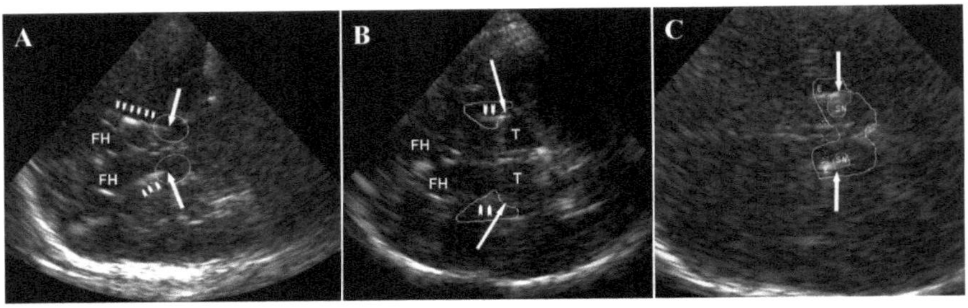

FIG. 4. Transcranial sonography (TCS) images obtained for the postoperative control of deep brain stimulation (DBS) electrode position in three patients. (A) Semi-axial TCS scan through the brain of a patient with severe essential tremor in whom DBS bilaterally of the nucleus ventralis intermedius (VIM) of the thalamus was performed. The thalami are surrounded for better recognition. The tips (arrows) of the bilateral DBS electrodes are visualized in the right and the left thalami. At the semi-axial scan, the more cranial parts of the DBS electrodes can be visualized simultaneously (arrow heads). FH denotes frontal horn of lateral ventricle. (B) Axial (transverse) TCS scan in a patient with idiopathic dystonia in whom DBS bilaterally of the globus pallidus interna (GPI) was performed. The bilateral lenticular nucleus is surrounded for better recognition. Considering the lower lateral compared to the axial image resolution on TCS (confer Fig. 1), the real position of the tips (arrows) of the bilateral DBS electrodes is exactly within the GPI, even though on the TCS image their position seems to be within the internal capsule. The more cranial parts of the electrodes are displayed within the GPI (arrow heads). FH denotes frontal horn of lateral ventricle; T, thalamus. (C) Axial (transverse) TCS scan in a patient with Parkinson's disease in whom DBS of the subthalamic nucleus (STN) was performed bilaterally. The midbrain and the bilateral markedly echogenic ("hyperechogenic") substantia nigra are surrounded for better recognition. Considering the lower lateral than axial image resolution on TCS (confer Fig. 1), the true position of the tips (arrows) of the bilateral DBS electrodes is exactly within the STN, even though on the TCS image their position seems to be within the substantia nigra.

VI. Current Limitations and Future Issues

It can be expected that the obvious advantages of TCS in comparison to other neuroimaging methods, such as high mobility, short investigation times, noninvasiveness, and less corruption by patients movements, will further promote the use of TCS for the intra- and postoperative monitoring of deep brain implants, especially in patients with movement disorders. The major current limitation of TCS application is, beside its dependence on the quality of transtemporal acoustic bone windows, the necessity of a highly qualified investigator. The investigator performing intraoperative TCS for guiding therapeutic decisions needs to be well trained beforehand in the pre- and postoperative routine setting. Moreover, the applied TCS

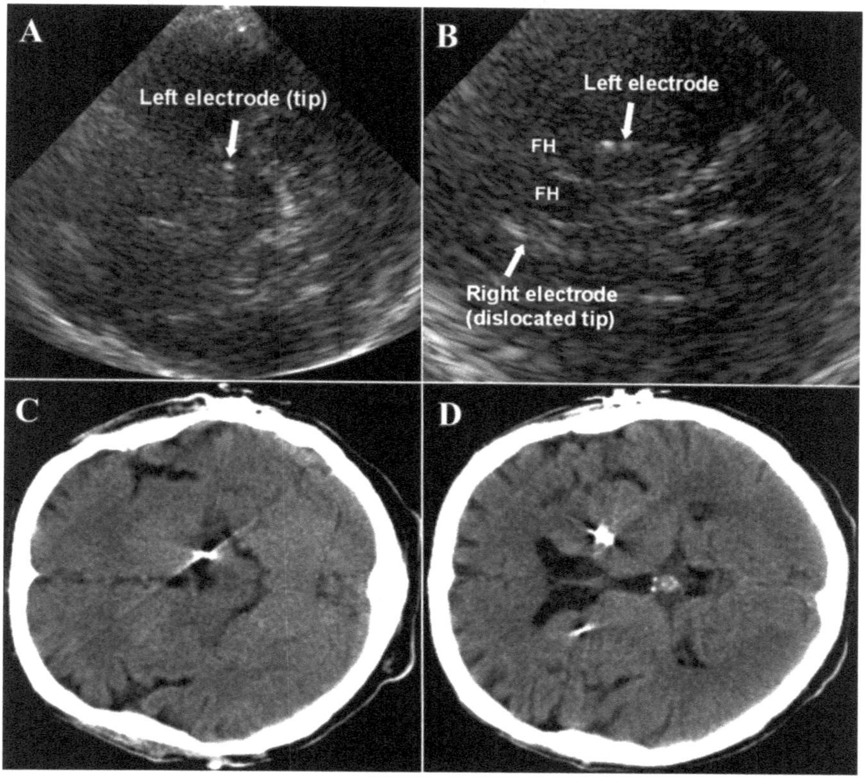

FIG. 5. Postoperative transcranial sonography (TCS) and corresponding computed tomography (CT) images of a patient in whom unilateral dislocation of deep brain stimulation (DBS) electrode was diagnosed. The DBS electrode dislocation on the right side was first detected with TCS and subsequently confirmed with CT. (A) TCS image showing the tip of the left-sided DBS electrode (arrow) correctly located within the subthalamic nucleus (STN) while the right-sided DBS electrode is not displayed in the STN due to dislocation. (B) TCS image showing the more cranial parts of the left-sided DBS electrode (arrow) and the tip of the dislocated right-sided electrode (arrow) lateral from the caudate nucleus. (C) CT image corresponding to the TCS image in (A). (D) CT image corresponding to the TCS image in (B).

system as well as the assessed brain implant should be studied in advance for the exact size of their imaging artifacts using a skull phantom as described earlier (Walter *et al.*, 2008, 2009). It may well be that novel technologies able to fuse preoperative MRI images and intraoperative TCS images may facilitate an easier and less investigator-dependent application of intraoperative TCS.

References

Babcock, D. S. (1995). Sonography of the brain in infants: Role in evaluating neurologic abnormalities. *AJR Am. J. Roentgenol.* **165**, 417–423.

Berg, D., Godau, J., and Walter, U. (2008). Transcranial sonography in movement disorders. *Lancet Neurol.* **7**, 1044–1055.

Claudon, M., Tranquart, F., Evans, D. H., Lefevre, F., and Correas M. (2002). Advances in ultrasound. *Eur. Radiol.* **12**, 7–18.

Clement, G. T. (2005). Spectral image reconstruction for transcranial ultrasound measurement. *Phys. Med. Biol.* **50**, 5557–5572.

Coenen, V. A., Krings, T., Weidemann, J., Hans, F. J., Reinacher, P., Gilsbach, J. M., and Rohde, V. (2005). Sequential visualization of brain and fiber tract deformation during intracranial surgery with three-dimensional ultrasound: An approach to evaluate the effect of brain shift. *Neurosurgery.* **56** (Suppl.), 133–141.

Comeau, R. M., Sadikot, A. F., Fenster, A., and Peters, T. M. (2000). Intraoperative ultrasound for guidance and tissue shift correction in image-guided neurosurgery. *Med. Phys.* **27**, 787–800.

Cosgrove, D. (2006). Developments in ultrasound. *Imaging* **18**, 82–96.

Hall, W. A., and Truwit, C. L. (2008). Intraoperative MR-guided neurosurgery. *J. Magn. Reson. Imaging* **27**, 368–375.

Khan, M. F., Mewes, K., Gross, R. E., and Skrinjar, O. (2008). Assessment of brain shift related to deep brain stimulation surgery. *Stereotact. Funct. Neurosurg.* **86**, 44–53.

Letteboer, M. M., Willems, P. W., Viergever, M. A., and Niessen, W. J. (2005). Brain shift estimation in image guided neurosurgery using 3-D ultrasound. *IEEE Trans. Biomed. Eng.* **52**, 268–276.

Lindner, D., Trantakis, C., Renner, C., Arnold, S., Schmitgen, A., Schneider, J., and Meixensberger, J. (2006). Application of intraoperative 3D ultrasound during navigated tumor resection. *Minim. Invasive Neurosurg.* **49**, 197–202.

Machado, H. R., Martelli, N., Assirati Júnior, J. A., and Colli, B. O. (1991). Infantile hydrocephalus: Brain sonography as an effective tool for diagnosis and follow-up. *Childs Nerv. Syst.* **7**, 205–210.

Moringlane, J. R., Fuss, G., and Becker, G. (2005). Peroperative transcranial sonography for electrode placement into the targeted subthalamic nucleus of patients with Parkinson disease: Technical note. *Surg. Neurol.* **63**, 66–69.

Mursch, K., Behnke, J., Christen, H. J., and Markakis, E (1996). Use of transcranial real-time ultrasonography for programming a shunt valve system. *Childs Nerv. Syst.* **12**, 392–395.

Rygh, O. M., Selbekk, T., Torp, S. H., Lydersen, S., Hernes, T. A., and Unsgaard, G. (2008). Comparison of navigated 3D ultrasound findings with histopathology in subsequent phases of glioblastoma resection. *Acta Neurochir. (Wien).* **150**, 1033–1041.

Schöning, M., Grunert, D., and Stier, B. (1988). Transcranial real-time sonography in children and adolescents, ultrasound anatomy of the brain. *Ultraschall. Med.* **9**, 286–292.

Walter, U. (2009). Imaging of the brain: Where is transcranial sonography superior to magnet resonance imaging? *Fortschr. Neurol. Psychiatr.* **77** (Suppl. 1), S39–41.

Walter, U., Behnke, S., Eyding, J., Niehaus, L., Postert, T., Seidel, G., and Berg, D. (2007). Transcranial brain parenchyma sonography in movement disorders: State of the art. *Ultrasound Med. Biol.* **33**, 15–25.

Walter, U., Kanowski, M., Kaufmann, J., Grossmann, A., Benecke, R., and Niehaus, L. (2008). Contemporary ultrasound systems allow high-resolution transcranial imaging of small echogenic deep intracranial structures similarly as MRI: A phantom study. *Neuroimage* **40**, 551–558.

Walter, U., Wolters, A., Wittstock, M., Benecke, R., Schroeder, H. W., and Müller, J. U. (2009). Deep brain stimulation in dystonia: Sonographic monitoring of electrode placement into the globus pallidus internus. *Mov. Disord.* **24**, 1538–1541.

White, P. J., Whalen, S., Tang, S. C., Clement, G. T., Jolesz, F., and Golby, A. J. (2009). An intraoperative brain shift monitor using shear mode transcranial ultrasound: Preliminary results. *J. Ultrasound Med.* **28**, 191–203.

CONCLUSIONS

Daniela Berg

Department of Neurodegeneration, Hertie Institute of Clinical Brain Research and German Center of Neurodegenerative Diseases (DZNE), Tübingen, Germany

I. General Remarks

When the first papers on transcranial sonography (TCS) in movement disorders were published in the mid- and end-1990s, no one could imagine the impact this method would have one day on both clinical routine and research. At that time, there were no means to detect abnormalities in the idiopathic forms of movement disorders such as Parkinson's disease or dystonia with routine structural neuroimaging methods. Therefore, it is well comprehensible that TCS as an easily applicable, quick-to-perform, and cheap method, which claimed to be able to detect disease-specific features, was encountered with skepticism, doubt, and sometimes also with a smile.

Since then, a large number of papers has been published on the value of TCS in the diagnosis and differential diagnosis of movement disorders as well as on its implications on the pathophysiological aspects of different diseases. This volume witnesses the development of the method and the spreading of its application to many diagnostic fields as well as to all continents of the world. Moreover, it gives an overview on the state-of-the-art technique of TCS and its contribution to our knowledge on the pathophysiology of the disorders investigated by this method.

Meanwhile, it is well known that TCS should be regarded as a supplementary tool to other neuroimaging methods. Because of specific underlying physical principles, information can be derived from this technique which differs from the aspects depicted, for example, by CCT or MRI. With this in mind, it also needs to be considered that the application of TCS entails limitations—as does every other neuroimaging technique as well. These limitations are named and carefully discussed in this volume, including, for example, an insufficient bone window in about 10% of the Caucasian population, nonspecificity of typical ultrasound features like substantia nigra or lentiform nucleus hyperechogenicity

INTERNATIONAL REVIEW OF NEUROBIOLOGY, VOL. 90
DOI: 10.1016/S0074-7742(10)90020-2

and difficulties in objectifying and quantifying observations. It is important to take the limitations seriously and stick to applications that are sound for clinical as well as research purposes. On the other hand, it should be considered and approached that many of the technical limitations can at least in part be overcome, as obviously testified by the development of the method in the past years. Therefore, further effort needs to be put into the improvement of the method. Doing so, new and even broader applications can be expected in the future.

How can the future of TCS be envisioned? Here, different fields of application need to be considered. To name a few:

1. The increase in application all over the world based on congruent results in a multitude of studies will lead to the implementation of TCS in the routine diagnostic workup of some movement disorders, especially Parkinsonian syndromes. Here, a diagnostic algorithm could be implemented, applying TCS routinely in different forms of movement disorders. Only in case of unclear results or discordance of clinical and TCS findings, the patient would be referred to other diagnostic measures—that is, functional neuroimaging. Moreover, countries with less access to functional neuroimaging methods will benefit from the possibility to "have a look into the brain" of their patients either to disclose specific features for, for example, Parkinson's disease or to exclude secondary causes mimicking the idiopathic forms of diseases (like hydrocephalus in the differential diagnosis of Parkinson's disease, or Wilson's disease in the differential diagnosis of dystonia).
2. A growing number of studies provide evidence that specific TCS findings may antecede the symptom manifestation of some diseases. The largest number and most convincing studies have been published concerning the premotor diagnosis of Parkinson's disease. Here an association of the "Parkinson sign" substantia nigra hyperechogenicity with a huge number of premotor markers has been shown. Moreover, first prospective longitudinal studies imply an increased risk for Parkinson's disease in yet healthy subjects with substantia nigra hyperechogenicity. Since the hypothesis that substantia nigra hyperechogenicity reveals a risk factor for Parkinson's disease is increasingly confirmed, application of the method as a screening instrument for a population at risk for Parkinson's disease in the nearer future is conceivable.
3. Although TCS findings have shed some light on the pathophysiological aspects of some movement disorders, many questions are still unsolved. Here, it can be hoped that with further improvement of the method and basic research in the field, further aspects can be elucidated.
4. New fields of applications have arisen ever since implementation of TCS in the diagnostic workup of movement disorders. Intra- and postoperative monitoring of the placement of brain electrodes is one promising field for

increasing application in the future, and also widening of the technique to other medical fields such as psychiatry (already done in depression or extrapyramidal symptoms induced by neuroleptic therapy) could mean an enrichment to the field.

In the past 15 years, TCS has proven to add to both diagnostic workup as well as research in different movement disorders. This volume shows that there is accumulating evidence that the value of the method can and will be implemented in a broader way.

II. Personal Remarks

Besides demonstrating the technique itself, its value, as well as the limitations of the method, this volume is also meant to honor Georg Becker, who was the first to detect the meanwhile well-known substantia nigra hyperechogenicity in Parkinson's disease and with whom I had the unique opportunity to pioneer the field of transcranial sonography until his tragic, unexpected death in 2003, which is still a great loss to the scientific community as well as to myself personally. Moreover, with this volume, I would like to thank those who dared to perform the method at a time in which it was not commonly used and was viewed with skepticism. Several of these colleagues contributed to this volume. Last but not least, we wish to thank the editors of *International Review of Neurobiology* serial, Peter Jenner and Adron Harris, who showed confidence in the ultrasound technique and encouraged us to compile and edit this volume.

INDEX

N

O

P

T

CONTENTS OF RECENT VOLUMES

Volume 39

Volume 40

Volume 41

Volume 42

Volume 43

Volume 44

Volume 45

Volume 46

Volume 47

Volume 48

Volume 49

Volume 50

Volume 51

Volume 52

Volume 53

Volume 54

Volume 55

Volume 56

Volume 57

Volume 58

Volume 59

Volume 60

Volume 61

Volume 62

Volume 67

Volume 68

Volume 69

Volume 70

Volume 71

Volume 72

Volume 73

Volume 74

Volume 75

Volume 76

Volume 77

Volume 78

Volume 79

Volume 80

Volume 81

Volume 82

Volume 83

Volume 84

Volume 85

Volume 86

Volume 87

Volume 88

Volume 89

FSC
www.fsc.org
MIX
Papier | Fördert gute Waldnutzung
FSC® C083411

Kolibri 360 GmbH
Ferdinand-Jühlke-Straße 7
99095 Erfurt, Deutschland
produktsicherheit@kolibri360.de